CISM COURSES AND LECTURES

The series presents lecture notes, monographs, edited works and proceedings in the field of Mechanics, Engineering, Computer Science and Applied Mathematics.
Purpose of the series is to make known in the international scientific and technical community results obtained in some of the activities organized by CISM, the International Centre for Mechanical Sciences.

INTERNATIONAL CENTRE FOR MECHANICAL SCIENCES

COURSES AND LECTURES - No. 471

PROBLEMS IN STRUCTURAL IDENTIFICATION AND DIAGNOSTICS: GENERAL ASPECTS AND APPLICATIONS

MURST Project n. MM08342598 - COFIN 2000

EDITED BY

CESARE DAVINI
UNIVERSITY OF UDINE, ITALY

ERASMO VIOLA
UNIVERSITY OF BOLOGNA, ITALY

This volume contains 131 illustrations

Originally published by Springer-Verlag Wien New York in 2003

SPIN 10970591

In order to make this volume available as economically and as rapidly as possible the authors' typescripts have been reproduced in their original forms. This method unfortunately has its typographical limitations but it is hoped that they in no way distract the reader.

ISBN 978-3-211-20492-4 ISBN 978-3-7091-2536-6 (eBook)
DOI 10.1007/978-3-7091-2536-6

PREFACE

The dynamic techniques as a tool for structural identification and diagnostics have a consolidated tradition. They owe their appeal to the ability to convey a great mass of information with relatively modest experimental burden and, overall, to the fact that this can be done without suspending machines or structures from service. Thus, a direct assessment of their integrity is possible. Of course, these techniques have become a viable tool of structural analysis because of the remarkable progress in the acquisition and treatment of experimental data that has characterized the final decades of the last century.

By all these reasons the dynamic techniques have long since attracted the interest of engineers in various areas, ranging from the aeronautical to civil applications, and with various purposes, including e.g. *model validation, failure detection, noise analysis, vibration control, etc.. Today there are many journals and conferences dedicated to the subject. Some, such as the* The International Seminar of Modal Analysis *organized by the Katholieke Universiteit of Leuven, have become institutional occasions where to periodically compare work and experience or to share expertise gained in doing things. In looking at the research activity which has been developed along the years, one is surprised by the huge amount of work done, but he also notices a progressive improvement in the level of awareness of the subtleties involved in this kind of problems. Scientific curiosity, interest for the power of the techniques and need of clarity, that at least some of us felt as a primary issue, were the motivation of the research project to which the present volume is dedicated and also of those that, in preceding years, have involved the same group of researchers.*

Attention was constantly addressed to the fact that the identification problems present pathology and difficulties typical of the inverse problems, and this obstructs a blind use of the above mentioned techniques. The presence of non-uniqueness, the lack of stability estimates, etc., make it difficult to renounce to aside information or assumptions in analyzing a specific problem, and to extend the analysis to general situations without discernment. It's not surprising that, especially at the beginning, in the national and international literature the attention was mainly devoted to the study of model problems and to data either simulated or coming from laboratory tests, where measures can be performed under well controlled conditions. Studies of complex structures are rare and often give results that are not easy to interpret. Thus, it's fair to say that the distance between the pieces of information that these techniques can provide and those expected by engineers still is noticeable and deeper understanding of both general and applied aspects of the discipline is needed. The long term objective of the project was in fact to reduce such a distance by bringing together researchers interested in the two aspects.

Our research has been developed along two lines. On one side, we cultivated the study of the general properties of the identification problems, with specific regard to the sensitivity of the natural frequencies to damage, the sensitivity of the identification techniques to the accuracy of the data and of the interpretation model, the determination of "minimal sets" of data needed for the reconstruction of localized damages (mainly for one dimensional structural elements), the determination of suitable damage indicators, etc.. On the other side, we tried to maintain the focus on the experimental aspects in the attempt to confront the objective difficulty of making and interpreting the measures, especially when performed on complex systems and in real conditions. In fact, our research broadened a little so as to include the identification of material properties and the use of both static and dynamic methods for the scope.

The present volume collects papers illustrating the work of the various research units with special emphasis on their most recent achievements. The papers were presented at a workshop held in Bologna on July 15-16, 2002, under the patronage of the Study and Research Centre for the Identification of Materials and Structures (CIMEST)-"M. Capurso". They cover the following themes: effects of the modeling errors on the identification techniques; damage identification from the measurement of a couple of frequencies in rods and beams; node shifting and damage detection; identification of damping and elastic constants in orthotropic plates; use of advanced techniques of signal theory in structural identification; structural identification of non linear dynamical systems under random excitation and applications to bridge structures; dynamical modeling, tests and property identification of building foundations on piles. As is seen, the papers encompass a rather broad spectrum. Some work includes experimentation on full scale structures and in situ *tests on monumental buildings (units of Catania and Torino), or tests on small scale models of bridge structures (unit of Palermo), all aiming at the study and monitoring of manufacts of great public interest. But we wish to stress that the experimental aspect, namely, the care that the theoretical results can be carried over to practical experimental techniques, is also present in other works that have a more methodological character. So, we believe that the project has kept its original goal of contributing to reduce the gap between theory and practice in the field.*

Cesare Davini
Erasmo Viola

CONTENTS

Identification of elastic constants by Bayesian approach.

Ivan Bartoli ‡ Antonio di Leo ‡ and Erasmo Viola‡

‡ Faculty of Engineering, DISTART Department, University of Bologna, Italy

Abstract A combined numerical-experimental procedure for the identification of the elastic material modulus of generally thin orthotropic plates is discussed in this paper. This method makes use of experimental plate response data, corresponding numerical predictions and Bayesian sensitivity analysis. The response data are a set of natural frequencies of flexural vibration of the plate. The numerical model is based on the Rayleigh Ritz method and the finite element method using a classical plate theory displacement field. Statistical Bayesian estimation is applied in an iterative scheme to direct the adjustments of material properties based on the discrepancies between the analytical and experimental responses. The confidence associated with frequencies and mechanical properties is incorporated into the revision procedure. The effects of the confidence associated with experimental and numerical data and parameter estimates are numerically investigated. The validity and efficiency of the present procedure is illustrated through several test cases.

1 Introduction

In this work, a numerical-experimental method for the identification of mechanical properties of generally orthotropic plate is described. For this purpose, the experimental results of the structure response are used. The dynamic free vibration behaviour of a structure made of anisotropic materials depends on its geometry, density and boundary conditions. The analytical model of the plate is obtained using Rayleigh-Ritz method and finite element method. Since the plates discussed are thin the classical lamination theory is employed to describe dynamic behaviour. It is assumed that geometry, material density and angle between edge and fibre direction are known in the identification approach. Generally in order to obtain good esteems of experimental vibration frequencies the plates are supposed to be with all boundary edges free. Introducing initial values of material properties into the mathematical model one can obtain analytical eigenvalues and eigenvectors. Frequently, this model does not initially produce mode shapes and natural frequencies, which concur with test results agree. The present approach is founded on an error function, which includes the differences between analytical and experimental frequencies. This error function also considers the deviations of the revised mechanical properties from their initial values based on some confidence limits. The unknown plate rigidities are evaluated at every step by an estimator, which is achieved by the minimization of the error function. The estimator is based on measured values and initial estimates

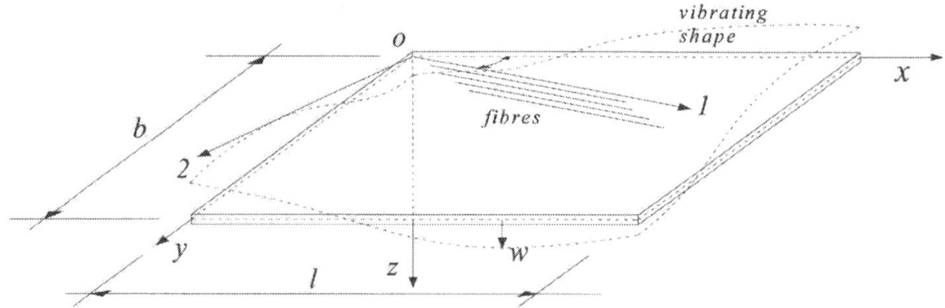

Figure 1. Orthotropic plate. Material $(1,2)$ and global (x,y) coordinate axis.

of the mechanical parameters. Sensitivity matrix is the most important operator of the estimator. Derivatives of the analytical eigenvalues with respect to the various constants constitute the elements of the matrix. The estimator also contains the uncertainties in the initial estimates, and those associated with modelling by two confidence matrices.

2 Numerical model

Figure 1 shows a rectangular plate of constant thickness h and plate dimensions l and b. A Cartesian coordinate system (x, y, z) is located at the middle plane as shown. θ represents the angle between material fibres and $x-$axis. The maximum potential energy expression for the 2-D laminated plate model, including only the bending effects, is (Bartoli, 2001)

$$V_{\max} = \frac{1}{2}\left[\int_{Area} D_{11}\left(\frac{\partial^2 W}{\partial x^2}\right)^2 + 2D_{12}\left(\frac{\partial^2 W}{\partial x^2}\right)\left(\frac{\partial^2 W}{\partial y^2}\right) + D_{22}\left(\frac{\partial^2 W}{\partial y^2}\right)^2 + \right.$$
$$\left. 4D_{16}\left(\frac{\partial^2 W}{\partial x^2}\right)\left(\frac{\partial^2 W}{\partial x \partial y}\right) + 4D_{26}\left(\frac{\partial^2 W}{\partial y^2}\right)\left(\frac{\partial^2 W}{\partial x \partial y}\right) + 4D_{66}\left(\frac{\partial^2 W}{\partial x \partial y}\right)^2\right] dS \tag{2.1}$$

where the D_{ij} are the conventional in-plane plate flexural stiffnesses from classical lamination theory (which in turn are related to the in-plane engineering constants E_1, E_2, G_{12} and v_{12}) (Jones, 1975). $W(x,y)$ is the plate deflection amplitude when an harmonic motion

$$w(x,y) = W(x,y)\, e^{i\omega t} \tag{2.2}$$

is considered in which ω is the frequency of vibration.

The maximum kinetic energy ignoring rotary inertia effects, is given by

$$K_{\max} = \frac{\rho h}{2}\omega^2 \int_0^l \int_0^b W^2 dx dy. \tag{2.3}$$

2.1 Rayleigh-Ritz modelling

Assuming the flexural deflection $W(x,y)$, to be represented as a linear series of assumed free-free characteristic beam functions

$$W(x,y) = \sum_{m=1}^{p} \sum_{n=1}^{q} a_{mn} \Phi_m(x)\, \Theta_n(y). \tag{2.4}$$

one can estimate the modal frequencies ω and the eigenvectors a_{ij} by solving the eigenvalue problem (Bartoli, 2002)

$$\sum_{m=1}^{p} \sum_{n=1}^{q} \left[C_{ik,mn} - \lambda \delta_{im} \delta_{kn}\right] a_{mn} = 0 \tag{2.5}$$

obtained from the Rayleigh-Ritz method. In (2.5)

$$\lambda = \rho h \omega^2 l^2 b^2, \qquad \delta_{rs} = \begin{cases} 1 & r = s \\ 0 & r \neq s \end{cases}$$

$$\begin{aligned}
C_{ik,mn} &= D_{11} J_{1,ik,mn} + D_{22} J_{2,ik,mn} + D_{12} J_{3,ik,mn} + \\
&\quad + D_{66} J_{4,ik,mn} + D_{16} J_{5,ik,mn} + D_{26} J_{6,ik,mn} = \\
&= \left[J_{1,ik,mn} \; J_{2,ik,mn} \; J_{3,ik,mn} \; J_{4,ik,mn} \; J_{5,ik,mn} \; J_{6,ik,mn}\right] \\
&\quad \cdot \left[D_{11} \; D_{22} \; D_{12} \; D_{66} \; D_{16} \; D_{26}\right]^{\mathrm{T}} = \\
&= \left[J_{1,ik,mn} \; J_{2,ik,mn} \; J_{3,ik,mn} \; J_{4,ik,mn} \; J_{5,ik,mn} \; J_{6,ik,mn}\right] \cdot \\
&\quad \cdot \begin{bmatrix} m^4 & n^4 & 2m^2n^2 & 4m^2n^2 \\ n^4 & m^4 & 2m^2n^2 & 4m^2n^2 \\ m^2n^2 & m^2n^2 & m^4+n^4 & -4m^2n^2 \\ m^2n^2 & m^2n^2 & -2m^2n^2 & (m^2-n^2)^2 \\ m^3n & -mn^3 & mn^3 - m^3n & 2mn^3 - 2m^3n \\ mn^3 & -m^3n & m^3n - mn^3 & 2m^3n - 2mn^3 \end{bmatrix} \begin{bmatrix} D_1 \\ D_2 \\ D_3 \\ D_4 \end{bmatrix}.
\end{aligned} \tag{2.6}$$

In equations (2.6) $m = cos\theta$, $n = sin\theta$, λ is the eigenvalue, ρ is the material density, $J_{r,ik,mn}$ are the integrals involving characteristic beam functions $\Phi_m(x)$ and $\Theta_n(y)$, D_i are the four independent bending stiffnesses of the plate with respect to the $(1,2)$ material coordinate axis (see Note 4).

2.2 Finite element modelling

Using Hamilton's theorem

$$\int_{t_1}^{t_2} \delta(K - V)\, dt = 0 \tag{2.7}$$

another form of the eigenvalue problem is given as follows with a finite element approach. In relation (2.7) K and V are the kinetic and strain energy. Variations of this two quantities lead to the expressions

$$\delta K = -\int_0^l \int_0^b \left[I \left(\frac{\partial \dot{w}}{\partial x}\right) \delta \left(\frac{\partial \dot{w}}{\partial x}\right) + I \left(\frac{\partial \dot{w}}{\partial y}\right) \delta \left(\frac{\partial \dot{w}}{\partial y}\right) + \rho h \dot{w} \delta \dot{w} \right] dx dy \tag{2.8}$$

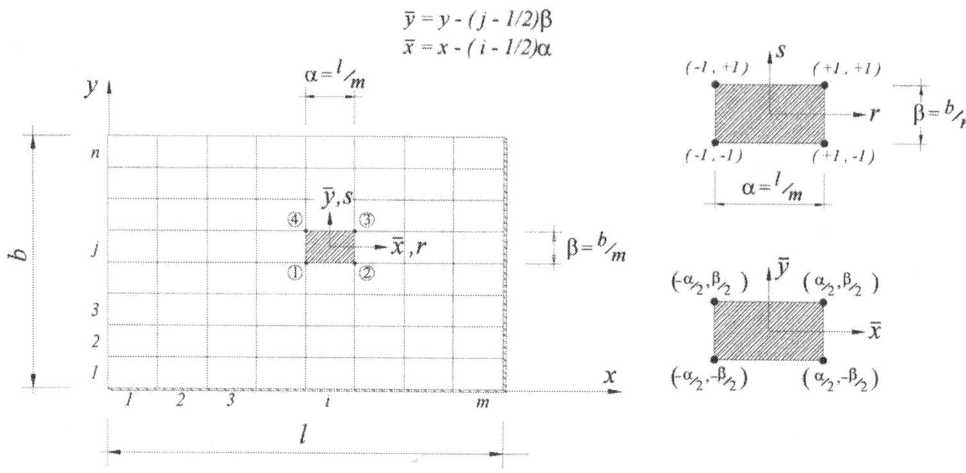

Figure 2. Discretization of the system.

$$\delta\Phi = \int_0^l \int_0^b \left\{ D_{11} \frac{\partial^2 w}{\partial x^2} \delta \left(\frac{\partial^2 w}{\partial x^2} \right) + D_{12} \left[\frac{\partial^2 w}{\partial y^2} \delta \left(\frac{\partial^2 w}{\partial x^2} \right) + \frac{\partial^2 w}{\partial x^2} \delta \left(\frac{\partial^2 w}{\partial y^2} \right) \right] + \right.$$
$$+ D_{22} \frac{\partial^2 w}{\partial y^2} \delta \left(\frac{\partial^2 w}{\partial y^2} \right) + 2D_{16} \left[\frac{\partial^2 w}{\partial x^2} \delta \left(\frac{\partial^2 w}{\partial x \partial y} \right) + \frac{\partial^2 w}{\partial x \partial y} \delta \left(\frac{\partial^2 w}{\partial x^2} \right) \right] +$$
$$\left. + 2D_{26} \left[\frac{\partial^2 w}{\partial y^2} \delta \left(\frac{\partial^2 w}{\partial x \partial y} \right) + \frac{\partial^2 w}{\partial x \partial y} \delta \left(\frac{\partial^2 w}{\partial y^2} \right) \right] + 4D_{66} \frac{\partial^2 w}{\partial x \partial y} \delta \left(\frac{\partial^2 w}{\partial x \partial y} \right) \right\} dx dy \quad (2.9)$$

In (2.8) the first two terms include the rotary inertia effects. The deformation field w is interpolated by using a four node quadrilateral element. Four degrees of freedom for every node

$$[w_i] = \left[w_i \quad \frac{\partial w_i}{\partial x} \quad \frac{\partial w_i}{\partial y} \quad \frac{\partial w_i}{\partial x \partial y} \right]^{\mathrm{T}} \qquad i = 1, 2, 3, 4 \quad (2.10)$$

are considered. Therefore the displacement variable inside the element can be approximated as[1]

$$w(x, y, t) = \sum_{i=1}^{16} h_i(\bar{x}, \bar{y})\, q_i(t) = \mathbf{N}^{\mathrm{T}} \mathbf{q} \quad (2.11)$$

[1] Shape functions $h_i(\bar{x}, \bar{y})$ are

$h_1 = N_1(\bar{x}) N_1(\bar{y})$	$h_2 = N_2(\bar{x}) N_1(\bar{y})$	$h_3 = N_1(\bar{x}) N_2(\bar{y})$	$h_4 = N_2(\bar{x}) N_2(\bar{y})$
$h_5 = N_3(\bar{x}) N_1(\bar{y})$	$h_6 = N_4(\bar{x}) N_1(\bar{y})$	$h_7 = N_3(\bar{x}) N_2(\bar{y})$	$h_8 = N_4(\bar{x}) N_2(\bar{y})$
$h_9 = N_3(\bar{x}) N_3(\bar{y})$	$h_{10} = N_4(\bar{x}) N_3(\bar{y})$	$h_{11} = N_3(\bar{x}) N_4(\bar{y})$	$h_{12} = N_4(\bar{x}) N_4(\bar{y})$
$h_{13} = N_1(\bar{x}) N_3(\bar{y})$	$h_{14} = N_2(\bar{x}) N_3(\bar{y})$	$h_{15} = N_1(\bar{x}) N_4(\bar{y})$	$h_{16} = N_2(\bar{x}) N_4(\bar{y})$

where

$$\begin{aligned} \mathbf{q}^{\mathrm{T}} &= \begin{bmatrix} q_1 & q_2 & q_3 & q_4 & .. & q_{16} \end{bmatrix} - \\ &= \begin{bmatrix} w_1 & \frac{\partial w_1}{\partial x} & \frac{\partial w_1}{\partial y} & \frac{\partial w_1}{\partial x \partial y} & .. & \frac{\partial w_4}{\partial x \partial y} \end{bmatrix} \end{aligned} \tag{2.12}$$

and $\bar{x}$-$\bar{y}$ is the local coordinate system. The assumption (2.11) conduces to the relation[2]

$$\int_{t_1}^{t_2} (\delta K - \delta \Phi)\, dt = -\int_{t_1}^{t_2} \delta \mathbf{w}^{\mathrm{T}} \left[\mathbf{M}\ddot{\mathbf{w}} + \mathbf{K}_e \mathbf{w}\right] dt = 0 \tag{2.21}$$

and considering a harmonic motion

$$\mathbf{w} = \mathbf{W} e^{i\omega t} \tag{2.22}$$

where

$$\begin{aligned} N_1(\bar{x}) &= \frac{1}{4}\left(1 - \frac{2\bar{x}}{\alpha}\right)^2 \left(2 + \frac{2\bar{x}}{\alpha}\right) \qquad N_2(\bar{x}) = \frac{1}{4}\left(1 - \frac{2\bar{x}}{\alpha}^2\right)\left(1 - \frac{2\bar{x}}{\alpha}\right) \\ N_3(\bar{x}) &= \frac{1}{4}\left(1 + \frac{2\bar{x}}{\alpha}\right)^2 \left(2 - \frac{2\bar{x}}{\alpha}\right) \qquad N_4(\bar{x}) = \frac{1}{4}\left(-1 + \frac{2\bar{x}}{\alpha}\right)^2 \left(1 + \frac{2\bar{x}}{\alpha}\right) \end{aligned}$$

[2] In order to express Hamilton's principle the following derivatives must be evaluated

$$\begin{aligned} w_{,x}(x,y,t) &= \frac{\partial w(x,y,t)}{\partial x} = \frac{\partial}{\partial x}\left[\mathbf{N}^{\mathrm{T}}(\bar{x},\bar{y})\,\mathbf{q}(t)\right] = \mathbf{D}_x^{\mathrm{T}}(\bar{x},\bar{y})\,\mathbf{q}(t) = \mathbf{q}^{\mathrm{T}}(t)\,\mathbf{D}_x(\bar{x},\bar{y}) \\ w_{,y}(x,y,t) &= \frac{\partial w(x,y,t)}{\partial y} = \frac{\partial}{\partial y}\left[\mathbf{N}^{\mathrm{T}}(\bar{x},\bar{y})\,\mathbf{q}(t)\right] = \mathbf{D}_y^{\mathrm{T}}(\bar{x},\bar{y})\,\mathbf{q}(t) = \mathbf{q}^{\mathrm{T}}(t)\,\mathbf{D}_y(\bar{x},\bar{y}) \\ w_{,xx}(x,y,t) &= \frac{\partial^2 w(x,y,t)}{\partial x^2} = \frac{\partial^2}{\partial x^2}\left[\mathbf{N}^{\mathrm{T}}(\bar{x},\bar{y})\,\mathbf{q}(t)\right] = \mathbf{B}_x^{\mathrm{T}}(\bar{x},\bar{y})\,\mathbf{q}(t) = \mathbf{q}^{\mathrm{T}}(t)\,\mathbf{B}_x(\bar{x},\bar{y}) \\ w_{,yy}(x,y,t) &= \frac{\partial^2 w(x,y,t)}{\partial y^2} = \frac{\partial^2}{\partial y^2}\left[\mathbf{N}^{\mathrm{T}}(\bar{x},\bar{y})\,\mathbf{q}(t)\right] = \mathbf{B}_y^{\mathrm{T}}(\bar{x},\bar{y})\,\mathbf{q}(t) = \mathbf{q}^{\mathrm{T}}(t)\,\mathbf{B}_y(\bar{x},\bar{y}) \\ w_{,xy}(x,y,t) &= \frac{\partial^2 w(x,y,t)}{\partial x \partial y} = \frac{\partial^2}{\partial x \partial y}\left[\mathbf{N}^{\mathrm{T}}(\bar{x},\bar{y})\,\mathbf{q}(t)\right] = \mathbf{B}_{xy}^{\mathrm{T}}(\bar{x},\bar{y})\,\mathbf{q}(t) = \mathbf{q}^{\mathrm{T}}(t)\,\mathbf{B}_{xy}(\bar{x},\bar{y}) \\ \ddot{w}(x,y,t) &= \frac{\partial^2 w(x,y,t)}{\partial t^2} = \frac{\partial^2}{\partial t^2}\left[\mathbf{N}^{\mathrm{T}}(\bar{x},\bar{y})\,\mathbf{q}(t)\right] = \mathbf{N}^{\mathrm{T}}(\bar{x},\bar{y})\,\ddot{\mathbf{q}}(t) = \ddot{\mathbf{q}}^{\mathrm{T}}(t)\,\mathbf{N}(\bar{x},\bar{y}) \end{aligned}$$

in which $\ddot{\mathbf{q}}$ is the nodal acceleration vector and $\mathbf{B}_{xy}$, $\mathbf{B}_x$, $\mathbf{B}_y$,$\mathbf{D}_x$, $\mathbf{D}_y$ are the matrices of the derivatives of the shape functions with respect to the global coordinates x, y.

By developing the equations (2.8) and (2.9) for each element

$$\delta K_i = -\delta \mathbf{q}^{\mathrm{T}} \left[\int_{-\frac{\alpha}{2}}^{+\frac{\alpha}{2}} \int_{-\frac{\beta}{2}}^{+\frac{\beta}{2}} \mathbf{D}_x\, I\, \mathbf{D}_x^{\mathrm{T}} + \mathbf{D}_y\, I\, \mathbf{D}_y^{\mathrm{T}} + \mathbf{N}\rho h \mathbf{N}^{\mathrm{T}} dx dy\right] \ddot{\mathbf{q}} \tag{2.14}$$

$$\begin{aligned} \delta \Phi_i = \delta \mathbf{q}^{\mathrm{T}} \Bigg\{ \int_{-\frac{\alpha}{2}}^{+\frac{\alpha}{2}} \int_{-\frac{\beta}{2}}^{+\frac{\beta}{2}} \Big[& D_{11} \mathbf{B}_x \mathbf{B}_x^{\mathrm{T}} + D_{12}\left[\mathbf{B}_x \mathbf{B}_y^{\mathrm{T}} + \mathbf{B}_y \mathbf{B}_x^{\mathrm{T}}\right] + \\ & + D_{22} \mathbf{B}_y \mathbf{B}_y^{\mathrm{T}} + 2 D_{16}\left(\mathbf{B}_x \mathbf{B}_{xy}^{\mathrm{T}} + \mathbf{B}_{xy} \mathbf{B}_x^{\mathrm{T}}\right) + \\ & + 2 D_{26}\left(\mathbf{B}_y \mathbf{B}_{xy}^{\mathrm{T}} + \mathbf{B}_{xy} \mathbf{B}_y^{\mathrm{T}}\right) + 4 D_{66} \mathbf{B}_{xy} \mathbf{B}_{xy}^{\mathrm{T}} \Big] dx dy \Bigg\} \mathbf{q} \end{aligned} \tag{2.15}$$

finally, from (2.21) one can obtain the eigenvalue problem

$$\left(\mathbf{K}_e - \omega^2 \mathbf{M}\right) \mathbf{W} = \mathbf{0}. \tag{2.23}$$

When geometry and mechanical properties are known, the two equation systems (2.23) and (2.5) allow us to calculate the eigenfrequencies ω and the eigenvectors $\mathbf{a}$ of the plate with Rayleigh-Ritz and finite element method.

Consequently the Hamilton's principle for the plate can be written as a sum of terms as follows

$$\int_{t_1}^{t_2} (\delta K - \delta\Phi)_i \, dt =$$

$$= -\int_{t_1}^{t_2} \delta\mathbf{q}^{\mathrm{T}} \left\{ \int_{-\frac{\alpha}{2}}^{+\frac{\alpha}{2}} \int_{-\frac{\beta}{2}}^{+\frac{\beta}{2}} \left[\mathbf{D}_x \, I \, \mathbf{D}_x^{\mathrm{T}} + \mathbf{D}_y \, I \, \mathbf{D}_y^{\mathrm{T}} + \mathbf{N}\rho h \mathbf{N}^{\mathrm{T}} \right] dx dy \ddot{\mathbf{q}} + \right.$$
$$+ \int_{-\frac{\alpha}{2}}^{+\frac{\alpha}{2}} \int_{-\frac{\beta}{2}}^{+\frac{\beta}{2}} \left[D_{11} \mathbf{B}_x \mathbf{B}_x^{\mathrm{T}} + D_{12} \left(\mathbf{B}_x \mathbf{B}_y^{\mathrm{T}} + \mathbf{B}_y \mathbf{B}_x^{\mathrm{T}} \right) + \right.$$
$$+ D_{22} \mathbf{B}_y \mathbf{B}_y^{\mathrm{T}} + 2D_{16} \left(\mathbf{B}_x \mathbf{B}_{xy}^{\mathrm{T}} + \mathbf{B}_{xy} \mathbf{B}_x^{\mathrm{T}} \right) +$$
$$\left. \left. + 2D_{26} \left(\mathbf{B}_y \mathbf{B}_{xy}^{\mathrm{T}} + \mathbf{B}_{xy} \mathbf{B}_y^{\mathrm{T}} \right) + 4D_{66} \mathbf{B}_{xy} \mathbf{B}_{xy}^{\mathrm{T}} \right] dx dy \mathbf{q} \right\} dt. \tag{2.16}$$

By introducing the equalities

$$\mathbf{m} = \int_{-\frac{\alpha}{2}}^{+\frac{\alpha}{2}} \int_{-\frac{\beta}{2}}^{+\frac{\beta}{2}} \left[\mathbf{D}_x \, I \, \mathbf{D}_x^{\mathrm{T}} + \mathbf{D}_y \, I \, \mathbf{D}_y^{\mathrm{T}} + \mathbf{N}\rho h \mathbf{N}^{\mathrm{T}} \right] dx dy \tag{2.17}$$

$$\mathbf{k}_e = \int_{-\frac{\alpha}{2}}^{+\frac{\alpha}{2}} \int_{-\frac{\beta}{2}}^{+\frac{\beta}{2}} \left[D_{11} \mathbf{B}_x \mathbf{B}_x^{\mathrm{T}} + D_{12} \left(\mathbf{B}_x \mathbf{B}_y^{\mathrm{T}} + \mathbf{B}_y \mathbf{B}_x^{\mathrm{T}} \right) + \right.$$
$$+ D_{22} \mathbf{B}_y \mathbf{B}_y^{\mathrm{T}} + 2D_{16} \left(\mathbf{B}_x \mathbf{B}_{xy}^{\mathrm{T}} + \mathbf{B}_{xy} \mathbf{B}_x^{\mathrm{T}} \right) +$$
$$\left. + 2D_{26} \left(\mathbf{B}_y \mathbf{B}_{xy}^{\mathrm{T}} + \mathbf{B}_{xy} \mathbf{B}_y^{\mathrm{T}} \right) + 4D_{66} \mathbf{B}_{xy} \mathbf{B}_{xy}^{\mathrm{T}} \right] dx dy \tag{2.18}$$

in which $\mathbf{m}$ is the element mass matrix and $\mathbf{k}_e$ is the element stiffness matrix, each term is given by

$$\int_{t_1}^{t_2} (\delta K - \delta\Phi)_i \, dt = -\int_{t_1}^{t_2} \delta\mathbf{q}^{\mathrm{T}} \left[\mathbf{m}\ddot{\mathbf{q}} + \mathbf{k}_e \mathbf{q} \right]_i dt. \tag{2.19}$$

By assembling

$$\int_{t_1}^{t_2} (\delta K - \delta\Phi) \, dt = -\int_{t_1}^{t_2} \delta\mathbf{w}^{\mathrm{T}} \left[\mathbf{M}\ddot{\mathbf{w}} + \mathbf{K}_e \mathbf{w} \right] dt \tag{2.20}$$

where

$\mathbf{w}$ is the global displacements vector

$\mathbf{M}$ is the plate mass matrix

$\mathbf{K}_e$ is the global stiffness matrix.

3 Bayesian estimation

The method of Bayesian estimation is employed in order to determine the suitable corrections of the mechanical properties. The present Bayesian procedure implies a minimization of the error function

$$e(\mathbf{r},\mathbf{p}) = (\hat{\mathbf{r}}-\mathbf{r})^{\mathrm{T}}\mathbf{C_R}(\hat{\mathbf{r}}-\mathbf{r}) + (\mathbf{p}-\mathbf{p_O})^{\mathrm{T}}\mathbf{C_P}(\mathbf{p}-\mathbf{p_O}). \tag{3.1}$$

where

$$\begin{aligned}\hat{\mathbf{r}} &= [\lambda_{\exp 1} \quad \lambda_{\exp 2} \quad \dots \quad \lambda_{\exp N}]^{\mathrm{T}} \qquad \mathbf{r} = [\lambda_1 \quad \lambda_2 \quad \dots \quad \lambda_N]^{\mathrm{T}} \\ \mathbf{p} &= [D_1 \quad D_2 \quad D_3 \quad D_4]^{\mathrm{T}} \qquad \mathbf{p_O} = [D_{1,0} \quad D_{2,0} \quad D_{3,0} \quad D_{4,0}]^{\mathrm{T}}\end{aligned} \tag{3.2}$$

In (3.2) $\hat{\mathbf{r}}$, $\mathbf{r}$ are the vectors of measured and predicted eigenvalues. $\mathbf{p}$ and $\mathbf{p_O}$ are the vectors of revised and initial bending stiffnesses of the plate. Weighting matrix[3]

$$\mathbf{C_R} = \begin{bmatrix} \frac{1}{\delta^2(\lambda_{exp1})} & 0 & .. & 0 \\ 0 & \frac{1}{\delta^2(\lambda_{exp2})} & .. & 0 \\ \vdots & \vdots & \vdots & \vdots \\ 0 & 0 & 0 & \frac{1}{\delta^2(\lambda_{expN})} \end{bmatrix}. \tag{3.5}$$

represents the confidence in the measured and predicted eigenvalues. Weighting matrix[4]

$$\mathbf{C_P} = \begin{bmatrix} \frac{1}{\delta^2(D_1)} & 0 & 0 & 0 \\ 0 & \frac{1}{\delta^2(D_2)} & 0 & 0 \\ 0 & 0 & \frac{1}{\delta^2(D_3)} & 0 \\ 0 & 0 & 0 & \frac{1}{\delta^2(D_4)} \end{bmatrix} \tag{3.7}$$

takes into consideration the reliability of the initial property estimates.

[3] Experimental eigenvalues are related to the measured natural frequencies f_{exp} by equation

$$\lambda_{exp} = \rho h \left(2\pi f_{exp}\right)^2 l^2 b^2. \tag{3.3}$$

Its variation leads to the following expression

$$\delta^2\left(\lambda_{exp}\right) = \left(\frac{\partial \lambda_{exp}}{\partial f_{exp}}\right)^2 \delta_f^2 = \left[8\rho h \left(\pi l b\right)^2 f_{exp}\delta_f\right]^2 \tag{3.4}$$

in which δ_f is the deviation of the natural frequency.

[4] The four independent flexural rigidities can be written as

$$D_1 = \frac{E_1 h^3}{12\left(1-\nu_{12}\nu_{21}\right)} \qquad D_3 = \frac{\nu_{12}E_2 h^3}{12\left(1-\nu_{12}\nu_{21}\right)} = \frac{\nu_{21}E_1 h^3}{12\left(1-\nu_{12}\nu_{21}\right)}$$

$$D_2 = \frac{E_2 h^3}{12\left(1-\nu_{12}\nu_{21}\right)} \qquad D_4 = \frac{G_{12}h^3}{12}.$$

A functional relationship between the modal characteristics and the structural parameters is expressible in terms of a Taylor series expansion as follows

$$\mathbf{r} = \tilde{\mathbf{r}} + \mathbf{S}|_{\tilde{\mathbf{p}}} (\mathbf{p} - \tilde{\mathbf{p}}) \tag{3.8}$$

where $\tilde{\mathbf{p}} = \begin{bmatrix} \tilde{D}_1 \; \tilde{D}_2 \; \tilde{D}_3 \; \tilde{D}_4 \end{bmatrix}^{\mathrm{T}}$ is the column matrix of prior mechanical properties. $\mathbf{r}$ and $\tilde{\mathbf{r}}$ (see 3.2) are the column matrices of the predicted and prior responses. The sensitive matrix

$$\mathbf{S}|_{\tilde{\mathbf{p}}} = \begin{bmatrix} \frac{\partial \lambda_1}{\partial D_1}|_{\tilde{\mathbf{p}}} & \frac{\partial \lambda_1}{\partial D_2}|_{\tilde{\mathbf{p}}} & \frac{\partial \lambda_1}{\partial D_3}|_{\tilde{\mathbf{p}}} & \frac{\partial \lambda_1}{\partial D_4}|_{\tilde{\mathbf{p}}} \\ \frac{\partial \lambda_2}{\partial D_1}|_{\tilde{\mathbf{p}}} & \frac{\partial \lambda_2}{\partial D_2}|_{\tilde{\mathbf{p}}} & \frac{\partial \lambda_2}{\partial D_3}|_{\tilde{\mathbf{p}}} & \frac{\partial \lambda_2}{\partial D_4}|_{\tilde{\mathbf{p}}} \\ \vdots & \vdots & \vdots & \vdots \\ \frac{\partial \lambda_N}{\partial D_1}|_{\tilde{\mathbf{p}}} & \frac{\partial \lambda_N}{\partial D_2}|_{\tilde{\mathbf{p}}} & \frac{\partial \lambda_N}{\partial D_3}|_{\tilde{\mathbf{p}}} & \frac{\partial \lambda_N}{\partial D_4}|_{\tilde{\mathbf{p}}} \end{bmatrix} \tag{3.9}$$

is the eigenvalue partial derivative matrix evaluated for the prior bending stiffnesses $\tilde{\mathbf{p}}$. The elements in (3.9) can be evaluated as (see Viola, 2001, p. 430)

$$\frac{\partial \lambda_{ik}}{\partial D_j} = \frac{1}{m_{ik}^*} \mathbf{a}_{ik}^{\mathrm{T}} \frac{\partial \mathbf{C}}{\partial D_j} \mathbf{a}_{ik} \tag{3.10}$$

where

$$m_{ik}^* = \mathbf{a}_{ik}^{\mathrm{T}} \mathbf{M} \mathbf{a}_{ik}$$

is the modal mass of the mode shape ik.

Using the above Rayleigh-Ritz method, $m_{ik}^* = 1$ because mass matrix $\mathbf{M}$ is an identity matrix. Analytical eigenvectors $\mathbf{a}_{ik}$ and eigenvalues λ_{ik} can be evaluated by solving the

---

The deviations of the stiffnesses D_1, D_2, D_3, D_4 can be expressed in terms of the deviations of elastic constants E_1, E_2, G_{12}, ν_{12} (Lai and Ip, 1996)

$$\delta(D_1) = \frac{\delta(E_1)\, h^3}{12(1-\nu_{12}\nu_{21})} \qquad \delta(D_2) = \frac{\delta(E_2)\, h^3}{12(1-\nu_{12}\nu_{21})} \qquad \delta(D_4) = \frac{h^3}{12}\, \delta(G_{12})$$

$$\begin{aligned} \delta(D_3) &= \sqrt{\left(\frac{\partial D_3}{\partial \nu_{12}}\right)^2 \delta^2(\nu_{12}) + \left(\frac{\partial D_3}{\partial E_2}\right)^2 \delta^2(E_2)} = \\ &= \frac{h^3}{12(1-\nu_{12}\nu_{21})} \sqrt{\left[\frac{\partial(\nu_{12}E_2)}{\partial \nu_{12}}\right]^2 \delta^2(\nu_{12}) + \left[\frac{\partial(\nu_{12}E_2)}{\partial E_2}\right]^2 \delta^2(E_2)} = \\ &= \frac{h^3}{12(1-\nu_{12}\nu_{21})} \sqrt{[E_2 \delta(\nu_{12})]^2 + [\nu_{12}\delta(E_2)]^2} \end{aligned} \tag{3.6}$$

Deviation of elastic constants are evaluated by relations (Lai and Ip, 1996)

$$\delta(E_1) = 12\frac{1-\nu_{12}\nu_{21}}{h^3}\delta(D_1) \qquad \delta(E_2) = 12\frac{(1-\nu_{12}\nu_{21})}{h^3}\delta(D_2)$$

$$\delta(G_{12}) = \frac{12}{h^3}\delta(D_4) \qquad \delta(\nu_{12}) = \sqrt{\left[\frac{\delta(D_3)}{D_2}\right]^2 - \left[\nu_{12}\frac{\delta(D_2)}{D_2}\right]^2}$$

eigenvalue problem (2.5) and

$$\frac{\partial \mathbf{C}}{\partial D_j} = \begin{bmatrix} \frac{\partial C_{11,11}}{\partial D_j} & \frac{\partial C_{11,12}}{\partial D_j} & .. & \frac{\partial C_{11,1q}}{\partial D_j} & \frac{\partial C_{11,21}}{\partial D_j} & .. & \frac{\partial C_{11,pq}}{\partial D_j} \\ \frac{\partial C_{12,11}}{\partial D_j} & \frac{\partial C_{12,12}}{\partial D_j} & .. & \frac{\partial C_{12,1q}}{\partial D_j} & \frac{\partial C_{12,21}}{\partial D_j} & .. & \frac{\partial C_{12,pq}}{\partial D_j} \\ \vdots & \vdots & & \vdots & \vdots & & \vdots \\ \frac{\partial C_{1q,11}}{\partial D_j} & \frac{\partial C_{1q,12}}{\partial D_j} & .. & \frac{\partial C_{1q,1q}}{\partial D_j} & \frac{\partial C_{1q,21}}{\partial D_j} & .. & \frac{\partial C_{1q,pq}}{\partial D_j} \\ \frac{\partial C_{21,11}}{\partial D_j} & \frac{\partial C_{21,12}}{\partial D_j} & .. & \frac{\partial C_{21,1q}}{\partial D_j} & \frac{\partial C_{21,21}}{\partial D_j} & .. & \frac{\partial C_{21,pq}}{\partial D_j} \\ \vdots & \vdots & & \vdots & \vdots & & \vdots \\ \frac{\partial C_{pq,11}}{\partial D_j} & \frac{\partial C_{pq,12}}{\partial D_j} & .. & \frac{\partial C_{pq,1q}}{\partial D_j} & \frac{\partial C_{pq,21}}{\partial D_j} & .. & \frac{\partial C_{pq,pq}}{\partial D_j} \end{bmatrix} \tag{3.11}$$

From equations (2.6) then

$$\begin{bmatrix} \frac{\partial C_{ik,mn}}{\partial D_1} \\ \frac{\partial C_{ik,mn}}{\partial D_2} \\ \frac{\partial C_{ik,mn}}{\partial D_3} \\ \frac{\partial C_{ik,mn}}{\partial D_4} \end{bmatrix} = \begin{bmatrix} m^4 & n^4 & 2m^2n^2 & 4m^2n^2 \\ n^4 & m^4 & 2m^2n^2 & 4m^2n^2 \\ m^2n^2 & m^2n^2 & m^4+n^4 & -4m^2n^2 \\ m^2n^2 & m^2n^2 & -2m^2n^2 & (m^2-n^2)^2 \\ m^3n & -mn^3 & mn^3-m^3n & 2mn^3-2m^3n \\ mn^3 & -m^3n & m^3n-mn^3 & 2m^3n-2mn^3 \end{bmatrix}^{\mathrm{T}} \begin{bmatrix} J_{1,ik,mn} \\ J_{2,ik,mn} \\ J_{3,ik,mn} \\ J_{4,ik,mn} \\ J_{5,ik,mn} \\ J_{6,ik,mn} \end{bmatrix}. \tag{3.12}$$

Using the finite element method, the different eigenvalue problem (2.23) must be solved and it leads to a different form of relation (3.10)

$$\frac{\partial \lambda_i}{\partial D_j} = \frac{1}{m_i^*}\mathbf{a}_i^{\mathrm{T}}\left(\frac{\partial \mathbf{K}}{\partial D_j}\right)\mathbf{a}_i. \tag{3.13}$$

From (2.18) one can see that the element stiffness matrix can be written as sum of six matrices

$$\begin{aligned} \mathbf{k}_e &= \mathbf{k}_{e_{11}} D_{11}(D_1,..D_4) + \mathbf{k}_{e_{12}} D_{12}(D_1,..D_4) + \mathbf{k}_{e_{22}} D_{22}(D_1,..D_4) + \\ &\quad + \mathbf{k}_{e_{16}} D_{16}(D_1,..D_4) + \mathbf{k}_{e_{26}} D_{26}(D_1,..D_4) + \mathbf{k}_{e_{66}} D_{66}(D_1,..D_4). \end{aligned} \tag{3.14}$$

For example, by derivation of $\mathbf{k}_e$ with respect to the first component of $\mathbf{p}$ it follows that

$$\begin{aligned} \mathbf{k}_{e,D_1} &= \mathbf{k}_{e_{11}} m^4 + \mathbf{k}_{e_{12}} m^2n^2 + \mathbf{k}_{e_{22}} n^4 + \mathbf{k}_{e_{16}} m^3n + \mathbf{k}_{e_{26}} mn^3 + \mathbf{k}_{e_{66}} m^2n^2 = \\ &= \int_{-\frac{\alpha}{2}}^{+\frac{\alpha}{2}} \int_{-\frac{\beta}{2}}^{+\frac{\beta}{2}} \left\{ \mathbf{B}_x\mathbf{B}_x^{\mathrm{T}} m^4 + \left(\mathbf{B}_x\mathbf{B}_y^{\mathrm{T}} + \mathbf{B}_y\mathbf{B}_x^{\mathrm{T}}\right) m^2n^2 + \right. \\ &\quad + \mathbf{B}_y\mathbf{B}_y^{\mathrm{T}} n^4 + 2\left(\mathbf{B}_x\mathbf{B}_{xy}^{\mathrm{T}} + \mathbf{B}_{xy}\mathbf{B}_x^{\mathrm{T}}\right) m^3n + \\ &\quad \left. + 2\left(\mathbf{B}_y\mathbf{B}_{xy}^{\mathrm{T}} + \mathbf{B}_{xy}\mathbf{B}_y^{\mathrm{T}}\right) mn^3 + 4\mathbf{B}_{xy}\mathbf{B}_{xy}^{\mathrm{T}} m^2n^2 \right\} dxdy. \end{aligned} \tag{3.15}$$

In the same manner $\mathbf{k}_{e,D_2}$ $\mathbf{k}_{e,D_3}$ $\mathbf{k}_{e,D_4}$ are evaluated. By assembling these matrices with the usual procedure, four global matrices $\mathbf{K}_{,D_j}$ can be obtained. Introducing these matrices into (3.13), the sensitivity matrix is finally calculated.

When $\mathbf{r}$ in equation (3.1), is substituted by equation (3.8), the error function depends solely on $\mathbf{p}$. The minimization of this function with respect to $\mathbf{r}$ leads to the following expression for the evaluation of the revised estimates[5]

$$\mathbf{p} = \mathbf{p_O} + \mathbf{C_P^{*-1}} \mathbf{S}|_{\tilde{\mathbf{p}}}^{\mathrm{T}} \mathbf{C_R} (\hat{\mathbf{r}} - \tilde{\mathbf{r}} - \mathbf{S}|_{\tilde{\mathbf{p}}} (\mathbf{p_O} - \tilde{\mathbf{p}})) \tag{3.20}$$

where $\mathbf{C_P^*} = \mathbf{C_P} + \mathbf{S}|_{\tilde{\mathbf{p}}}^{\mathrm{T}} \mathbf{C_R} \mathbf{S}|_{\tilde{\mathbf{p}}}$ is the revised parameters confidence matrix. The diagonal elements of its inverse gives the deviations of the revised parameters $D_1..D_4$ at every step of the iterative procedure.

4 Determination of elastic constants with Rayleigh-Ritz model

In order to test the estimation procedure, the modal characteristics of several orthotropic plates discussed in the literature were considered.

4.1 Generally orthotropic plates

The physical properties of three thin glass-epoxy plates discussed by Lai and Ip (1996) are written below.

- Plate 1: $a = 0.2003m$ $b = 0.2003m$ $h = 9.423mm$ $\rho = 2.77g/cm^3$ $\theta = 60°$
- Plate 2: $a = 0.3490m$ $b = 0.3000m$ $h = 7.946mm$ $\rho = 2.77g/cm^3$ $\theta = 60°$
- Plate 3: $a = 0.3000m$ $b = 0.1980m$ $h = 7.946mm$ $\rho = 2.77g/cm^3$ $\theta = 30°$

In this work the first plate is considered. The experimental natural frequencies and deviations δ_f are shown in Table 1. Analytical frequencies are evaluated by introducing in the Rayleigh-Ritz model the initial esteems of the mechanical properties shown in Table

[5] When (3.8) is substituted in (3.1), by rearranging this error function, the only terms that depend on $\mathbf{p}$ are summarized as

$$\tilde{e}(\mathbf{p}) = -2\mathbf{p}^{\mathrm{T}}\mathbf{S}^{\mathrm{T}}\mathbf{C_R}\hat{\mathbf{r}} + 2\mathbf{p}^{\mathrm{T}}\mathbf{S}^{\mathrm{T}}\mathbf{C_R}\tilde{\mathbf{r}} + \mathbf{p}^{\mathrm{T}}\mathbf{S}^{\mathrm{T}}\mathbf{C_R}\mathbf{S}\mathbf{p} - 2\mathbf{p}^{\mathrm{T}}\mathbf{S}^{\mathrm{T}}\mathbf{C_R}\mathbf{S}\tilde{\mathbf{p}} + \mathbf{p}^{\mathrm{T}}\mathbf{C_P}\mathbf{p} - 2\mathbf{p}^{\mathrm{T}}\mathbf{C_P}\mathbf{p_O}. \tag{3.16}$$

The minimization of (3.16) and consequently the minimization of the error function can be obtained as

$$\frac{\partial \tilde{e}(\mathbf{p})}{\partial \mathbf{p}} = \frac{\partial e(\mathbf{p})}{\partial \mathbf{p}} = 2\left(-\mathbf{S}^{\mathrm{T}}\mathbf{C_R}\hat{\mathbf{r}} + \mathbf{S}^{\mathrm{T}}\mathbf{C_R}\tilde{\mathbf{r}} + \mathbf{S}^{\mathrm{T}}\mathbf{C_R}\mathbf{S}\mathbf{p} - \mathbf{S}^{\mathrm{T}}\mathbf{C_R}\mathbf{S}\tilde{\mathbf{p}} + \mathbf{C_P}\mathbf{p} - \mathbf{C_P}\mathbf{p_O}\right) = \mathbf{0}. \tag{3.17}$$

Further developments lead to the equality

$$\left(\mathbf{C_P} + \mathbf{S}^{\mathrm{T}}\mathbf{C_R}\mathbf{S}\right)\mathbf{p} = \left(\mathbf{C_P} + \mathbf{S}^{\mathrm{T}}\mathbf{C_R}\mathbf{S}\right)\mathbf{p_O} - \mathbf{S}^{\mathrm{T}}\mathbf{C_R}\tilde{\mathbf{r}} + \mathbf{S}^{\mathrm{T}}\mathbf{C_R}\hat{\mathbf{r}} - \mathbf{S}^{\mathrm{T}}\mathbf{C_R}\mathbf{S}\mathbf{p_O} + \mathbf{S}^{\mathrm{T}}\mathbf{C_R}\mathbf{S}\tilde{\mathbf{p}}. \tag{3.18}$$

Finally the expression of the following estimator is obtained

$$\begin{aligned}\mathbf{p} = \mathbf{p_O} &+ \left(\mathbf{C_P} + \mathbf{S}^{\mathrm{T}}\mathbf{C_R}\mathbf{S}\right)^{-1}\mathbf{S}^{\mathrm{T}}\mathbf{C_R}(\hat{\mathbf{r}} - \tilde{\mathbf{r}}) - \left(\mathbf{C_P} + \mathbf{S}^{\mathrm{T}}\mathbf{C_R}\mathbf{S}\right)^{-1}\mathbf{S}^{\mathrm{T}}\mathbf{C_R}\mathbf{S}(\mathbf{p_O} - \tilde{\mathbf{p}}) \\ &= \mathbf{p_O} + \mathbf{C_P^*}^{-1}\mathbf{S}^{\mathrm{T}}\mathbf{C_R}(\hat{\mathbf{r}} - \tilde{\mathbf{r}}) - \mathbf{C_P^*}^{-1}\mathbf{S}^{\mathrm{T}}\mathbf{C_R}\mathbf{S}(\mathbf{p_O} - \tilde{\mathbf{p}}) = \\ &= \mathbf{p_O} + (\mathbf{C_P^*})^{-1}\mathbf{S}^{\mathrm{T}}\,\mathbf{C_R}\left[(\hat{\mathbf{r}} - \tilde{\mathbf{r}}) - \mathbf{S}(\mathbf{p_O} - \tilde{\mathbf{p}})\right].\end{aligned} \tag{3.19}$$

2. The deviations are the difference between the frequencies obtained by the Rayleigh-Ritz method and an accurate finite element method (see Lai and Ip, 1996). Under this assumption, deviations of frequencies δ_f depend only on modelling error. Generally this assumption is not correct because one should also consider the measurement errors of the natural frequencies. In the paper by Lai and Ip (1996) this error was considered negligible. Initial esteems of the elastic constants and their deviations shown in Table 2 were obtained by adopting the rule of mixture (see Jones, 1975, and Lai and Ip, 1996). By relations (3.6) – (3.7) the weighting matrix of mechanical properties can be evaluated. All possible combinations utilizing four and more experimental frequencies are used in the inverse identification method. For example, as shown in Table 2, there are seven possible combinations when six modes are chosen. For every set of experimental frequencies the procedure converges. Introducing deviations δ_f in the expressions (3.4) – (3.5) the confidence matrix of vibration frequencies can be obtained.

At each step of the iterative procedure, the sensitivity matrix must be calculated introducing prior properties estimates $\tilde{\mathbf{p}}$ (at first step $\tilde{\mathbf{p}} = \mathbf{p_0}$) by equations (3.9) – (3.11). By solving the eigenvalue problem (2.5), analytical eigenvectors and eigenfrequencies are known. Eigenfrequencies are used to form the revised vector $\mathbf{r}$ (3.2).

Table 1. Natural frequencies of Plate 1

	Natural frequencies (Hz)			
Mode	Experimental[a]	F.E.M.[a]	Rayleigh-Ritz[b]	δ_f (Hz)
1 (1,1)	301.3	273.5	279.3	5.8
2 (2,0)	405.0	363.7	369.2	5.5
3 (0,2)	615.9	597.6	605.9	8.3
4 (2,1)	717.9	652.6	656.3	3.7
5 (1,2)	913.4	829.1	850.8	21.7
6 (3,0)	1291.4	1142.7	1152.5	9.8
7 (1,1)	1319.8	1259.8	1272.9	13.1

[a]Lai and Ip (1996) [b]Present work

Table 2. Estimated elastic constants of Plate 1.

	$E_1(GPa)$	$E_2(GPa)$	$G_{12}(GPa)$	v_{12}
Initial esteems [a]	10.05 {1.35}	3.25 {0.91}	1.21 {0.34}	0.331 {0.007}
Static values [a]	9.49	4.21	1.25	0.284
All modes [b]	9.58 {0.25}	3.71 {0.041}	1.73 {0.046}	0.405 {0.024}
6 out of 7 [b] (7/7)	9.84	3.66	1.76	0.382
5 out of 7 [b] (21/21)	10.06	3.66	1.76	0.361
4 out of 7 [b] (35/35)	10.23	3.72	1.71	0.338

[a]Lai and Ip (1996) [b]Present work

Eigenvectors are employed in equation (3.10) to calculate the sensitivity matrix elements.

Finally by equation (3.20), a new set of elastic constants $\mathbf{p}$ is obtained. The iteration scheme generally takes less than ten iteration steps for convergence.

Values in the braces in Table 2 (see second row) represent the initial deviations for each elastic constant. The final deviations, which are obtained when seven frequencies are employed in iterative procedure, are shown in the braces in the fourth row. These values are evaluated by diagonal elements of matrix $\mathbf{C_P^*} = \mathbf{C_P} + \mathbf{S}|_{\tilde{\mathbf{p}}}^{T} \mathbf{C_R} \mathbf{S}|_{\tilde{\mathbf{p}}}$ and consequently they depend on sensitivity matrix $\mathbf{S}$. A significant reduction of the deviation for a parameter means that high sensitivities are associated with the chosen modes and the characterization of the constants is accurate. Generally estimation of Poisson's ratio denotes great difficulties and this problem is due to the low sensitivities that flexural modes show with respect this parameter.

In this estimation procedure, identification values depend on initial esteems of the elastic constants and associated confidence to mechanical and vibration characteristics. In this work, some considerations about this question are reported. In order to study the influence of all uncertainties in the identification, several simulations were produced with different values of confidence for the initial elastic constants and experimental frequencies.

4.2 Sensitivity analysis

Let's consider an orthotropic plate whose physical characteristics are described as follow (Larsson, 1997):

plate OSB2: $l = 2.44m$ $b = 1.22m$ $h = 10mm$ $\rho = 659kg/m^3$. The mechanical properties of the plate are evaluated for different initial esteems of elastic constants and different confidence for this quantities. In Figure 3 are illustrated the values of elastic constants obtained. Different initial values were considered. $\bar{\mu}$ represents the values proposed by Larsson (1997), which are

$$E_1 = 7.14GPa \qquad E_2 = 3.50GPa \qquad G_{12} = 1.97GPa \qquad v_{12} = 0.31.$$

For example, the curves named μ_1 show the estimates of elastic constants when initial values of the properties equal $\mu_1 = \bar{\mu}/1.5$. A different choice for the set of initial estimates produces different final values for all four parameters. Nevertheless by assuming high variation coefficient δ/μ, one can observe a decreasing influence of a different choice for initial values. The modal characteristics shown in Table 3 were identified by Larsson (1997). No statistical data are known for the experimental frequencies. Here the uncertainties of experimental data are considered by a constant variation coefficient $C_v = \delta f_i/\mu f_{\text{exp},i} = 0.025$ for each experimental frequency (see Figure 3). δf_i is the deviation of ith frequency and $\mu f_{exp,i}$ is ith frequency. In this manner the confidence associated to experimental data is established. Numerous numerical simulations with different variation coefficients for experimental data were performed. Generally it was observed that a decrease of this variation coefficient $C_v = \delta f_i/\mu f_{exp,i}$ is associated with an increase of the weight of experimental data and with a consequent decrease of the weight of initial property estimates in the error function (3.1). For this reason, estimates gradually become independent from the uncertainties of the initial values of mechanical parameters. Figure 3 also shows that when correct initial estimates of mechanical prop-

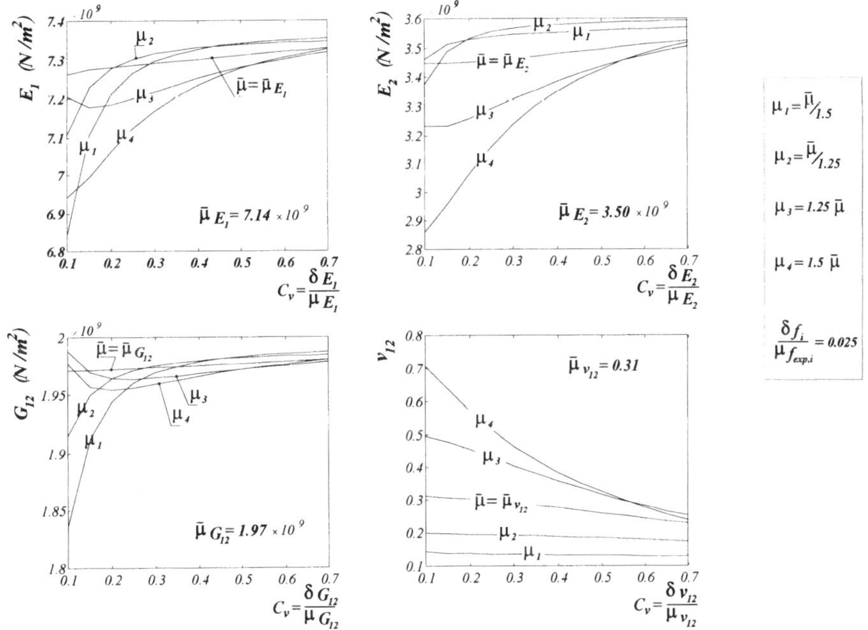

Figure 3. Estimates for various variation coefficients and different initial property values.

erties are introduced in the procedure (see curves named $\bar{\mu}$), final estimates are somewhat independent from the confidence associated to all the parameters.

5 Determination of elastic constants by finite element method

By using finite element modelling the elastic constants of plate OSB2 have again been identified. In this procedure, the evaluation of the sensitivity matrix is obtained by relations (3.11) – (3.15). Different values for variation coefficients of frequencies and initial property estimates have been used. Table 3 illustrates the experimental frequencies identified by Larsson (1997). Figure 4 shows convergence of the iteration method. In Table 4 summarizes the esteems of elastic constants obtained with a variation coefficient of experimental data $C_v = \delta f_i / \mu f_{exp,i} = 0.02$. It can be observed from the above table that if a smaller number of frequencies is introduced, the procedure shows some drawbacks, especially in the evaluation of Poisson's ratio.

Table 3. Experimental frequencies of OSB2 plate.

Mode	1	2	3	4	5	6	12
Frequency [a] (Hz)	5.57	5.63	13.4	15.6	16.7	20.7	46.30

[a]Larsson (1997)

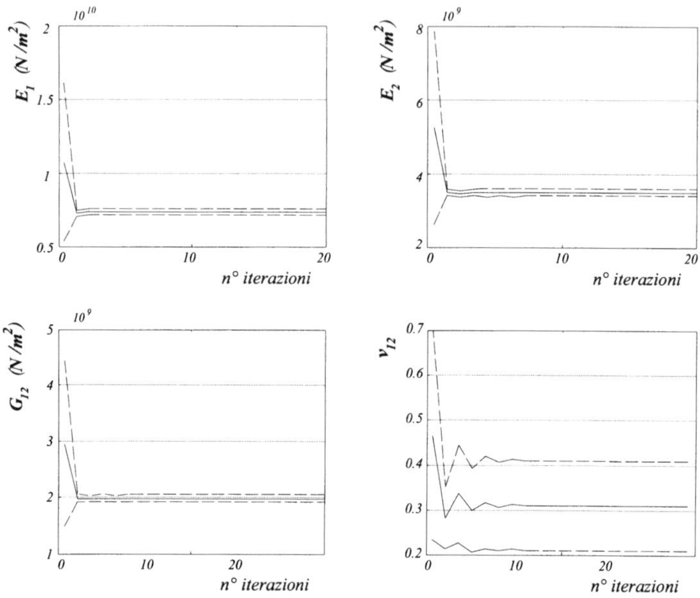

Figure 4. Convergence of elastic constants for OSB2 plate.

Table 4. Estimated elastic constants of OSB2 plate.

	$E_1(GPa)$		$E_2(GPa)$		$G_{12}(GPa)$		v_{12}	
Initial esteems [b]	10.7	{5.35}	5.25	{2.62}	2.95	{1.45}	0.46	{0.23}
Reference values [a]	7.14		3.5		1.97		0.31	
All modes [b]	7.33	{0.20}	3.47	{0.091}	1.98	{0.064}	0.31	{0.099}
6 out of 7 [b]	7.33		3.44		1.99		0.29	
5 out of 7 [b]	7.17		2.94		1.96		0.77	

[a]Larsson (1997) [b]Present work

6 Conclusions

In this paper, several numerical simulations to identify the mechanical parameters of generally orthotropic thin plates are illustrated. The inverse identification procedure is based on a Bayesian analysis using the Rayleigh-Ritz and the finite element method. In the Bayesian approach, an error function is used in order to include confidence in the initial property estimates, the uncertainties of the experimental measurements and approximations of the analitycal model. The input data for the inverse analysis are experimental natural frequencies obtained by vibration tests of free boundary plates. The procedure minimizes the above error function and a set of final esteems is obtained with a small number of iteration steps. In this work, a small part of the numerical simulations carried out has been reported. Good results have been obtained for all the plates considered. Therefore, this estimation procedure can be an alternative method for evaluation of mechanical properties and can be employed also for other systems and probably for thick composite plates.

Bibliography

I. Bartoli. Identificazione dinamica dei parametri di piastre ortotrope. *Tesi di laurea (Relatore Prof. E. Viola)*, Facoltà di Ingegneria, Università di Bologna A.A. 2000/2001.

I. Bartoli, A. di Leo, E. Viola. Identificazione delle costanti elastiche attraverso l'approccio Bayesiano per materiali generalmente ortotropi. *Nota Tecnica 83, Anno 2002*, D.I.S.T.A.R.T., Facoltà di Ingegneria, Università di Bologna.

G. Caldersmith, T.D. Rossing. Determination of modal coupling in vibrating rectangular plates. *Applied Acoustics*, 17:33–44, 1984.

L.R. Deobald, R.F. Gibson. Determination of elastic constants of orthotropic plates by a modal analysis/Rayleigh Ritz technique. *Journal of Sound and Vibration*, 124(2): 269–283, 1988.

R.M. Jones. Mechanics of Composite Materials, Scripta Book Company, Washington D.C., 1975.

P. Laarson, O. Laarson. Determination of Young's and shear moduli from flexural vibrations of beams. *Journal of Sound and Vibration*, 146(1):111–123, 1991.

D. Larsson. Using modal analysis for estimation of anisotropic material constants. *Journal of Engineering Mechanics*, 3(123):222–229, 1997.

T.C. Lai, K.H. Ip. Parameter estimation of orthotropic plates by Bayesian sensitivity analysis. *Composite Stuctures*, 34:29–42, 1996.

T.C. Lai, T.C. Lau. Determination of elastic constants of a generally orthotropic plate by modal analysis. *The international Journal of Analitical and Experimental Modal Analysis*, 1(8):15–33, 1993.

M.E. McIntyre, J. Woodhouse. The influence of geometry on linear damping. *Acustica*, 39(4):209–224, 1988.

M.E. McIntyre, J. Woodhouse. On measuring the elastic and damping constants of orthotropic sheet materials. *Acta Metallurgica*, 6(36):1397–1416, 1988.

F. Moussu, M. Nivoit. Determination of elastic constants of orthotropic plates by a modal analysis/method of superposition. *Journal of Sound and Vibration*, 165(1):149–163, 1993.

J.N. Reddy. Energy and Variational Methods in Applied Mechanics, Wiley Interscience, 1984.

E. Viola. Fondamenti di dinamica e vibrazione delle strutture, Volume II, Pitagora Editrice, 2001.

E. Viola. Scienza delle costruzioni, Volume I, Teoria dell'elasticità, Pitagora Editrice, 2000.

Problems concerning the elastic constants determination of isotropic and orthotropic plates.

Ivan Bartoli [‡] and Erasmo Viola[‡]

[‡] Faculty of Engineering, DISTART Department, University of Bologna, Italy

Abstract In this paper a procedure for estimating the four in-plane elastic constants from measured natural frequencies of a completely free plate is described. The parameter estimation problem is solved by a least square approach. The numerical model for calculating the natural frequencies comprises a Rayleigh-Ritz formulation. The performance of the identification approach is illustrated by numerical simulations. The results of the procedure of identification are also compared with the esteems obtained in several articles for various examples of isotropic and orthotropic plates.

1 Introduction

The determination of stiffness parameters for composite materials is much more complicated than for isotropic materials. This is not only due to the high number of independent parameters but also because the anisotropy of these materials produces specific problems when traditional tests are employed in an attempt to evaluate the desired quantities in a direct way. Boundary effects, sample size dependencies and difficulties in obtaining homogeneous stress-strain fields are some of the more serious problems. As an alternative to tensile testing, mixed numerical experimental techniques have recieved increasing attention.

In this paper a method for estimating elastic constants of isotropic and orthotropic plates is described. The material parameters are deduced from an overdetermined number of natural frequencies of completely free plates by a least square approach. The classical lamination theory is employed in the vibration model, therefore only the four in-plane elastic constants can be determined. The Rayleigh Ritz technique is chosen to model the dynamic behaviour of rectangular plates.

2 Governing equations

The small deflection theory of bending of thin plates is known as the classical lamination theory and is based on assunptions similar to those used in the Euler-Bernoulli theory (or Classical beam theory). Under these hypotheses one can introduce the following displacement field (see Viola, 2001, p. 220, and Reddy, 1984, p. 310)

$$u = -\frac{\partial w}{\partial x} z, \qquad v = -\frac{\partial w}{\partial y} z, \qquad w = w(x, y, t) \tag{2.1}$$

where u, v are the displacements along the $x-$ and $y-$ directions of the reference system $oxyz$ (Figure 1). The xy is assumed to coincide with the middle plane of the plate before deflection. The z axis is perpendicular to the middle plane. The transverse deflection w is small in comparison with the plate thickeness h and does not vary over the thickness itself. Using assumption (2.1) the components of the strain tensor are

$$\begin{aligned}
\varepsilon_x &= \frac{\partial u}{\partial x} = -z\frac{\partial^2 w}{\partial x^2} = z\chi_x \\
\varepsilon_y &= \frac{\partial v}{\partial y} = -z\frac{\partial^2 w}{\partial y^2} = z\chi_y \\
\varepsilon_z &= \frac{\partial w}{\partial z} = 0 \\
\gamma_{xy} &= \frac{\partial u}{\partial y} + \frac{\partial v}{\partial x} = -2z\frac{\partial^2 w}{\partial x \partial y} = z\chi_{xy} \\
\gamma_{xz} &= \frac{\partial u}{\partial z} + \frac{\partial w}{\partial x} = -\frac{\partial w}{\partial x} + \frac{\partial w}{\partial x} = 0 \\
\gamma_{yz} &= \frac{\partial v}{\partial z} + \frac{\partial w}{\partial y} = -\frac{\partial w}{\partial y} + \frac{\partial w}{\partial y} = 0
\end{aligned} \tag{2.2}$$

where

$$\chi_x = -\frac{\partial^2 w}{\partial x^2} \qquad \chi_y = -\frac{\partial^2 w}{\partial y^2} \qquad \chi_{xy} = -2\frac{\partial^2 w}{\partial x \partial y}. \tag{2.3}$$

Eqs (2.3) can be written in the following symbolic form

$$\boldsymbol{\chi} = \mathbf{D}w \tag{2.4}$$

where $\boldsymbol{\chi}$ is the curvatures vector. The strain components in matrix notation become

$$\boldsymbol{\varepsilon} = z\boldsymbol{\chi} = z\mathbf{D}w \tag{2.5}$$

To derive the equilibrium equations of the classical plate theory, it is necessary to introduce the following inplane stress resultants

$$\begin{bmatrix} N_x \\ N_y \\ N_{xy} \end{bmatrix} = \int_{-h/2}^{h/2} \begin{bmatrix} \sigma_x \\ \sigma_y \\ \tau_{xy} \end{bmatrix} dz \quad \text{or synthetically} \quad \mathbf{N} = \int_{-h/2}^{h/2} \boldsymbol{\sigma} dz \tag{2.6}$$

bending and twisting moment resultants

$$\begin{bmatrix} M_x \\ M_y \\ M_{xy} \end{bmatrix} = \int_{-h/2}^{h/2} z \begin{bmatrix} \sigma_x \\ \sigma_y \\ \tau_{xy} \end{bmatrix} dz \quad \text{or synthetically} \quad \mathbf{M} = \int_{-h/2}^{h/2} z\boldsymbol{\sigma} dz \tag{2.7}$$

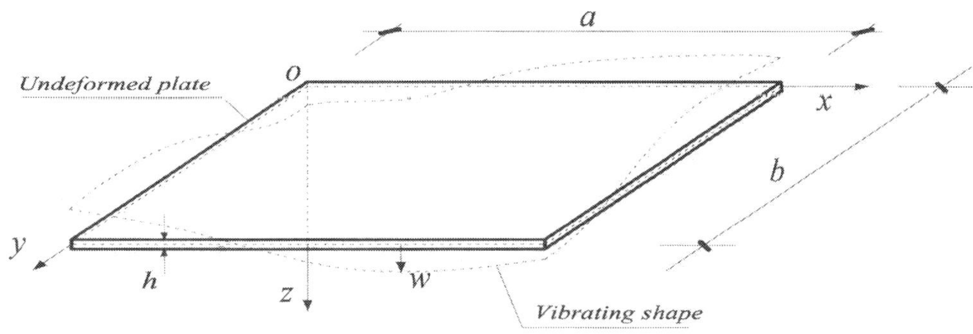

Figure 1. Transverse vibration of plate.

and shear force resultants

$$\begin{bmatrix} T_x \\ T_y \end{bmatrix} = \int_{-h/2}^{h/2} \begin{bmatrix} \tau_{xz} \\ \tau_{yz} \end{bmatrix} dz \quad \text{or synthetically} \qquad \mathbf{T} = \int_{-h/2}^{h/2} \boldsymbol{\tau} dz \tag{2.8}$$

The second order differential equation of the free motion in the $z-$ direction is

$$\frac{\partial^2 M_x}{\partial x^2} + 2\frac{\partial^2 M_{xy}}{\partial x \partial y} + \frac{\partial^2 M_y}{\partial y^2} = \rho h \frac{\partial^2 w}{\partial t^2} \quad \text{or synthetically} \qquad \mathbf{D}^* \mathbf{M} = \rho h \tfrac{\partial^2 w}{\partial t^2} \tag{2.9}$$

in which μ is the density of plate material and

$$f_I - -\rho h \ddot{w}(x, y, t) - -\rho h \tfrac{\partial^2 w}{\partial t^2} \tag{2.10}$$

is the term of the inertia force.

To express the flexural equation of motion (2.9) in terms of displacement $w = w(x, y, t)$, the constitutive relations have to be employed. For a specially orthotropic plate, the moments vector $\mathbf{M} = [M_x \ \ M_y \ \ M_{xy}]^{\mathrm{T}}$ only depends on the bending curvatures vector $\boldsymbol{\chi} = [\chi_x \ \ \chi_y \ \ \chi_{xy}]^{\mathrm{T}}$ (Jones, 1975)

$$\mathbf{M} = \mathbf{C_f} \boldsymbol{\chi} = \mathbf{C_f} \mathbf{D} w \tag{2.11}$$

In (2.11) $\mathbf{C_f}$ is the bending stiffnesses matrix

$$\mathbf{C_f} = \begin{bmatrix} D_{11} & D_{12} & 0 \\ D_{12} & D_{22} & 0 \\ 0 & 0 & D_{66} \end{bmatrix} = \frac{h^3}{12} \begin{bmatrix} \frac{E_x}{\mu} & \nu_{xy}\frac{E_x}{\mu} & 0 \\ \nu_{yx}\frac{E_y}{\mu} & \frac{E_y}{\mu} & 0 \\ 0 & 0 & G_{xy} \end{bmatrix} \tag{2.12}$$

Substitution of (2.11) into the right end of (2.9) gives the fundamental equation

$$\mathbf{D}^* \mathbf{C_f} \mathbf{D} w = D_{11} \frac{\partial^4 w}{\partial x^4} + 2(D_{12} + 2D_{66}) \frac{\partial^4 w}{\partial x^2 \partial y^2} + D_{22} \frac{\partial^4 w}{\partial y^4} = \mu h \frac{\partial^2 w}{\partial t^2} \tag{2.13}$$

that is differential equation of the deflection surface of the plate.

For specially orthotropic rectangular thin plate in free vibration as shown in Figure 1 the bending strain energy due to transverse vibration is given by

$$V = \tfrac{1}{2}\int_S \mathbf{M}^{\mathrm{T}}\chi dS = \tfrac{1}{2}\int_S \chi^{\mathrm{T}}\mathbf{C}_{\mathbf{f}}\chi dS = \tfrac{1}{2}\int_S^{\mathrm{T}} (\mathbf{D}w)^{\mathrm{T}}\,\mathbf{C}_{\mathbf{f}}\mathbf{D}w dS =$$
$$= \frac{1}{2}\int_S \left[D_{11}\left(\tfrac{\partial^2 w}{\partial x^2}\right)^2 + 2D_{12}\tfrac{\partial^2 w}{\partial x^2}\tfrac{\partial^2 w}{\partial y^2} + D_{22}\left(\tfrac{\partial^2 w}{\partial y^2}\right)^2 + 4D_{66}\left(\tfrac{\partial^2 w}{\partial x \partial y}\right)^2\right] dS \qquad (2.14)$$

The term of kinetic energy is

$$K = \frac{\rho h}{2}\int_0^a \int_0^b \dot{w}^2 dxdy \qquad (2.15)$$

where a, b, h are the plate dimensions.

3 Rayleigh-Ritz model

The derivation of the Rayleigh-Ritz model for orthotropic plate vibrations is briefly summarized here. For a conservative system, the maximum potential energy $V_{\max}$ equals the maximum kinetic energy $K_{\max}$. After assuming harmonic motion at frequency ω

$$w(x,y) = W(x,y)\cos\omega t. \qquad (3.1)$$

the maximum kinetic energy is given by

$$K_{\max} = \frac{\rho h}{2}\omega^2 \int_0^a \int_0^b W^2 dxdy = \omega^2 T. \qquad (3.2)$$

Setting $K_{\max} = V_{\max}$ allows to form the Rayleigh's quotient as

$$\omega^2 = \frac{V_{\max}}{T} = \frac{2}{\rho h}\frac{V_{\max}}{\int_0^a \int_0^b w^2 dxdy}. \qquad (3.3)$$

This quotient provides an estimate of the fundamental frequency, ω, in terms of the maximum potential energy, $V_{\max}$, material density, ρ, assumed mode shape, $W(x,y)$, and plate dimensions.

The Rayleigh-Ritz method is applied by assuming the maximum deflection, $W(x,y)$, to be represented as a linear series of assumed free-free characteristic beam functions

$$W(x,y) = \sum_{m=1}^{p}\sum_{n=1}^{q} A_{mn}\Phi_m(x)\,\Theta_n(y) \qquad (3.4)$$

that satisfy the natural boundary condition along the edges of a plate having zero Poisson's ratio. Substituting (3.4) into (3.3) gives ω^2 as a function of the coefficients A_{mn}, and beam functions $\Phi_m(x), \Theta_n(y)$. Searching the stationary value of the frequency minimize ω^2 with respect to the coefficients A_{ik} is possible to obtain the eigenvalue problem describing free vibration as shown in Appendix A

$$\sum_{j=1}^{p\times q} [C_{ij} - \lambda\delta_{ij}]\,A_j = 0 \qquad (3.5)$$

where

$$C_{ij} = D_{11}Q_{1ij} + D_{22}Q_{2ij} + D_{12}Q_{3ij} + D_{66}Q_{4ij}. \tag{3.6}$$

In (3.6) Q_{1ij}, Q_{2ij}, Q_{3ij}, Q_{4ij} are integral matrices and D_{ij} are the in-plane flexural stiffnesses.

4 Determination of elastic constants

The present paper is a development and an extension of the work by Deobald and Gibson (1988). It should be noted that, measuring four natural frequencies of an orthotropic plate from modal analysis, the possibility exists for determining four independent elastic constants.

The first step is to form four integral matrices. Only one calculation is necessary for these matrices. After a prior property estimate of elastic constants four bending stiffnesses can be calculated. A first evaluation of analytical eigenvalues λ_i and eigenvectors $\mathbf{A}^{(i)}$ can be obtained solving the eigenvalue problem (3.5).

The experimental natural frequencies f_i^e can be converted to four eigenvalues λ_1^e, λ_2^e, λ_3^e, λ_4^e through the expressions

$$\lambda_i^e = \rho h \left(2\pi f_i^e ab\right)^2. \tag{4.1}$$

When the prior estimates of elastic constants are reasonable and the eigenvalues are isolated the frequencies can be used as input for that inverse method. In Deobald and Gibson (1988) the identification procedure adopts a four by four matrix relating the four measured eigenvalues to the four bending stiffnesses. For each eigenpair $\lambda_i, \mathbf{A}^{(i)}$, the $h-th$ row in equations (3.5) is isolated:

$$\begin{aligned} \lambda_i A_h^{(i)} &= A_1^{(i)}\left(D_{11}Q_{1h1} + D_{22}Q_{2h1} + D_{12}Q_{3h1} + D_{66}Q_{4h1}\right) + \\ &\quad + A_2^{(i)}\left(D_{11}Q_{1h2} + D_{22}Q_{2h2} + D_{12}Q_{3h2} + D_{66}Q_{4h2}\right) + \\ &\quad + \ldots + A_n^{(i)}\left(D_{11}Q_{1hn} + D_{22}Q_{2hn} + D_{12}Q_{3hn} + D_{66}Q_{4hn}\right), \\ i &= 1,2,3,4, \qquad 1 < h \leq n, \qquad n = p \times q \end{aligned} \tag{4.2}$$

One equation from the set of n equations is necessary for each eigenpair. The hth equation can be selected, in which $A_h^{(i)}$ is the maximum value of the ith eigenvector. Replacing the four approximate eigenvalues λ_i in equations (4.2) by the eigenvalues λ_i^e calculated from the natural frequencies (4.1), one can reduce equations (4.2) to the following equations:

$$\begin{Bmatrix} \lambda_1^e \\ \lambda_2^e \\ \lambda_3^e \\ \lambda_4^e \end{Bmatrix} = \begin{bmatrix} H_{11} & H_{12} & H_{13} & H_{14} \\ H_{21} & H_{22} & H_{23} & H_{24} \\ H_{31} & H_{32} & H_{33} & H_{34} \\ H_{41} & H_{42} & H_{43} & H_{44} \end{bmatrix} \begin{Bmatrix} D_{11} \\ D_{22} \\ D_{12} \\ D_{66} \end{Bmatrix}. \tag{4.3}$$

in which

$$H_{ij} = \sum_{l=1}^{n} \frac{A_l^{(i)}}{A_h^{(i)}} Q_{jhl}, \qquad i,j = 1,2,3,4, \qquad n = p \times q, \qquad 1 \leq h \leq n \tag{4.4}$$

are the coefficients of the "solution" matrix that is a function of the approximate eigenvectors, $\mathbf{A}^{(i)}$, and the integral matrices Q_{rmn}.

Even when the estimated bending stiffnesses are considerably different from the true values, the analytical eigenvectors from the approximate eigenvalue problem are generally a good representation of the true eigenvectors. Solving equations (4.3) one can obtain new estimates of material rigidities. This is first step of the iterative revision method of the constants. The iteration procedure is repeated until all stiffnesses fall near previous values.

Finally, the four elastic constants are determined by using the following relations:

$$\begin{aligned} \nu_{xy} &= \frac{D_{12}}{D_{22}}, \qquad \nu_{yx} = \nu_{xy}\frac{D_{22}}{D_{11}}, \qquad G_{xy} = \frac{12D_{66}}{h^3}, \\ E_x &= \frac{12D_{11}\left(1-\nu_{xy}\nu_{yx}\right)}{h^3}, \qquad E_y = \frac{12D_{22}\left(1-\nu_{xy}\nu_{yx}\right)}{h^3}. \end{aligned} \tag{4.5}$$

It is possible to change the method just described enlarging the matrix H. Using a greater number of experimental frequencies equations (4.3), it becomes an overdetermined system

$$\left\{ \begin{array}{c} \lambda_1^e \\ \lambda_2^e \\ \cdot \\ \cdot \\ \lambda_i^e \\ \cdot \\ \cdot \\ \lambda_{N-1}^e \\ \lambda_N^e \end{array} \right\} = \left[\begin{array}{cccc} H_{11} & H_{12} & H_{13} & H_{14} \\ H_{21} & H_{22} & H_{23} & H_{24} \\ \cdot & \cdot & \cdot & \cdot \\ \cdot & \cdot & \cdot & \cdot \\ H_{i1} & H_{i2} & H_{i3} & H_{i4} \\ \cdot & \cdot & \cdot & \cdot \\ \cdot & \cdot & \cdot & \cdot \\ H_{N-1,1} & H_{N-1,2} & H_{N-1,3} & H_{N-1,4} \\ H_{N,1} & H_{N,2} & H_{N,3} & H_{N,4} \end{array} \right] \left\{ \begin{array}{c} D_{11} \\ D_{22} \\ D_{12} \\ D_{66} \end{array} \right\}. \tag{4.6}$$

where $\mathbf{H}$ is a rectangular matrix. Therefore a greater number of experimental frequencies is necessary. The unknown coefficients $D_{11}, D_{22}, D_{12}, D_{66}$ of the overdetermined system of linear equation (4.6) can be computed by doing a least square fit.

With this modification various advantages are obtained. As described in Deobald and Gibson (1988), the square 4×4 solution matrix frequently turns out to be ill conditioned and convergence problems arise in the iterative scheme. Only for some sets of four frequencies reasonable esteems are possible. Moreover, introducing more experimental frequencies in the vector $\boldsymbol{\lambda}^e$ (4.6) the identification method becomes richer in information. Therefore one can introduce all the measured frequencies obtained from the vibration testing. To verify this inverse method for determining the four coefficients $D_{11}, ..., D_{66}$ several models of plates discussed in some papers are considered in this work.

4.1 Aluminium plate

The dimensions and mass density of the square aluminium plate discussed in Deobald and Gibson (1988) are:

$a = 25.4cm \qquad b = 25.4cm \qquad h = 0.316cm \qquad \rho = 2.77g/cm^3$.

The Young's modulus of this isotropic plate is $E_x = E_y = 72.4\ GPa$, the shear modulus $G_{xy} = 28\ GPa$ and the Poisson's ratio is $v_{xy} = 0.33$. These properties were obtained by static tests. Only the F-F-F-F plate boundary condition (all edges free) is considered. The natural frequencies achieved by impulse technique and modal analysis are shown in Table 1. In addition, it also shows the natural frequencies obtained by both the Rayleigh-Ritz technique and finite element method. In both methods the actual elastic constants have been used. The first eight vibration modes obtained with the Rayleigh-Ritz method, are illustrated in Figure 2.

Table 1. Natural frequencies of aluminium plate

Mode	Natural frequencies (Hz)			
	Experimental[a]	Sap F.E.M.[a]	Rayleigh-Ritz[a]	Rayleigh-Ritz[b]
1 (1,1)	156,7	161,4	163,2	163,3
2 (0,2)	232,5	230,0	237,6	237,6
3 (2,0)	300,4	292,3	299,9	299,9
4 (1,2)	411,7	412,9	424,3	424,5
5 (2,1)	411,7	412,9	424,3	424,5
6 (0,3)	744,9	719,3	749,4	749,4
7 (3,0)	744,9	719,3	749,4	749,4
8 (2,2)	755,7	755,9	780,5	780,6
9 (3,1)	821,8	803,2	843,4	843,5
10 (1,3)	936,5	917,6	949,8	951,1

[a]Deobald and Gibson (1988) [b]Present work

Table 2. Estimated elastic constants of aluminium plate.

First set of initial values	No. of frequencies	$E_x(GPa)$	$E_y(GPa)$	$G_{xy}(GPa)$	ν_{xy}
$E_x = 109\ GPa$	5	69,82	69,82	25,47	0,362
$E_y = 109\ GPa$	6	69,76	69,83	25,47	0,363
$G_{xy} = 39\ GPa$	7	69,92	69,92	25,45	0,361
$\nu_{xy} = 0,39$	8	69,98	69,98	25,41	0,360
Second set of initial values	No. of frequencies	$E_x(GPa)$	$E_y(GPa)$	$G_{xy}(GPa)$	ν_{xy}
$E_x = 48\ GPa$	5	69,82	69,82	25,47	0,362
$E_y = 48\ GPa$	6	69,92	69,92	25,45	0,361
$G_{xy} = 20\ GPa$	7	69,76	69,83	25,47	0,363
$\nu_{xy} = 0,28$	8	69,82	69,82	25,47	0,361
Average value		69,85	69,87	25,46	0,362
Paper[a] value		69,5	69,9	25,6	0,361
(static test)[a]		72,4	72,4	28	0,33

[a]Deobald and Gibson (1988)

Two different sets of initial values are put in the present inverse identification method by using 5,6,7 and 8 experimental frequencies. As shown on Table 2, large initial values are employed in the first set and small initial values are used in the second set. Therefore eight esteems of elastic constants are possible.

The iteration scheme generally takes less than five iteration steps for convergence.

4.2 Graphite epoxy plate

The dimensions of the graphite epoxy plate discussed in Deobald and Gibson (1988) are $25.4cm \times 25.4cm \times 0.1483cm$ and the density is $1.584g/cm^3$. By static tests Deobald and Gibson (1988) obtained the following properties: the major Young's modulus is $E_x = 127.9\ GPa$ and $E_y = 10.27\ GPa$, the shear modulus $G_{xy} = 7,312\ GPa$ and the major Poisson's ratio $v_{xy} = 0.22$. The natural frequencies identified by vibration tests and the frequencies calculated by Rayleigh-Ritz method and finite element method, are shown on Table 3. The number of frequencies employed in the method described is greater than the total number of experimental frequencies. Comparing on Table 3 the experimental natural frequencies with the analytical frequencies, important differences between the corresponding values can be observed. These errors probably cause some convergence problems in the iteration scheme that produce incorrect Poisson's ratio evaluations. In order to solve these problems some analytical frequencies have to be added to experimental data.

Table 3. Natural frequencies of graphite-epoxy plate.

	Natural frequencies (Hz)			
Mode	Experimental[a]	Sap F.E.M.[a]	Rayleigh-Ritz[a]	Rayleigh-Ritz[b]
1 (1,1)	49,37	51,33	51,81	51,84
2 (0,2)	78,89	60,54	60,19	60,20
3 (1,2)	113,2	120,7	122,0	122,0
4 (0,3)	188,3	161,8	165,9	165,9
5 (2,0)	210,5	208,6	212,7	212,7
6 (1,3)	222,4	225,2	229,0	229,2
7 (2,1)	231,6	231,6	236,7	236,7
8 (2,2)	295,2	299,2	306,5	306,5
9 (0,4)	350,0	316,2	326,1	326,1
10 (1,4)	381,3	378,8	386,2	386,3

[a]Deobald and Gibson (1988) [b]Present work

It's well known which higher modes of orthotropic plates are much more heavily influenced by shear deformation effect. The higher mode frequencies are not well predicted with classical lamination theory. Therefore the utilization of too many frequencies in the classical lamination theory can lead to incorrect results. Table 4 shows all the frequencies used in this procedure. In the first and second row there are only experimental data, while all the data in the third and fourth row were obtained with Rayleigh Ritz approach.

Two different sets of initial values of elastic constants are put in the inverse identification method with an increasing number of frequencies (see Table 5). Large initial values are used in the first set and small initial values are used in the second set. Eight different esteems of elastic constants are possible. It is clear that different initial values still cause some scatters on the results.

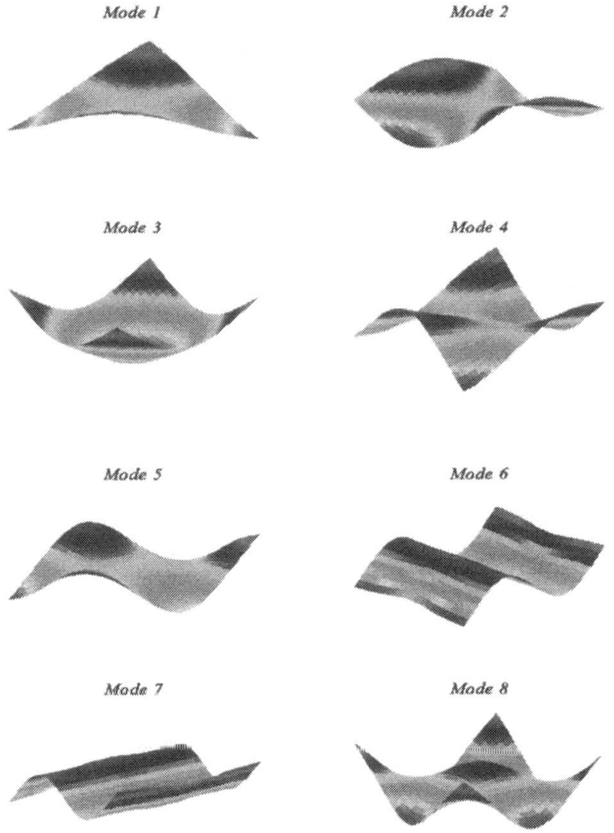

Figure 2. The first eight vibration modes of the square aluminium plate.

Table 4. Experimental and numerical frequencies of graphite-epoxy plate.

Experimental [a] frequencies	Mode	1	2	3	4	5
	Frequency (Hz)	49,37	78,89	113,2	188,3	210,5
	Mode	6	7	8	9	10
	Frequency (Hz)	222,4	231,6	295,2	350,0	381,3
Numerical frequencies (Rayleigh-Ritz method)[b]	Mode	11	12	13	14	15
	Frequency (Hz)	424	538,4	586	590	593
	Mode	16	17	18	19	20
	Frequency (Hz)	609	671	775	794	942

[a]Deobald and Gibson (1988) [b]Present work

Table 5. Estimated elastic constants of graphite-epoxy plate.

First set of initial values	No. of frequencies	$E_x(GPa)$	$E_y(GPa)$	$G_{xy}(GPa)$	ν_{xy}
$E_x = 180\ GPa$	12	126,7	10,38	7,19	0,373
$E_y = 14\ GPa$	13	127,4	10,35	7,14	0,378
$G_{xy} = 9,5\ GPa$	14	127,2	10,34	7,17	0,374
$\nu_{xy} = 0,26$	15	126,9	10,35	7,15	0,373
Second set of initial values	No. of frequencies	$E_x(GPa)$	$E_y(GPa)$	$G_{xy}(GPa)$	ν_{xy}
$E_x = 90\ GPa$	12	125,3	10,38	7,30	0,393
$E_y = 7,3\ GPa$	13	125,8	10,38	7,33	0,399
$G_{xy} = 5,6\ GPa$	14	126,1	10,38	7,27	0,365
$v_{xy} = 0,18$	15	125,9	10,38	7,29	0,370
Average value		126,41	10,36	7,22	0,374
Paper[a] value		125,2	10,3	6,6	-0,249
(static test)[a]		127,9	10,27	7,312	0,22

[a]Deobald and Gibson (1988)

4.3 Other plates

In Larsson (1997) some dynamic properties are obtained from vibration testing and modal analysis for three wood-based panels (called OSB) modelled as orthotropic thin plate. In this paper the first of the three structural panels (OSB1) is considered. The dimensions of the OSB1 panel are $244cm \times 122cm \times 10cm$ and the density is $650.7kg/m^3$. The final estimates of the elastic in-plane constants obtained in Larsson (1997) are: $E_x = 6.79\ GPa$, $E_y = 3.68\ GPa$, $G_{xy} = 1.72\ GPa$ and $v_{xy} = 0.32$.

In Table 6 all the frequencies used in this procedure are shown. In the first two rows there are only experimental natural frequencies, while all the data in the third row were obtained with the Rayleigh Ritz approach using elastic constants predicted by Larsson (1997).

Table 7 shows eight different esteems of the four elastic constants obtained as a function of the number of frequencies when large and small initial values are considered. These results are all close enough to the values obtained by Larsson (1997).

Table 6. Experimental and numerical frequencies of OSB1 plate.

Experimental[a] frequencies	Mode	1	2	3	4	5
	Frequency (Hz)	5,57	5,63	13,4	15,6	16,7
	Mode	6	12			
	Frequency (Hz)	20,7	46,30			
Numerical frequencies[b] (Rayleigh Ritz)	Mode	7	8	9	10	11
	Frequency (Hz)	23,63	29,32	31,34	38,39	42,76

[a]Larsson (1997) [b]Present work

Table 7. Estimated elastic constants of OSB1 plate.

First set of initial values	No. of frequencies	$E_x(GPa)$	$E_y(GPa)$	$G_{xy}(GPa)$	ν_{xy}
$E_x = 10,5\ GPa$	9	6,91	3,69	1,81	0,309
$E_y = 5,90\ GPa$	10	6,70	3,64	1,82	0,353
$G_{xy} = 2,80\ GPa$	11	6,82	3,69	1,79	0,326
$\nu_{xy} = 0,42$	12	6,81	3,65	1,80	0,330
Second set of initial values	No. of frequencies	$E_x(GPa)$	$E_y(GPa)$	$G_{xy}(GPa)$	ν_{xy}
$E_x = 4,70\ GPa$	9	6,89	3,69	1,81	0,32
$E_y = 2,60\ GPa$	10	6,70	3,69	1,81	0,35
$G_{xy} = 1,20\ GPa$	11	6,82	3,69	1,81	0,33
$v_{xy} = 0,18$	12	6,81	3,69	1,81	0,33
Average value		6,81	3,67	1,80	0,33
Paper[a] value		6,79	3,68	1,72	0,32

[a]Larsson (1997)

Table 8. Experimental natural frequencies of the fibre-reinforced plate.

	Mode	1	2	3	4	5
Experimental frequencies	Frequency (Hz)	243	396	600	632	760
	Mode	6	7	8	9	10
	Frequency (Hz)	1113	1244	1310	1624	1780

Table 9. Estimated elastic constants of the fibre-reinforced plate.

First set of initial values	No. of frequencies	$E_x(GPa)$	$E_y(GPa)$	$G_{xy}(GPa)$	ν_{xy}
$E_x = 39,0\ GPa$	7	31,38	27,96	7,40	0,285
$E_y = 31,6\ GPa$	8	30,99	27,38	7,34	0,353
$G_{xy} = 8,20\ GPa$	9	30,51	27,30	7,38	0,326
$\nu_{xy} = 0,31$	10	31,36	28,29	7,38	0,330
Second set of initial values	No. of frequencies	$E_x(GPa)$	$E_y(GPa)$	$G_{xy}(GPa)$	ν_{xy}
$E_x = 27,1\ GPa$	7	31,38	27,96	7,40	0,285
$E_y = 23,9\ GPa$	8	30,99	27,38	7,34	0,318
$G_{xy} = 5,70\ GPa$	9	30,51	27,30	7,38	0,320
$v_{xy} = 0,21$	10	31,31	28,25	7,37	0,246
Average value		31,06	27,73	7,37	0,33
Paper[a] value		32,51	27,45	6,82	0,24

[a]McIntyre and Woodhouse (1988)

The measuring of the elastic and damping constants of other models of rectangular thin orthotropic plates was discussed by McIntyre and Woodhouse (1988). A procedure was presented whereby all four elastic constants may be determined. The method was illustrated with measurements on a range of different sheet materials: wooden plates and two very different fibre reinforced composites. The inverse method proposed in this paper was applied to all these models of plate discussed in the work of the above mentioned authors with good results.

In this paper the results achieved for second model of plate and the esteems obtained by McIntyre and Woodhouse (1988) are presented. The dimensions of the glass-fiber-reinforced plate are $15.0cm \times 17.7cm \times 0.323cm$ and the density is $2081 kg/m^3$. The final estimates of the elastic in-plane constants obtained in McIntyre and Woodhouse (1988) are: $E_x = 32,51\ GPa$, $E_y = 27,45\ GPa$, $G_{xy} = 6,82\ GPa$ and $v_{xy} = 0.24$.

Table 8 shows all the experimental natural frequencies used in this iterative procedure. Table 8 shows the eight different esteems for each elastic constant obtained with large and small initial values. These results are all in agreement with the values evaluated by McIntyre and Woodhouse (1988).

5 Conclusions

A method for measuring four in-plane elastic constants for a specially orthotropic thin plate has been presented. The approach is based on the analysis of frequencies of vibration modes which are measured by vibration testing. A Rayleigh-Ritz technique has been employed in order to model the dynamic behaviour of free edge plates. When only four experimental frequencies are used, the identification procedure can give an ill-conditioned problem. Using a major number of experimental data and introducing a least square method, one can extract the elastic constants. The procedure presented here provides a fast and accurate identification method for isotropic materials. Moreover, the values obtained for the mechanical properties of orthotropic plates are in good agreement with the values reported in the literature. It should be noted that some convergence problems in the iteration procedure could be found when available data are insufficient. These difficulties have been solved adding some numerical frequencies to the measured natural frequencies in the iterative scheme. It has been demonstrated that reasonably identification is possible for orthotropic material plates by starting from an increasing number of experimental data.

Bibliography

I. Bartoli. Identificazione dinamica dei parametri di piastre ortotrope. *Tesi di laurea (Relatore Prof. E. Viola)*, Facoltà di Ingegneria, Università di Bologna A.A. 2000/2001.

I. Bartoli, E. Viola. Problemi di stima delle costanti elastiche di piastre isotrope ed ortotrope. *Nota Tecnica 57, Anno 2002*, D.I.S.T.A.R.T., Facoltà di Ingegneria, Università di Bologna.

G. Caldersmith, T.D. Rossing. Determination of modal coupling in vibrating rectangular plates. *Applied Acoustics*, 17:33–44, 1984.

L.R. Deobald, R.F. Gibson. Determination of elastic constants of orthotropic plates by a modal analysis/Rayleigh Ritz technique. *Journal of Sound and Vibration*, 124(2): 269–283, 1988.

R.M. Jones. Mechanics of Composite Materials, Scripta Book Company, Washington D.C., 1975.

P. Laarson, O. Laarson. Determination of Young's and shear moduli from flexural vibrations of beams. *Journal of Sound and Vibration*, 146(1):111–123, 1991.

D. Larsson. Using modal analysis for estimation of anisotropic material constants. *Journal of Engineering Mechanics*, 3(123):222–229, 1997.

T.C. Lai, K.H. Ip. Parameter estimation of orthotropic plates by Bayesian sensitivity analysis. *Composite Stuctures*, 34:29–42, 1996.

T.C. Lai, T.C. Lau. Determination of elastic constants of a generally orthotropic plate by modal analysis. *The international Journal of Analitical and Experimental Modal Analysis*, 1(8):15–33, 1993.

A.W. Leissa. The free vibration of rectangular plates. *Journal of Sound and Vibration*, 31(3):257–293, 1973.

M.E. McIntyre, J. Woodhouse. On measuring the elastic and damping constants of orthotropic sheet materials. *Acta Metallurgica*, 6(36):1397–1416, 1988.

M.E. McIntyre, J. Woodhouse. The influence of geometry on linear damping. *Acustica*, 39(4):209–224, 1988.

R.D. Mindlin, M.A. Medick. Extensional vibrations of elastic plates. *Journal of Applied Mechanics*, 26:561–569, 1959.

F. Moussu, M. Nivoit. Determination of elastic constants of orthotropic plates by a modal analysis/method of superposition. *Journal of Sound and Vibration*, 165(1):149–163, 1993.

J.N. Reddy. Energy and Variational Methods in Applied Mechanics, Wiley Interscience, 1984.

E. Viola. Fondamenti di dinamica e vibrazione delle strutture, Volume II, Pitagora Editrice, 2001.

E. Viola. Scienza delle costruzioni, Volume I, Teoria dell'elasticità, Pitagora Editrice, 2000.

Appendix A

The function ω^2 (3.3) have to be derived with respect to A_{ik} parameters.

$$\begin{aligned}\frac{\partial\omega^2}{\partial A_{ik}} &= \frac{2}{\rho h}\frac{\partial}{\partial A_{ik}}\frac{V_{\max}}{\int_0^a\int_0^b W^2dxdy} = \frac{2}{\rho h}\left[\frac{\partial V_{\max}}{\partial A_{ik}}\frac{1}{\iint W^2dxdy} - \frac{\partial}{\partial A_{ik}}\iint W^2dxdy\frac{V_{p,\max}}{\left(\iint W^2dxdy\right)^2}\right] = \\ &= \frac{2}{\rho h}\left[\frac{\partial V_{\max}}{\partial A_{ik}}\frac{1}{\iint W^2dxdy} - \frac{\partial}{\partial A_{ik}}\iint W^2dxdy\frac{\rho h\omega^2}{2}\frac{1}{\iint W^2dxdy}\right] = \\ &= \frac{2}{\rho h}\frac{1}{\iint W^2dxdy}\left[\frac{\partial V_{p,\max}}{\partial A_{ik}} - \frac{\rho h\omega^2}{2}\frac{\partial}{\partial A_{ik}}\iint W^2dxdy\right].\end{aligned} \tag{A.1}$$

By equating to zero the derivative (A.1), we obtain

$$\frac{\partial\omega^2}{\partial A_{ik}} = \frac{\partial V_{\max}}{\partial A_{ik}} - \frac{\rho h\omega^2}{2}\frac{\partial}{\partial A_{ik}}\iint W^2dxdy = 0. \tag{A.2}$$

The maximum strain energy can be expressed as

$$V_{\max} = \frac{1}{2}\int\int(D_{11}W_{xx}^2 + 2D_{12}W_{xx}W_{yy} + D_{22}W_{yy}^2 + 4D_{66}W_{xy}^2)dxdy \tag{A.3}$$

where

$$W_{xx} = \frac{\partial^2 W}{\partial x^2}, \qquad W_{yy} = \frac{\partial^2 W}{\partial y^2}, \qquad W_{xy} = \frac{\partial^2 W}{\partial x \partial y}. \tag{A.4}$$

By using (A.3), the first term on the right-hand side of equation (A.2) can be expressed as

$$\frac{\partial}{\partial A_{ik}} \frac{1}{2} \int \int (D_{11} W_{xx}^2 + 2D_{12} W_{xx} W_{yy} + D_{22} W_{yy}^2 + 4D_{66} W_{xy}^2) dxdy - \frac{\rho h \omega^2}{2} \frac{\partial}{\partial A_{ik}} \int\int W^2 dxdy = 0 \tag{A.5}$$

In order to solve (A.5) the following expressions have to be worked out

$$W(x,y)^2 = \left[\sum_{m=1}^{p} \sum_{n=1}^{q} A_{mn} \Phi_m(x) \Theta_n(y) \right]^2, \qquad \frac{\partial}{\partial A_{ik}} W^2 = 2 \sum_{m=1}^{p} \sum_{n=1}^{q} A_{mn} \Phi_m \Theta_n (\Phi_i \Theta_k)$$

$$\frac{\partial}{\partial A_{ik}} \int\int W^2 dxdy = 2 \sum_{m=1}^{p} \sum_{n=1}^{q} \left[\int_0^a \Phi_m \Phi_i dx \int_0^b \Theta_n \Theta_k dy \right] A_{mn}$$

$$W_{xx} = \sum_{m=1}^{p} \sum_{n=1}^{q} A_{mn} \Phi''_m \Theta_n, \qquad W_{xx}^2 = \left[\sum_{m=1}^{p} \sum_{n=1}^{q} A_{mn} \Phi''_m \Theta_n \right]^2$$

$$\frac{\partial}{\partial A_{ik}} \int\int W_{xx}^2 dxdy = 2 \sum_{m=1}^{p} \sum_{n=1}^{q} \left[\int_0^a \Phi''_m \Phi''_i dx \int_0^b \Theta_n \Theta_k dy \right] A_{mn}$$

$$W_{yy} = \sum_{m=1}^{p} \sum_{n=1}^{q} A_{mn} \Phi_m \Theta''_n, \qquad W_{yy}^2 = \left[\sum_{m=1}^{p} \sum_{n=1}^{q} A_{mn} \Phi_m \Theta''_n \right]^2$$

$$\frac{\partial}{\partial A_{ik}} \int\int W_{yy}^2 dxdy = 2 \sum_{m=1}^{p} \sum_{n=1}^{q} \left[\int_0^a \Phi_m \Phi_i dx \int_0^b \Theta''_n \Theta''_k dy \right] A_{mn}$$

$$W_{xx} W_{yy} = \left(\sum_{m=1}^{p} \sum_{n=1}^{q} A_{mn} \Phi''_m \Theta_n \right) \left(\sum_{r=1}^{p} \sum_{s=1}^{q} A_{rs} \Phi_r \Theta''_s \right)$$

$$\frac{\partial}{\partial A_{ik}} \int\int W_{xx} W_{yy} dxdy = \sum_{m=1}^{p} \sum_{n=1}^{q} \left[\int_0^a \Phi''_m \Phi_i dx \int_0^b \Theta_n \Theta''_k dy + \int_0^a \Phi''_i \Phi_m dx \int_0^b \Theta_k \Theta''_n dy \right] A_{mn}$$

$$W_x = \sum_{m=1}^{p} \sum_{n=1}^{q} A_{mn} \Phi'_m \Theta_n, \qquad W_{xy} = \sum_{m=1}^{p} \sum_{n=1}^{q} A_{mn} \Phi'_m \Theta'_n$$

$$\frac{\partial}{\partial A_{ik}} \int\int W_{xy}^2 dxdy = \frac{\partial}{\partial A_{ik}} \int\int \left[\sum_{m=1}^{p} \sum_{n=1}^{q} A_{mn} \Phi'_m \Theta'_n \right]^2 dxdy =$$

$$= 2 \sum_{m=1}^{p} \sum_{n=1}^{q} \left[\int_0^a \Phi'_m \Phi'_i dx \int_0^b \Theta'_n \Theta'_k dy \right] A_{mn}. \tag{A.6}$$

Using expressions (A.6) one can write the equation $(A.1)$ as

$$\sum_{m=1}^{p} \sum_{n=1}^{q} \left\{ D_{11} \left[\int_0^a \Phi''_m \Phi''_i dx \int_0^b \Theta_n \Theta_k dy \right] + \right.$$

$$+ D_{12} \left[\int_0^a \Phi''_m \Phi_i dx \int_0^b \Theta_n \Theta''_k dy + \int_0^a \Phi''_i \Phi_m dx \int_0^b \Theta_k \Theta''_n dy \right] +$$

$$D_{22} \left[\int_0^a \Phi_m \Phi_i dx \int_0^b \Theta''_n \Theta''_k dy \right] + 4D_{66} \left[\int_0^a \Phi'_m \Phi'_i dx \int_0^b \Theta'_n \Theta'_k dy \right] +$$

$$\left. - \rho h \omega_S^2 \left[\int_0^a \Phi_m \Phi_i dx \int_0^b \Theta_n \Theta_k dy \right] \right\} A_{mn} = 0$$

$$i = 1, 2..p \qquad k = 1, 2..q. \tag{A.7}$$

The following integrals

$$\int_0^a \Phi''_m \Phi''_i dx \quad \int_0^b \Theta''_n \Theta''_k dy \quad \int_0^a \Phi_m \Phi_i dx \quad \int_0^b \Theta_n \Theta_k dy \tag{A.8}$$

are non zero only for $m=i$ or for $n=k$. The above integrals satisfy orthogonality relationships. This is due to the particular assumed beam functions $\Phi(x)$, $\Theta(y)$, which are the solutions of characteristic equations of vibrating beams

$$\begin{aligned} V_1 &= 1 \quad V_2 = \sqrt{3} - 2\sqrt{3}\frac{x}{a} \\ V_i &= \cos(\varepsilon_i \tfrac{x}{a}) + \cosh(\varepsilon_i \tfrac{x}{a}) - \sin(\varepsilon_i \tfrac{x}{a}) - \sinh(\varepsilon_i \tfrac{x}{a}) \\ \varepsilon_3 &= 4.73 \quad \varepsilon_4 = 7.853 \quad \varepsilon_5 = 10.996 \quad \varepsilon_6 = 14.137 \end{aligned} \tag{A.9}$$

where $V_1..V_6$ are beam eigenfunctions and ε_i are beam eigenvalues.

The four remaining integrals are named

$$\begin{aligned} F^x_{mi} &= a \int_0^a \Phi_m \Phi''_i dx \,, \quad F^y_{nk} = b \int_0^b \Theta_n \Theta''_k dy, \\ H^x_{mi} &= a \int_0^a \Phi'_m \Phi'_i dx \,, \quad H^y_{nk} = b \int_0^b \Theta'_n \Theta'_k dy \end{aligned} \tag{A.10}$$

All the integrals were non-dimensionalized so that only one calculation is necessary.

Finally, considering relations (A.8)-(A.10), equations (A.7) reduce to the eigenvalue problem

$$\sum_{m=1}^{p} \sum_{n=1}^{q} \left[C_{ik,mn} - \lambda \delta_{im} \delta_{kn}\right] A_{mn} = 0 \tag{A.11}$$

in which

$$\lambda = \rho h \omega_S^2 a^2 b^2 \tag{A.12}$$

$$C_{ik,mn} = D_{11} Q_{1,ik,mn} + D_{22} Q_{2,ik,mn} + D_{12} Q_{3,ik,mn} + D_{66} Q_{4,ik,mn} \tag{A.13}$$

$$Q_{1,ik,mn} = \int_0^a \Phi''_m \Phi''_i dx \int_0^b \Theta_n \Theta_k dy = (\varepsilon^x_i)^4 \left(\frac{b^2}{a^2}\right) \delta_{im} \delta_{nk} \tag{A.14}$$

$$\begin{aligned} Q_{3,ik,mn} &= \int_0^a \Phi''_m \Phi_i dx \int_0^b \Theta_n \Theta''_k dy + \int_0^a \Phi''_i \Phi_m dx \int_0^b \Theta_k \Theta''_n dy \\ &= F^x_{mi} F^y_{kn} + F^x_{im} F^y_{nk} \end{aligned} \tag{A.15}$$

$$Q_{2,ik,mn} = \int_0^a \Phi_m \Phi_i dx \int_0^b \Theta''_n \Theta''_k dy - (\varepsilon^y_k)^4 \left(\frac{a^2}{b^2}\right) \delta_{im} \delta_{nk} \tag{A.16}$$

$$Q_{4,ik,mn} = 4 \int_0^a \Phi'_m \Phi'_i dx \int_0^b \Theta'_n \Theta'_k dy = 4 H^x_{mi} H^y_{nk} \tag{A.17}$$

$$i = 1,2,3..,p, \qquad k = 1,2,3..,q, \qquad \delta_{rs} = \left\{ \begin{smallmatrix} 1 & \text{for } r=s \\ 0 & \text{for } r\neq s \end{smallmatrix} \right\}.$$

Effects of damage on the response of Euler-Bernoulli beams traversed by a moving mass

Cristiano Bilello[‡]

[‡] Dipartimento di Ingegneria Strutturale e Geotecnica, Universitá degli Studi di Palermo, Viale delle Scienze, 90128, Palermo, Italy.

Abstract The perturbation induced by damage in the dynamic response of Euler-Bernoulli beams traversed by a moving mass is investigated. The structure is discretized into segments of constant bending stiffness, connected together by elastic hinges representing damaged sections. The beam-moving mass interaction force is modelled in the most accurate way by taking into account the effective structural mass distribution and the convective acceleration terms, often omitted in similar studies. The analytical response is obtained through a series expansion of the unknown deflection in a basis of the beam eigenfunctions. The results of experimental tests, performed on a small-scale model of a prototype bridge structure, are presented for a validation of the proposed solution procedure, and to provide data of practical engineering relevance. A good agreement with theoretical predictions is observed, moreover, the nature of the applied load seems to increase the structural damage sensitivity.

1 Introduction

The analysis of continuous elastic systems subjected to moving sub-systems has been a subject of interest in many fields but it is especially in bridge engineering that this problem finds its natural application. Nowadays, advances in material technologies push the designers towards larger and more flexible structures. From one side this determines a reduction of the construction costs, on the other side it requires more complex structural models and a deeper understanding of their fundamental source of excitation: the dynamic interaction between the continuous system (bridge) and the sub-system (vehicle) moving on it.

However, this problem is not only crucial for design purposes. Bridge structures, in fact, are often subjected to aggressive environmental conditions that can lead to corrosion, material loss, supports deterioration, cracks, joint failures etc. It is clear that the presence of localized or diffused damage, combined with the dynamic nature of the vehicle excitation can dramatically reduce the useful life of the structure.

For these reasons, and in the light of the potential developments in the structural health monitoring, analysis tools that account for the presence of existing structural damage are required. For the undamaged case, a large number of theoretical investigations have been based on the use of a beam model; despite its simplicity, it provides useful information for investigating structural response. The complexity of the solution depends on the moving vehicle model. For the simplest moving force model, closed form solutions are available primarily based on the series expansion of the unknown displacement function (Meirovitch, 1967; Fryba, 1972; Pesterev et al., 2002). On the other hand, the so-called moving mass problem, where the inertia of the moving sub-system is taken into account, does not have a closed form solution. Thus, proposed approximate solutions have been based on series expansions (Stanisic et al., 1974; Pesterev and Bergman, 1998), iterative solution of integral equation (Sadiku and Leipholz, 1987), and finite element discretization (Akin and Mofid, 1997; Mofid and Shadam, 2000). A more complex and realistic model is represented by a moving oscillator (Pesterev and Bergman, 1997a,b; Yang et al., 2000), from which more advanced vehicle models originate (Swa; Humar and Kashif, 1993; Chatterjee et al., 1994; Yang and Yau, 1997).

Interest in the dynamic analysis of damaged beams has been ongoing for the past forty years, particularly because of its relationship to the automatic monitoring of structural integrity (see Dimarogonas, 1996 for a thorough literature survey). Damage models presented in the literature can be cast into two categories. The most common, and widely used to study cracked beams, is the so-called "spring hinge model" (Irwin, 1957; Rice and Levy, 1972; Dimarogonas, 1976), in which the effect of structural damage is modelled through a local compliance which quantifies, in a macroscopic way, the relation between the applied load and the strain surrounding the crack area . In the most general case a 6 x 6 local flexibility matrix is introduced whose elements are evaluated using linear elastic fracture mechanics theory (Anifantis and Dimarogonas, 1983; Papadopoulus and Dimarogonas). Recently, starting from the work of Christides and Barr (1984), new continuous vibration theories for cracked beams (Shen and Pierre, 1990; Chondros and Dimarogonas, 1998; Chondros et al., 1997) have been developed employing so-called *crack disturbance functions* that modify the stress, strain, and displacement fields in the entire damaged element. These represent a more advanced and accurate damage model; however, experimental tests and numerical simulations (Chondros and Dimarogonas, 1998; Chondros et al., 1997) have shown that the elastic hinge model, despite its simplified assumptions (localized damage effect), provides a good approximation of the beam response.

In spite of the large number of publications focused on the two fields, moving loads on continuous elastic systems and vibration of damaged structures, to the authors' knowledge, only two (Parhi and Behera, 1997; Mahamoud, 2001) have addressed these two topics simultaneously. In Parhi and Behera (1997), the problem of a mass moving on cantilever beam with a single damage location is investigated, while in Mahamoud (2001), a procedure is presented for determining stress intensity factors for single and double edge cracks in a simply supported undamped Euler-Bernoulli beam under a moving force. Linear elastic fracture mechanics is used to calculate the compliance of the equivalent elastic hinge. In both of these papers, however, a number of undocumented assumptions are made, either in the structural model (lumped masses in Mahamoud, 2001) or in the evaluation of the beam-moving mass interaction force (convective acceleration terms are omitted in Parhi and Behera, 1997; inertia of the moving load is not considered in Mahamoud, 2001). Moreover, the experimental data in Parhi and Behera (1997), due to the selected range of variation of parameters governing the problem, do not provide information of significant engineering relevance.

The first objective of this work, treated in section 2, is to provide an analytical tool for the analysis of a damaged Euler-Bernoulli beam under a moving mass, with an arbitrary number of damaged sections, and generic boundary conditions. An elastic hinge damage model is used (Section 3). Both the structural model and the beam-moving mass interaction force are modelled in the most accurate way by taking into account the effective structural mass distribution and the convective acceleration terms. Though relatively simple, this model can be used to study the behavior of a damaged bridge structure subjected to moving vehicles. In the second part of this work (Section 4), an experimental investigation of a small-scale model of a prototype bridge structure is presented. The objective of this section is not only to validate the analytical tools developed, but also to provide data of practical engineering relevance for analyzing the structural damage sensitivity under a moving mass.

2 Theory

2.1 Formulation of the Problem

Let us discretize the given beam into N segments of constant linear mass density ρ, bending stiffness EI (undamaged beam stiffness), and length l_i (Figure 1). Damping is not considered. The segments are connected together through elastic hinges (damaged sections) whose compliances are denoted by c_i. The equation of motion for the ith segment can then be written as

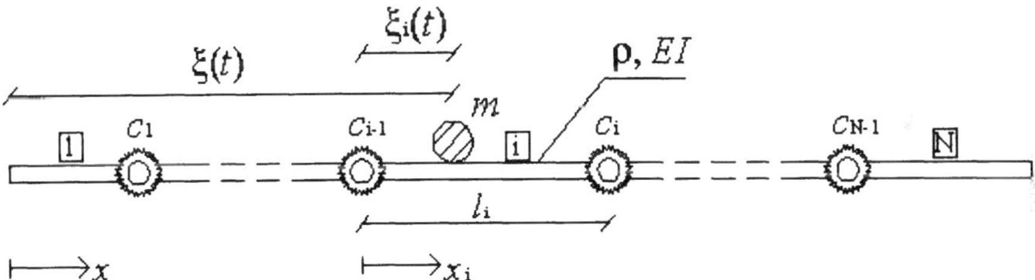

Figure 1. Beam with rotational springs representing damaged sections traversed by a moving mass.

$$\rho\ddot{w}\left(x_i,t\right)+EI\partial^4 w\left(x_i,t\right)/\partial x^4=f_i\left(t\right)W\left[\xi_i\left(t\right),0,l_i\right]\delta\left[x_i-\xi_i\left(t\right)\right],\ 0<x_i<l_i \quad (2.1)$$

where $w\left(x_i,t\right)$ is the unknown deflection, overdots denote derivatives with respect to the time variable, $f_i\left(t\right)$ is the beam-moving mass interaction force, $W\left(\cdot\right)$ is a window function, $\delta\left(\cdot\right)$ is the Dirac delta function, and $\xi_i\left(t\right)$ denotes the location of the moving load at time t (in the local coordinate system). The window function is defined as

$$W\left[\xi_i\left(t\right),0,l_i\right]=U\left[\xi_i\left(t\right),0\right]-U\left[\xi_i\left(t\right),l_i\right] \quad (2.2)$$

in which $U(\cdot)$ is the unit step function; thus, the right-hand side of Eq. (2.1) is different from zero $\forall t$: $0 \leqslant \xi_i(t) \leqslant l_i$ (the mass transits over the ith segment).

The expression for $f_i(t)$ depends on the analytical model used to represent the moving load; in the case of a moving mass model it is given by

$$f_i(t) = m\left\{g - T\left[w(x_i, t)\right]\big|_{x_i=\xi_i(t)}\right\} \tag{2.3}$$

where m is the moving mass, g is the acceleration of gravity, and $T[\cdot]$ is a linear differential operator for the acceleration of the mass

$$T[\cdot] = \left(\frac{\partial^2}{\partial t^2} + 2\dot{\xi}\frac{\partial^2}{\partial x \partial t} + \dot{\xi}^2\frac{\partial^2}{\partial x^2} + \ddot{\xi}\frac{\partial}{\partial x}\right)[\cdot]\,. \tag{2.4}$$

In this expression $\dot{\xi}$ and $\ddot{\xi}$ are the speed and acceleration of the moving mass, respectively. In a number of previous studies (Stanisic et al., 1974; Akin and Mofid, 1997; Parhi and Behera, 1997) the convective terms in (2.4) were omitted $\left(T[\cdot] \cong \partial^2[\cdot]/\partial t^2\right)$; however, this approximation is not generally reasonable unless the mass moves at very low speed, and it may lead to significant errors in the evaluation of the system response.

From Eq. (2.3) it is apparent that the interaction force depends on the beam response itself; moreover, using Eqs. (2.3-2.4) in (2.1), it is straightforward to show that the governing partial differential equation is characterized by time-varying singular coefficients. Besides the initial conditions, $w_0(x) = w(x,0)$ and $\dot{w}_0(x) = \dot{w}(x,0)$ for $0 \leqslant x \leqslant l$ and the external boundary conditions, the interior compatibility and equilibrium relationships must be taken into account in the formulation of the problem; for the case under study they are given by

$$w_i(0,t) = w_{i-1}(l_{i-1},t) \tag{2.5}$$

$$\vartheta_i(0,t) = \vartheta_{i-1}(l_{i-1},t) + c_{i-1}M_{i-1}(l_{i-1},t) \tag{2.6}$$

$$M_i(0,t) = M_{i-1}(l_{i-1},t) \tag{2.7}$$

$$V_i(0,t) = V_{i-1}(l_{i-1},t) \quad \text{for} \quad i = 2,3,....,N-1 \tag{2.8}$$

where $\vartheta(x,t)$, $M(x,t)$ and $V(x,t)$ denote the slope, bending moment and shear force, respectively. An exact solution of this problem is not, in general, possible. An approximate solution can be obtained by expanding the unknown function $w(x,t)$ in a series of the beam eigenfunctions.

2.2 Approximate Solution

The eigenvalues ω_r and eigenfunctions $\varphi_r(x)$ of the damaged beam are calculated through the *Transfer Matrix Method*. This approach, first introduced by Pestel and Leckie (1983), is still an attractive tool for the solution of the eigenvalue problem for one-dimensional systems with non-uniform mechanical properties. Details on the use of this method for the problem under investigation can be found in Bilello and Bergman (2002). Here, it should be stressed that, with respect to the work of Mahamoud (2001), where a lumped-mass transfer matrix is used, the effective structural mass distribution has be taken into account through the use of a proper transfer matrix that significantly improves the accuracy of the solution. Moreover, due to the nature of the problem, the beam eigenfunctions are piecewise continuous and it can be easily shown that they satisfy the orthonormality relationships (Hayashikawa and Watanabe, 1981; Lee and Bergman, 1994; Bilello, 2001)

$$\sum_{i=1}^{N}\int_{0}^{l_i} \varphi_{is}(x_i)\,\rho\varphi_{ir}(x_i)\,dx_i = \delta_{sr} \tag{2.9}$$

$$\sum_{i=1}^{N}\int_{0}^{l_i} \varphi_{is}(x_i)\,EI\varphi_{ir}^{IV}(x_i)\,dx_i = \omega_s^2\delta_{sr} \quad \text{for} \quad s,r = 1,2.....,\infty \tag{2.10}$$

where δ_{sr} is the Kronecker delta.

These functions can be used as basis for a series expansion of the unknown deflection

$$w(x_i,t) \cong \sum_{r=1}^{n} \varphi_{ir}(x_i)\,q_r(t) \quad \text{for} \quad i = 1,2,....,N \tag{2.11}$$

in which the approximation is due to the series truncation, and the $q_r(t)$ are unknown time coefficients to be calculated. Using Eqs. (2.11) and (2.4) in Eq. (2.3), the beam-moving mass interaction force is rewritten as (with explicit time omitted)

$$f_i(t) = m\left\{g - \sum_{r=1}^{n} \varphi_{ir}(\xi_i)\,\ddot{q}_r + 2\dot{\xi}\varphi'_{ir}(\xi_i)\,\dot{q}_r + \left[\dot{\xi}^2\varphi''_{ir}(\xi_i) + \ddot{\xi}\varphi'_{ir}(\xi_i)\right]q_r\right\} \tag{2.12}$$

where the apex prime denotes differentiation with respect to x.

Operating on Eq. (2.1) by $\int_0^{l_i} \varphi_{is}(x_i)\,[\cdot]\,dx_i$, summing over the N segments, and taking into account the orthonormality relationships (2.9, 2.10) yields

$$\ddot{q}_s(t) + \omega_s^2 q_s(t) = F_s(t) \quad \text{for} \quad s = 1,2,....,n \tag{2.13}$$

where

$$\begin{aligned} F_s(t) &= \sum_{i=1}^{N}\int_{0}^{l_i} \varphi_{is}(x_i)\,f_i(t)\,W[\xi_i(t),0,l_i]\,\delta[x_i - \xi_i(t)]\,dx_i = \\ &= \sum_{i=1}^{N} \varphi_{is}[\xi_i(t)]\,f_i(t)\,W[\xi_i(t),0,l_i] \end{aligned} \tag{2.14}$$

The right-hand side of Eqs. (2.13) depends on the functions $f_i(t)$ (Eq. (2.14)) which, in turn, depend on the coefficients $q_r(t)$ (Eq. (2.12)); it is clear, then, that Eqs. (2.13) are a set of coupled second order linear differential equations. However, within all the terms of the summation in Eq. (2.14), only one will be different from zero, i.e., the one for which $W[\xi_i(t), 0, l_i] = 1$. Thus, to calculate $F_s(t)$ at any time t, one must scan the entire beam to find out which is the loaded segment.

In order to simplify the program implementation and allow for a more succinct presentation of the equations involved, introduce the functions

$$\phi_{r,p}(x) = \sum_{i=1}^{N} \varphi_{ir}^{(p)}(x - L_{i-1}) W(x, L_{i-1}, L_i) \quad \text{for} \quad r = 1, 2....n \tag{2.15}$$

where $L_i = \sum\limits_{k=1}^{i} l_k$ and $L_0 = 0$. These will be called *computational eigenfunctions* (note that $\phi_{r,p}(x) \neq \phi_r^{(p)}(x)$). This approach is similar to the one used in Pesterev et al. (2000) to study the dynamic response of a beam traversed by a stream of oscillators. Substituting Eq. (2.12) into (2.14) and following the notation (2.15) yields

$$\begin{aligned} F_s(t) &= m\phi_{s,0}[\xi(t)] \left\langle g - \sum_{r=1}^{n} \phi_{r,0}[\xi(t)] \ddot{q}_r + 2\dot{\xi}\phi_{r,1}[\xi(t)] \dot{q}_r + \right. \\ &+ \left. \left\{ \dot{\xi}^2 \phi_{r,2}[\xi(t)] + \ddot{\xi}\phi_{r,1}[\xi(t)] \right\} q_r \right\rangle \end{aligned} \tag{2.16}$$

Finally, using Eq. (2.16) one can rewrite the set of Eqs. (2.13) in a compact matrix form as

$$\mathbf{M}(t)\ddot{\mathbf{q}}(t) + \mathbf{D}(t)\dot{\mathbf{q}}(t) + \mathbf{K}(t)\mathbf{q}(t) = mg\mathbf{\Phi}_0[\xi(t)] \tag{2.17}$$

where $\mathbf{q}(t)$ is an n-dimensional vector collecting the unknowns $q_r(t)$, the system matrices are given by

$$\mathbf{M}(t) = \mathbf{I}_n + m\mathbf{\Phi}_0^T[\xi(t)]\mathbf{\Phi}_0[\xi(t)] \tag{2.18}$$

$$\mathbf{D}(t) = 2m\dot{\xi}(t)\mathbf{\Phi}_0^T[\xi(t)]\mathbf{\Phi}_1[\xi(t)] \tag{2.19}$$

$$\mathbf{K}(t) = \mathbf{\Omega}^2 + m\mathbf{\Phi}_0^T[\xi(t)] \left\{ \dot{\xi}^2(t)\mathbf{\Phi}_2[\xi(t)] + \ddot{\xi}(t)\mathbf{\Phi}_1[\xi(t)] \right\} \tag{2.20}$$

and

$$\mathbf{\Phi}_p(x) = [\phi_{1,p}(x), \phi_{2,p}(x), ...\phi_{n,p}(x)] \tag{2.21}$$

In Eq. (2.18) $\mathbf{I}_n$ is an n-dimensional identity matrix while $\mathbf{\Omega}^2$ in (2.20) is a diagonal matrix collecting the beam eigenvalues. The initial conditions associated with Eq. (2.17) are written as

$$\mathbf{q}(0) = \int_0^l \mathbf{\Phi}_0^T(x)\,\rho w_0(x)\,dx; \quad \dot{\mathbf{q}}(0) = \int_0^l \mathbf{\Phi}_0^T(x)\,\rho \dot{w}_0(x)\,dx. \tag{2.22}$$

Due to the interaction between the beam and the mass moving on it, all of the system matrices in (2.17) are time dependent. As already mentioned, in a number of previous studies, an approximation to the interaction force has been used so that $\mathbf{D} \equiv \mathbf{0}$ and $\mathbf{K} \equiv \boldsymbol{\Omega}^2$; however, as can be seen from Eqs. (2.19, 2.20), these matrices depend on the speed and the acceleration of the moving mass. It can be shown that, at high speeds or under significant acceleration, the approximation leads to an underestimation of the structural response. Thus, for certain types of structures (for example bridges for high-speed railways, decks for aircraft carriers), especially in presence of damage, the correct interaction force is necessary for accurate response prediction.

Once the vector $\mathbf{q}(t)$ has been obtained, one can calculate the beam response according to the expressions

$$w(x,t) = \mathbf{\Phi}_0^T(x)\mathbf{q}(t) \tag{2.23}$$

$$\vartheta(x,t) = \mathbf{\Phi}_1^T(x)\mathbf{q}(t) \tag{2.24}$$

$$M(x,t) = -EI\mathbf{\Phi}_2^T(x)\mathbf{q}(t) \tag{2.25}$$

$$V(x,t) = -EI\mathbf{\Phi}_3^T(x)\mathbf{q}(t) \tag{2.26}$$

It is worth noting that, for $\mathbf{M} \equiv \mathbf{I}_n$, $\mathbf{D} \equiv \mathbf{0}$ and $\mathbf{K} \equiv \boldsymbol{\Omega}^2$ in Eq. (2.17), the moving force problem is recovered. The procedure just described can easily be extended to include the case in which each segment of the beam differs in stiffness and mass density.

3 Damage model

The rotational spring model has been shown to lead to reasonable predictions of the static as well as the dynamic behavior of damaged beams (Irwin, 1957; Rice and Levy, 1972; Dimarogonas, 1976; Papadopoulus and Dimarogonas; Chondros et al., 1997; Chondros and Dimarogonas, 1998). Though it has been used mainly to model cracked beams, experimental tests (Bilello, 2001) have shown that, for simpler problems (beam under a bending load, stable damage size), the spring model can also be applied to model saw-cuts. The damage configuration used for the experiments presented in the next section is shown in Figure 2.

Damage is realized through symmetric saw-cuts with a depth equal to a. The beam cross section has height h and width $2b$. To the authors' knowledge, the equivalent compliance corresponding to this damage and load configuration was not available in the literature. Thus, following the classical approach based on linear elastic fracture mechanics and taking into account the effective stress distribution throughout the cut

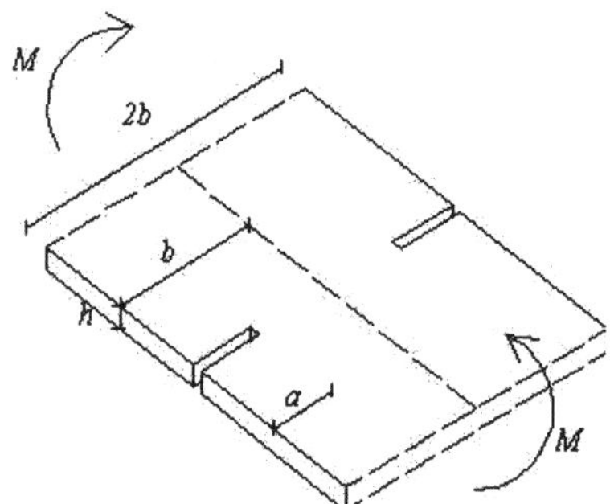

Figure 2. Damage and load configuration.

depth (Anifantis and Dimarogonas, 1983; Tada et al., 1985), the equivalent compliance has been calculated as (Bilello, 2001)

$$c = \frac{6\pi}{Eh^3} \int_0^{\beta} \bar{y} F^2(\bar{y})\, d\bar{y}, \; \bar{y} = y/b, \; \beta = a/b \tag{3.1}$$

where

$$F(\bar{y}) = \frac{1.122 - 0.561\bar{y} - 0.205\bar{y}^2 + 0.471\bar{y}^3 - 0.190\bar{y}^4}{\sqrt{1-\bar{y}}} \tag{3.2}$$

Eqs. (3.1, 3.2) have been used for all numerical simulations herein.

4 Experimental Validation

The analytical solution has been validated through experimental tests. To provide results of practical engineering relevance, a small-scale model of a prototype bridge structure has been designed according to the similitude requirements set forth by a proper dimensional analysis (Bilello et al., 2002).

4.1 Experimental Model

The model used in the experimental investigation consists of a 6061 T-6 aluminum beam with dimensions 1071.5 x 105.25 x 6.35 mm. Two steel rods were glued to the model to form a rail that is used to guide the mass moving on it (Figure 3).

The weight of the beam, including the rail, is 19.7927 N, corresponding to a linear mass density $\rho = 184.7\text{x}10^8\text{Ns}^2\text{mm}^{-2}$. The bending stiffness, estimated from preliminary static tests, is $EI = 162.6 \text{ x } 10^6\text{Nmm}^2$. The beam is simply supported at both ends along two pre-machined lines 10 mm from the end edges. A plywood ramp is placed next to the model for the acceleration of the moving mass. A thin aluminum layer forms its rolling surface on which the same rail as in the model is glued. The inclination of the surface is sixty degrees. The ramp stands on three adjustable feet that allow levelling

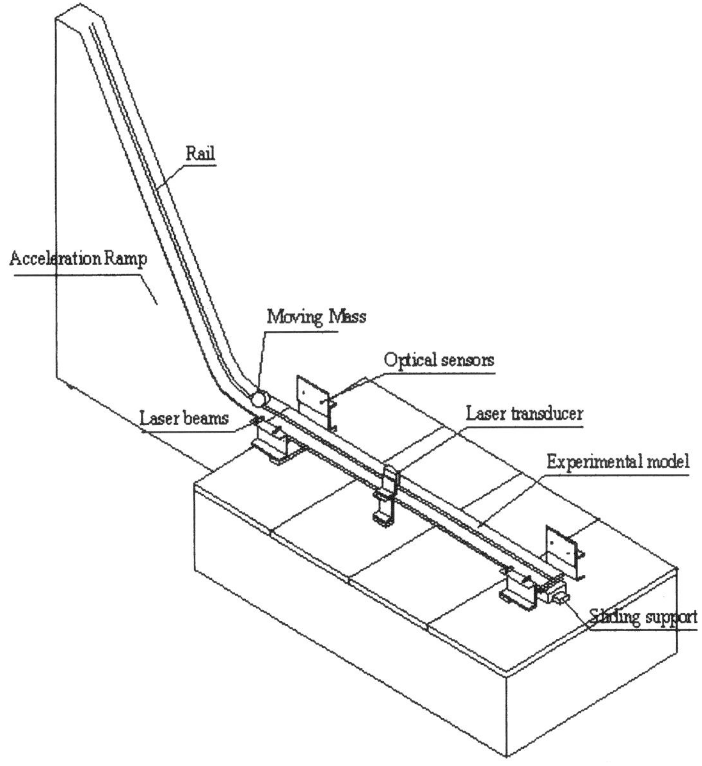

Figure 3. Experimental model and set-up.

with the beam. The moving mass is a 4.952 N steel disk, whose diameter and thickness are, respectively, 50.8 mm and 38.1 mm.

4.2 Experimental set-up

Four optical diodes are used to measure the entrance and the exit speed of the moving disk. The beam deflection is measured through a laser displacement transducer focused on a point 460 mm($7/16l$) from the left-end support. The output signals are acquired through an A/D board integrated with a Tektronix 2630 spectrum analyzer. No windows are applied to the signals. Four model configurations have been investigated: undamaged model, damaged Level 1, damaged Level 2, and damaged Level 3 (Table 1).

The number of tests performed, for each speed range and each model configuration, varied from 30 to 100 depending on the difficulties encountered in the repeatability of the experiments. The tests were labelled V1 through V8 to distinguish between the eight entrance speeds. The average and standard deviation of the entrance speeds and decelerations are reported in Table 2. It is worth noting that the disk deceleration is reasonably small. The speed range goes from $v \approx 3$Km/h ($v \approx 23$Km/h in the prototype

Table 1. Damaged model configurations.

Notation	Location	$\beta = a/b$
Level 1	6/16l	0.3
Level 2	6/16l	0.6
Level 3 = Level 2 +	12/16l	0.6

scale) to a maximum of $v \approx 7.6$Km/h ($v \approx 60$Km/h in the prototype scale). Larger values could not be obtained without risking damage to the experimental set-up, nevertheless, this range is sufficient to represent realistic vehicle speeds.

Table 2. Averages and standard deviations of the disk entrance speeds and decelerations.

Test	Exp. Model		Prot. Structure		Exp. and Prot.	
	E[v] Km/h	s[v] Km/h	E[v] Km/h	s[v] Km/h	E[a] m/s2	s[a] m/s2
V1	3.06	0.05	23.10	0.37	-0.07	0.012
V2	4.12	0.08	31.08	0.59	-0.10	0.022
V3	4.63	0.06	34.93	0.48	-0.10	0.025
V4	5.18	0.07	39.11	0.56	-0.12	0.033
V5	5.72	0.09	43.16	0.65	-0.14	0.046
V6	6.35	0.09	47.92	0.65	-0.16	0.052
V7	6.92	0.10	52.25	0.78	-0.17	0.075
V8	7.59	0.13	57.31	1.01	-0.17	0.099

4.3 Results

In Table 3 the first six theoretical and experimental natural frequencies (obtained from impact hammer tests) are compared. Note that, for the model configurations Level 1 and Level 2, damage is located in the vicinity of a node of the third mode shape ($5.33/16l$); thus, the corresponding natural frequency is nearly unchanged. The theoretical results agree very well with the experimental ones. The maximum absolute error is equal to 1.726 % (f_6 for damage Level 2) while the maximum difference between the undamaged and damaged measured fundamental frequencies is equal to 4.52%.

Table 3. Comparison between theoretical and experimental frequencies (Hz).

	Undam.		Dam. Level 1		Dam. Level 2		Dam. Level 3	
	Theor.	Exp.	Theor.	Exp.	Theor.	Exp.	Theor.	Exp.
f_1	13.322	13.507	13.231	13.233	12.963	13.007	12.743	12.896
f_2	53.291	52.965	53.078	52.560	52.478	52.147	50.623	50.840
f_3	119.90	118.41	119.76	118.44	119.36	117.87	117.49	116.53
f_4	213.16	210.58	211.48	209.60	206.97	205.54	206.95	205.17
f_5	333.07	334.41	332.68	330.99	331.68	328.34	325.82	324.85
f_6	479.62	479.60	477.73	471.60	472.68	464.52	459.71	453.98

Some results of the moving mass tests corresponding to damage Level 3 are reported in Figure 4; the results of the entire experimental investigation can be found in Bilello (2001). The theoretical solution is compared with the average measured one. It has been observed that the experimental response is basically due only to the first mode; thus, a digital Butterworth low-pass filter with a cut-off frequency $f_c = 30$Hz has been applied to the data to remove unwanted high frequency content which does not affect the maximum response. In the light of this consideration, the theoretical response has been calculated retaining only the first term of the series expansion (2.11).

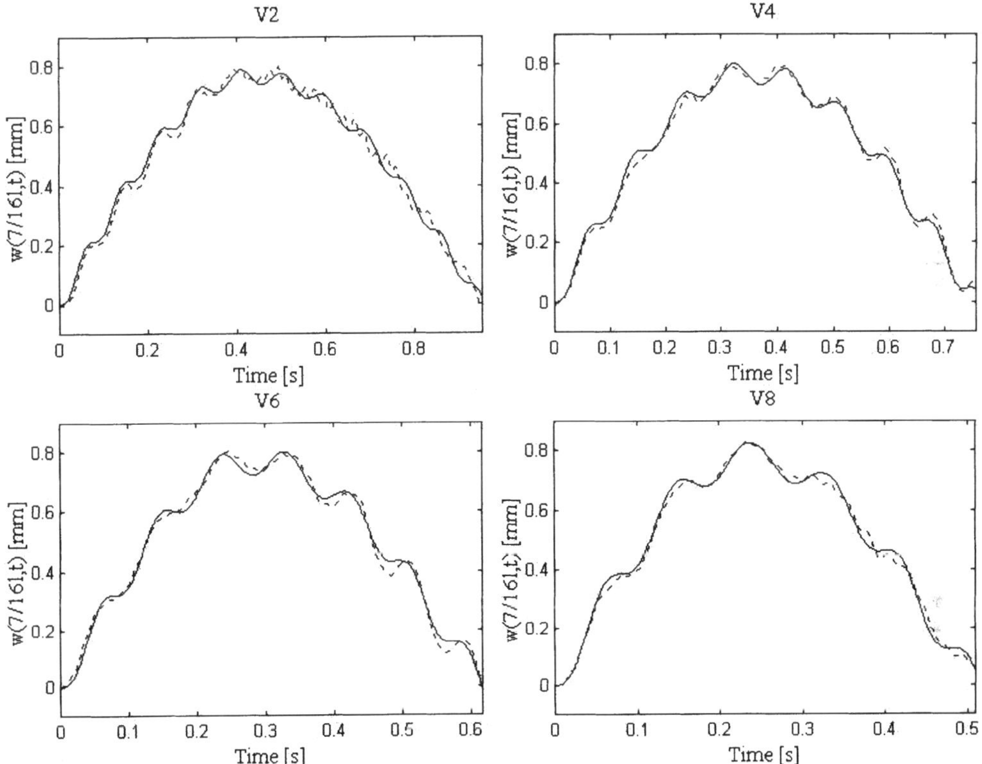

Figure 4. Theoretical (solid) and experimental (dashed) beam deflection at $x = 7/16l$ versus crossing time: damage Level 3.

The experimental results agree very well with the theoretical ones in terms of trend and maximum deflection (Table 4). In general the experimental response is larger than the calculated one and, after the peak, it shows higher response than the theoretical one. This may be caused by out-of-plane oscillations of the moving disk that, in turn, can be due to rolling surface irregularities or small deviations of the rail-guide from a straight line. The maximum absolute error between theoretical and the experimental results is 5.32%, while the maximum difference between the damaged and undamaged measured deflection is 11.37% (damage level 3 V6). It should be noted that the deflection increases with damage level but not always with the moving disk's speed (for the experimental as well as the theoretical data); moreover, the percentages of variation in the beam response to the moving mass are larger than those induced in the beam dynamic parameters. Thus, a higher structural sensitivity is observed under the effect of a moving interacting load. Finally, it should be noted that these experimental observations can be directly extended, through proper scale factors (Bilello et al., 2002), to the prototype structure.

Table 4. Comparison between maximum theoretical and experimental deflection (mm).

	Undam.		Dam. Level 1		Dam. Level 2		Dam. Level3	
	Theor.	Exp.	Theor.	Exp.	Theor.	Exp.	Theor.	Exp.
V1	0.729	0.747	0.730	0.746	0.764	0.777	0.784	0.799
V2	0.726	0.756	0.738	0.763	0.773	0.801	0.783	0.787
V3	0.730	0.736	0.741	0.737	0.779	0.791	0.799	0.797
V4	0.729	0.770	0.742	0.767	0.775	0.814	0.790	0.796
V5	0.736	0.739	0.747	0.751	0.783	0.794	0.802	0.775
V6	0.743	0.712	0.754	0.733	0.788	0.764	0.806	0.793
V7	0.732	0.743	0.742	0.747	0.789	0.812	0.808	0.801
V8	0.749	0.787	0.761	0.779	0.804	0.832	0.821	0.821

5 Conclusions

The effects of damage in Euler-Bernoulli beams traversed by a moving mass has been investigated. Damage is modelled through elastic hinges whose compliance is calculated using fracture mechanics theory. To evaluate the beam-moving mass interaction force the convective acceleration terms, often omitted in similar studies, have been taken into account improving the accuracy of the solution. An eigenfunction expansion of the unknown deflection has been used to obtain an approximate solution. By introducing a set of computational eigenfunction, a compact mathematical formulation, adaptable to computer implementation, has been provided.

The analytical solution has been validated through experimental tests on a small-scale model of prototype bridge structure. In this way data of practical engineering relevance can be obtained and the experimental observation are extendable, through the use of proper scale factors, to the real scale structure. The experimental results are in good agreement with theoretical predictions, however, they often appears to be slightly larger than the calculated one. It is remarkable that damage causes larger perturbations (in percentage) to the dynamic response under the moving mass rather than to the system frequencies. Thus, the nature of the dynamic loads seems to increase the structural damage sensitivity. This observation appears to be very important in light of the potential developments in the field of structural health monitoring and damage identification in bridge structures. Indeed, enormous economic advantages could arise from the use of simple moving load tests for integrity checks in highway bridges.

6 Acknowledgements

This work is part of a larger national research project coordinated by Prof. C. Davini and supported by M.U.R.S.T through grant 2000-2002 number MM08342598-003. The financial support is gratefully acknowledged.
Many thanks go to Prof. L. A. Bergman for his continuous support and valuable suggestions. The financial support of the U.S. National Science Foundation, through grant number CMS-9800136, which made the experimental investigation reported in this paper possible, is also gratefully acknowledged.

Bibliography

J. E. Akin and M. Mofid. Numerical solution for response of beam with moving mass. *Journal of Structural Engineering*, 115:120–131, 1997.

N. Anifantis and A. Dimarogonas. Stability of columns with a single crack subjected to follower and vertical loads. *International Journal of Solids Structures*, 19:281–291, 1983.

C. Bilello. *Theoretical and experimental Investigation on Damaged Beams under Moving Systems.* PhD. Thesis, Dip. Ing. Strut. e Geotec., Università degli Studi di Palermo, 2001.

C. Bilello and L. A. Bergman. Vibration of damaged beams under a moving mass: theory and experimental validation. *Journal of Sound and Vibration*, under review, 2002.

C. Bilello, L. A. Bergman, and D. Kuchma. Experimental investigation of a small scale bridge model under a moving mass. *Journal of Structural Engineering*, under review, 2002.

P.K. Chatterjee, T. K. Datta, and C. S. Surana. Vibration of continuous bridges under moving vehicles. *Journal of Sound and Vibration*, 169:619–632, 1994.

T. G. Chondros and A. D. Dimarogonas. Vibration of a cracked cantilever beam. *Journal of Vibration and Acoustics*, 120:742–746, 1998.

T. G. Chondros, A. D. Dimarogonas, and J. A. Yao. A continuous cracked beam vibration theory. *Journal of Vibration and Acoustics*, 217:17–34, 1997.

S. Christides and A. D. S. Barr. One-dimensional theory of cracked Bernoulli-Euler beams. *International Journal of Mechanical Science*, 26:639–648, 1984.

A. D. Dimarogonas. *Vibration Engineering.* West Publisher, 1976.

A. D. Dimarogonas. Vibration of cracked structures: a state of the art. *Engineering Fracture Mechanics*, 55:831–857, 1996.

L. Fryba. *Vibration of Solids and Structures under Moving Mass.* Noordhoff International Publishing, 1972.

T. Hayashikawa and N. Watanabe. Dynamic behavior of continuous beams with moving loads. *Journal of the Engineering Mechanics Division*, 107:229–246, 1981.

J.L. Humar and A.M. Kashif. Dynamic response of bridges under traveling loads. *Canadian Journal of Civil Engineers*, 20:287–298, 1993.

G. R. Irwin. Analysis of stresses and strains near the end of a crack traversing a plate. *Journal of Applied Mechanics*, 24:361–364, 1957.

J. Lee and L. A. Bergman. The vibration of stepped beams and rectangular plates by an elemental dynamic flexibility method. *Journal of Sound and Vibration*, 171:617–640, 1994.

M. A. Mahamoud. Stress intensity factors for single and double edge cracks in a simple beam subjected to a moving load. *International Journal of Fracture*, 111:151–161, 2001.

L. Meirovitch. *Analytical Methods in Vibrations*. MacMillan, 1967.

M. Mofid and M. Shadam. On the response of beams with internal hinges under moving mass. *Advances in Engineering Software*, 31:323–328, 2000.

C. A. Papadopoulus and A. D. Dimarogonas. Coupled longitudinal and bending vibrations of a rotating shaft with an open crack. *Journal of Sound and Vibration*.

D. R. Parhi and A. K. Behera. Dynamic deflection of a cracked shaft subjected to moving mass. *Transactions of the CSME*, 21:295–316, 1997.

E. C. Pestel and F. A. Leckie. *Matrix Methods in ElastoMechanics*. McGraw and Hill, 1983.

A. V. Pesterev and L. A. Bergman. Vibration of elastic continuum carrying moving linear oscillator. *Journal of Engineering Mechanics*, 123:878–884, 1997a.

A. V. Pesterev and L. A. Bergman. Vibration of elastic continuum carrying accelerating oscillator. *Journal of Engineering Mechanics*, 123:886–889, 1997b.

A. V. Pesterev and L. A. Bergman. A contribution to the moving mass problem. *Journal of Vibration Acoustics*, 120:824–826, 1998.

A. V. Pesterev, L. A. Bergman, and C. A. Tan. Response and stress calculation of an elastic continuum carrying multiple oscillators. In *Proceedings of Advance in Structural Dynamics Vol. I*, pages 545–552, 2000.

A. V. Pesterev, L. A. Bergman, and C. A. Tan. Revisiting the moving force problem. *Journal of Sound and Vibration*, in press, 2002.

J. R. Rice and N. Levy. The part-through a surface crack in an elastic plate. *Journal of Applied Mechanics*, March:185–195, 1972.

S. Sadiku and H.H.E. Leipholz. On the dynamics of elastic systems with moving concentrated mass. *Ingenieur Archive*, 57:223–242, 1987.

M.-H. H. Shen and C. Pierre. Natural modes of Bernoulli-Euler beams with symmetric cracks. *Journal of Sound and Vibration*, 138:115–134, 1990.

M. M. Stanisic, J. A. Euler, and S. T. Montgomery. On a theory concerning the dynamic behavior of structures carrying moving mass. *Ingenieur Archives*, 43:295–305, 1974.

H. Tada, P. C. Paris, and G. R. Irwin. *The Stress Analysis and Cracks Handbook*. Del Research Corporation, 1985.

B. Yang, C. A. Tan, and L. A. Bergman. Direct numerical procedure for solution of moving oscillator problems. *Journal of Engineering Mechanics*, 126:62–69, 2000.

Y. B. Yang and J. D. Yau. Vehicle-bridge interaction element for dynamic analysis. *Journal of Structural Engineering*, 123:1512–1518, 1997.

DYNAMIC BEHAVIOUR AND MODELLING OF RC BEAMS RETROFITTED WITH CFRP SHEETS

Barbara Bonfiglioli *‡ and Giovanni Pascale †‡

* PhD, Research Fellow

† Associated Professor

‡ DISTART - University of Bologna, Viale Risorgimento, 2 - 40136 BOLOGNA ITALY

Abstract A procedure for assessing stiffness variations due to damaging and retrofitting of reinforced concrete beams is presented in this paper. This procedure is based on modal analysis and testing. Two beams were subjected to cyclic loading at increasing load levels, in order to induce in them increasing cracking patterns. The beams were then retrofitted by externally bonded CFRP sheets. Modal tests were carried out after each loading-unloading cycle. A theoretical model was applied, in order to simulate the behaviour of the beams in each situation. This model is based on the concepts of Fracture Mechanics to evaluate the stiffness of cracked beam elements, with and without external strengthening. The theoretical results are in good agreement with the experimental ones. The research suggests that dynamic testing can be used to obtain useful information about the effectiveness of the strengthening system.

1 INTRODUCTION

This research deals with the dynamic identification of structures damaged and subsequently retrofitted with Fibre Reinforced Polymers,or FRP, sheets externally bonded. This technique has been recently used in civil engineering, especially as external strengthening of reinforced concrete structural elements. For example, bending and shear strengthening of beams and wrapping of columns can be realised with FRP sheets externally bonded, due to their inviting features like good mechanical properties, lightness, ease of handling and application, versatility and resistance against corrosion.

A typical application of FRPs in civil engineering deals with the rehabilitation of beams damaged by overloading or accidental events, such as impact of trucks against beams of highway overpasses. The structural behaviour of RC beams strengthened with FRP was studied by several researchers (Arduini and Nanni, 1997) and a guideline on concrete structures externally reinforced with FRP was recently published (fib, 2001).

New experimental techniques were developed in order to perform effective quality control and long term monitoring of external FRP strengthening (Pascale et al , 2000). Nevertheless, these techniques can only provide a control of the application, aimed at detecting local defects or detachments, but they are not capable to evaluate the stiffness recovery and the global behaviour of the structure after the rehabilitation. Damaged areas in structures can be detected and localised by modal testing. Many researchers have given their contributions in this field as (Eilbracht, 1995, Zonta, 2000) and (Pimentel et al , 1999).

The aim of this research is to evaluate the effectiveness of modal testing for assessing the overall stiffness variations in RC beams extensively cracked and subsequently repaired with FRP sheets. Both an experimental investigation and a theoretical approach are presented in this paper.

The research is based on the assumption that stiffness changes of a structural element can be estimated by variations of its dynamic behaviour. Dynamic tests were already evaluated by the authors as a method capable of assessing the effectiveness of the strengthening (Pascale and Bonfiglioli, 2001): a RC beam strengthened with GFRP was tested in that work.

In this paper, further developments of that research are presented. Two RC beams, strengthened with CFRP, were tested in laboratory and their dynamic behaviour was observed in several phases: undamaged, cracked at different damage levels and, finally, repaired.

The theoretical model developed in (Pascale and Bonfiglioli, 2001), based on previous models reported in literature (Massabó and Carpinteri, 1997, Viola and Federici, 1998), was improved using the Fracture Mechanics concepts, to simulate the flexural behaviour of beam elements containing cracks. Different cases were modelled: both undamaged and cracked elements, with and without external strengthening, were studied. For each case the compliance matrix was derived and subsequently the stiffness one. Based on this model, a finite element analysis was carried out in order to evaluate both the static and dynamic behaviour of damaged beams.

2 EXPERIMENTAL

2.1 Testing specimens and set up

The shape and dimensions of the two beams tested are shown in Figure 1.

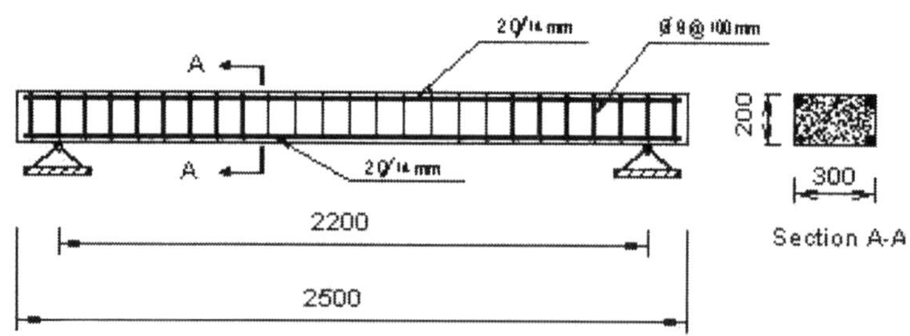

Figure 1. Shape and dimensions (in mm) of the beams

Both beams, identified as C1 and C2, were repaired with unidirectional single-layer CFRP sheets. All FRP sheets were impregnated with epoxy resin. The same resin was used as adhesive between FRP and concrete. The length and width of the sheets were respectively 2000 and 260 mm. The fibre resisting thickness was 0,165 mm.
The mechanical properties of the concrete and the CFRP were previously determined in laboratory on specimens and are summarised in Table 1. The static tests were carried out in three point bending. Two displacement transducers were used to measure the deflection at mid-span section. The dynamic tests were performed with shock excitation, applied by an impact hammer. Accelerometers were used as receiving transducers. Both the excitation point and the position of the accelerometers was varied during testing, to determine the different vibration modes, up to the third one.

Table 1. Mechanical properties of the materials

–	–	–	Concrete	CFRP
Young's modulus	E	MPa	28500	273500
Compressive strength	fc	MPa	64.5	–
Tensile strength	ft	MPa	3.2	2600
Mass density	m	kg/m3	2397	–

The following experimental apparatus was used:

- displacement transducers LVDT HBM TKS 5 mm;
- accelerometers DYTRAN 3100A (sensitivity 100, 2 mV/g, range ±50 g, 1-3500 Hz);
- impact hammer DYTRAN with hard rubber head;
- acquisition card National Instruments AT-MIO 64F-5 200 KS/s 12 bit A/D.

The time history free response was filtered and transferred in Frequency Response Function (FRF) by the Fast Fourier Transform (FFT). Several acquisitions were done for each test, varying the hitting point along the entire span. For each dynamic test, also the damping ratio was determined by evaluating the decay in oscillation amplitude. To do this, the time history free response experimental plots were replaced by exponential curves obtained by fitting the peaks.

2.2 Procedure

Each beam was placed on a loading frame. A first dynamic test was performed, to characterise the undamaged beam. Then some loading-unloading cycles were carried out at increasing load levels. After each cycle, the cracking pattern was recorded, and a dynamic test was performed. Then, the beam was retrofitted with CFRP and other loading-unloading cycles were carried out up to about the same load levels previously reached. After each cycle, a dynamic test was performed. Finally, the beam was loaded up to failure.
Slightly different procedures were used for retrofitting the two beams: the beam C2 was turned off before applying the composite; the beam C1 was just turned on one side. In the first case, this caused a partial closure of the cracks, while in the second case the

cracks were still quite open, when the composite was applied. Due to this difference, only for the beam C2 the primer was placed with great care before applying the FRP.

2.3 Results and comments

The load-deflection curves are shown in Figures 2 and 3 for the beams C1 and C2, respectively, before the application of the CFRP strengthening, and in Figures 4 and 5, after repairing. For both beams the increase of damage and the benefit resulting from FRP strengthening, after its application, is observed. The slopes of the descending branches of each loading cycle are shown in all the figures, allowing us to calculate the stiffness of the beams during unloading.

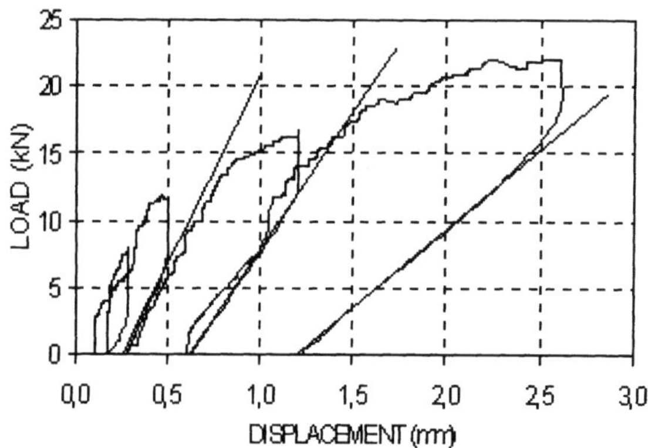

Figure 2. Load vs. mid-span deflection curves for the beam C1 before strengthening

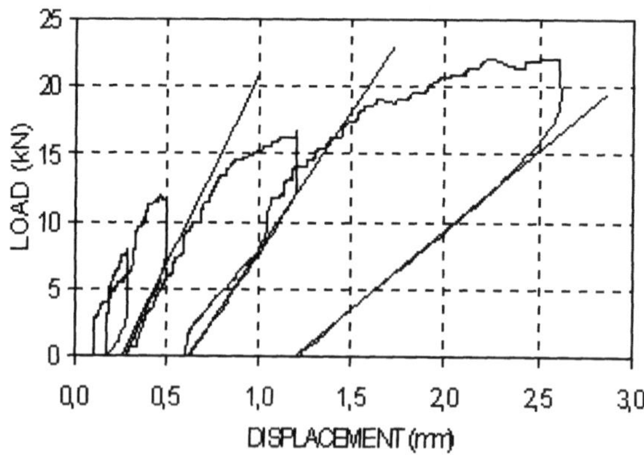

Figure 3. Load vs. mid-span deflection curves for the beam C2 before strengthening

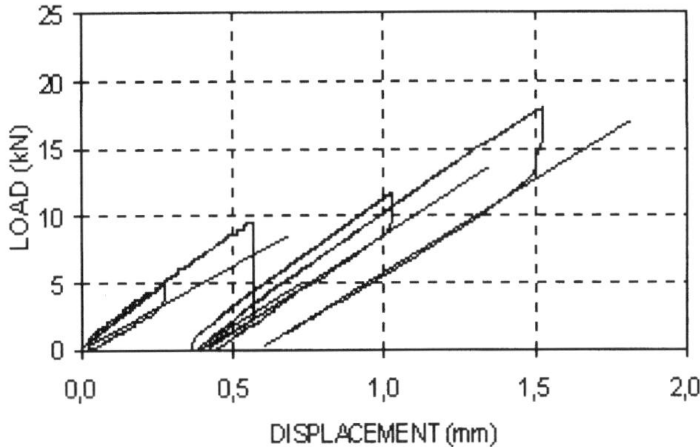

Figure 4. Load vs. mid-span deflection curves for the beam C1 after strengthening

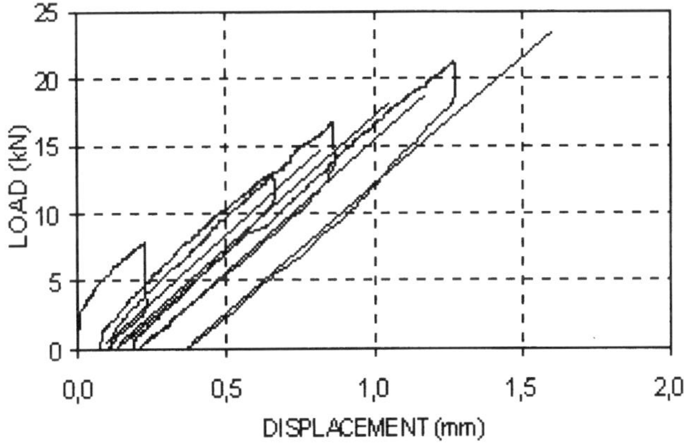

Figure 5. Load vs. mid-span deflection curves for the beam C2 after strengthening

The values of the stiffness, as well as the values of both the frequency and the damping ratio of the first vibration mode, are summarised in Table 2 and in Table 3.
After retrofitting, a partial recovery of the initial stiffness is observed. The recovery is more pronounced for the beam C2 than for the beam C1. This is probably due to the different strengthening procedure, but it has also to be noted that the beam C2 was more damaged, before strengthening.
After strengthening, the stiffness is slightly affected by the loading-unloading cycles. This is due to the action of the strengthening, which gives to the beam quite elastic behaviour.
The results obtained by dynamic tests show similar trend: cracking results in a reduction of the frequency, whereas the application of the composite makes it increasing.
The increase is greater for the beam C2. For both beams the frequency is quite constant after retrofitting.

Table 2. Values of stiffness, frequency and damping ratio for the beams in different situations of the beam C1

Cycle	Max load	Situation	Stiffness	1st Frequency	Damping Ratio
–	kN	–	kN/mm	Hz	–
1	6.00	Undamaged	29.19	57.8	0.029
2	9.50	1 crack	19.63	56.7	0.029
3	11.80	2 crack	15.64	53.8	0.032
4	17.96	3 crack	11.50	50.2	0.033
5	4.90	Retrofitted	18.90	52.0	0.027
6	9.50	Retrofitted	13.18	51.9	0.029
7	11.36	Retrofitted	15.12	51.9	0.026
8	17.96	Retrofitted	13.47	52.0	0.025

Table 3. Values of stiffness, frequency and damping ratio for the beams in different situations of the beam C2

Cycle	Max load	Situation	Stiffness	1st Frequency	Damping Ratio
–	kN	–	kN/mm	Hz	–
1	8.17	Undamaged	–	60.0	0.024
2	12.16	1 crack	29.45	60.0	0.024
3	17.00	2 cracks	20.23	59.0	0.022
4	22.50	5 cracks	11.92	54.3	0.035
5	8,03	Retrofitted	20,88	57,0	0,019
6	13,00	Retrofitted	20,52	58.0	0.021
7	17,00	Retrofitted	19,00	58.0	0.024
8	21,55	Retrofitted	20,20	58.0	0.022

The damping ratio does not appear to be very sensitive to stiffness variations for the beam C1. Instead, it shows a pronounced increase for the beam C2 at the maximum damage level. This is in accordance with the previous remarks: the beam C2 was more damaged than the beam C1, before strengthening. In fact, for the beam C2 strong reduction of the damping ratio can be observed after retrofitting, whereas the variations are very low for the beam C1, which was less damaged.

In general, stiffness variations caused by cracking and retrofitting with FRP can be evaluated by means of both frequency and damping variations, but the sensitivity is different.

Frequency is well correlated to stiffness, but its sensitivity to stiffness variations is not very high.

On the other hand, damping is not very well correlated to stiffness, but it is very sensitive to strong damage and stiffness recovery due to retrofitting.

In order to utilize in practice the experimental results obtained, the following procedure can be adopted. A dynamic test is performed on the structural element to be retrofitted. Another dynamic test is carried out after retrofitting, in the same conditions, and the variations of both frequency and damping ratio are recorded. In Figure 6 the results obtained by applying this procedure to the two beams tested in this research are shown. The percent variations of damping and, in particular, of frequency are lower than those of stiffness, but the trend is similar: the stiffness recovery of the beam C1 is smaller than for the beam C2, and this is revealed by the variations of both dynamic parameters.

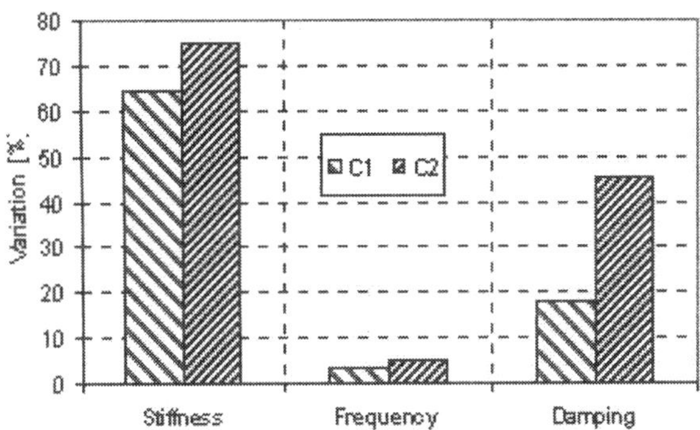

Figure 6. Percent variation of stiffness, frequency and damping ratio after beam retrofitting

3 THEORETICAL ANALYSIS

3.1 Analytical model

An analytical model based on Fracture Mechanics was developed to simulate both the static and dynamic behaviour of a RC beam, with and without cracks. This theoretical approach is based on the following hypotheses :

- the behaviour of all the materials is assumed linear elastic; sections remain plane;
- both the concrete-steel bond and the concrete-FRP bond are assumed perfect;
- the tensile strength of the concrete is taken into account. This hypothesis allows us to evaluate the cracking growth at increasing load levels.

Four cases were studied, which represent actual conditions in RC structures:

1. beam undamaged and not strengthened;
2. beam cracked and not strengthened;
3. beam undamaged and strengthened;
4. beam cracked and strengthened.

First the Castigliano's theorem is used to determine the compliance matrix and subsequently the stiffness matrix is determined by the principle of virtual displacements. According to the Castigliano's theorem, the compliance matrix elements can be calculated by deriving twice the total work with respect to the forces and the bending moments. For the undamaged beam element the total work is equal to the strain energy, while for a cracked beam element it is obtained by summing up the strain energy and the additional energy due to crack propagation. More details can be found in (Pascale and Bonfiglioli, 2001).

3.2 Numerical results and comparison with the experimental data

The analytical model was tested, reproducing the beam used in the experimental tests. The beam was modelled by dividing it in beam elements.
In Tables 3 and 4, the calculated values of the static stiffness and of the first frequency for each cycle are summarised for the beams C1 and C2, respectively.

Table 4. Static stiffness for the beams C1 and C2 in different situations

Cycle	C1-Exp	C1-Num	C2-Exp	C2-Num
–	kN/mm	kN/mm	kN/mm	kN/mm
Before the stengthening				
1	29.19	27.27	–	29.18
2	19,63	18,75	29,45	25,87
3	15,64	16,44	20,23	17,00
4	11,50	15,09	11,92	10,07
After strengthening				
8	13,47	19,52	20,20	13,86

Table 5. 1st frequency for the beams C1 and C2 in different situations

Cycle	C1-Exp	C1-Num	C2-Exp	C2-Num
–	Hz	Hz	Hz	Hz
Before the stengthening				
1	57.8	66.0	60.0	66.0
2	56.7	58.0	60.0	52.8
3	53.8	52.2	59.0	51.5
4	50.2	47.8	54.2	40.1
After strengthening				
8	52.0	53.9	58.0	43.1

All calculated values are compared with the corresponding experimental ones. Both in static and dynamic fields, the results show good agreement between numerical and experimental data, when the beam is undamaged or slightly damaged.
On the contrary, considerable difference is observed, when the beam is strongly damaged. This can be noted in particular for the beam C2 with five cracks.
To explain this discrepancy, we can note that the cracks hypothesised in the theoretical model are supposed to be plane and smooth, and to have height known and constant along the beam width. The height was evaluated by observing the beams during testing, after each cycle. The actual situation differs considerably from this assumption. In fact, the real cracks are neither plane nor smooth, nor of constant height.
In addition, some cracks appeared on both lateral surfaces and consequently were as-

sumed to cross the entire section, but actually cracking area could not involve the internal portion of the section. Remarkable difference is observed also between numerical and experimental results for the second mode frequency.
Somc hypotheses were done, dealing with the experimental procedure and the support condition of the beam during dynamic testing. Further investigation is needed, to better explain this difference.
A comparison between theoretical and experimental in terms of load-deflection curves for the beam C2 is shown in Figure 7 , before the strengthening. To allow a comparison in analogous situations, the experimental curve was modified, by eliminating the unloading-reloading cycles and the hysteretic components of displacement due to these cycles. Also, the curves show good agreement at low load levels, whereas the difference increases when the damage becomes strong, according to the previous remarks.

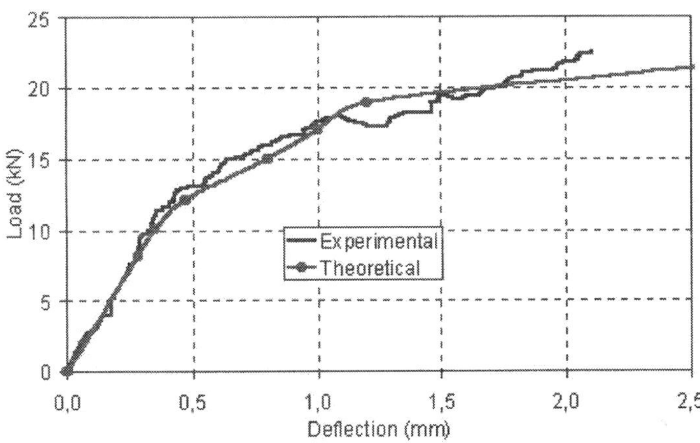

Figure 7. Experimental and theoretical curves of the beam C2 before strengthening

4 CONCLUSION

In this paper, experimental and theoretical research is presented, aimed at evaluating the capability of dynamic testing to give useful information about the stiffness recovery due to external CFRP strengthening of RC beams which were previously damaged. The research showed that the stiffness variations caused by cracking and retrofitting with FRP can be evaluated by means of both frequency and damping variations. Frequency is well correlated to stiffness, but its sensitivity to stiffness variations is not very high. Damping is very sensitive to strong damage and stiffness recovery due to retrofitting. The results suggest that modal testing can be used to obtain useful information not only about the stiffness variations due to damage, but also on the effectiveness of the strengthening system. The theoretical model developed show good agreement between numerical and experimental data, when the beam is undamaged or slightly damaged. On the contrary, considerable difference is observed, when the beam is strongly damaged.

The difference between theoretical and experimental results is probably due to the difficulty to detect the actual cracking pattern, instead of to the model adopted. Further developments are needed to identify and evaluate the damage area of a beam, starting from the experimental data.

5 ACKNOWLEDGEMENTS

The experimental tests were carried out at Experimental Laboratory for the Strength of Construction Materials of the Bologna University. The authors are grateful to the technical staff of the laboratory, namely Mr. Roberto Carli, Mr. Davide Betti and Mr. Gregorio Bartolotta. Prof. Erasmo Viola is acknowledged for his suggestions and encouragement. Ms. Rita Trentini and Mr. Filippo Bastianini are acknowledged for their collaboration. The Italian company MAC Italia S.p.A. is acknowledged for having supplied CFRP and epoxy resin for retrofitting the beams. The research program was funded by ConFibreCrete, a European Union network for the Training and Mobility of Researchers (TMR) and by the Italian Ministry of School and University MIUR.

Bibliography

M. Arduini and A. Nanni, *Behavior of Precracked RC Beams Strengthened with Carbon FRP Sheets*,ournal of Composites for Construction, 1(2), 1997.

G. Eilbracht and M. Link, *Identification of Crack Parameters in Concrete Beams using Modal Test Data*, In roceeding of International Symposium on Non-Destructive Testing in Civil Engineering, pages 327-334 , 1995.

Bulletin 14, *Externally bonded FRP reinforcement for RC structures*, echnical report on the "Design and use of externally bonded fibre reinforced polymer reinforcement (FRP EBR) for reinforced concrete structures" by 'EBR' working party of fib TG 9.3, ISBN 2-88394-054-1 , July 2001.

R. Massabó and A. Carpinteri, *Continuous Vs. Discontinuous bridged-crack model for fiber-reinforced materials in flexure*,nternational Journal of Solids Structures, 34(18):2321-2338, 1997.

G. Pascale and B. Bonfiglioli, *Reinforced concrete beams damaged and repaired with GFRP: dynamic testing and modelling*, In roceeding of International Conference on FRP composites in civil engineering, pages 441-448 , 2001.

G. Pascale, B. Bonfiglioli and F. Bastianini, *Metodologie di indagine per la valutazione della qualità dei rinforzi in FRP applicati a strutture in calcestruzzo e altri materiali*, In roceedings of the XXIX Convegno Nazionale dell'Associazione Italiana per l'Analisi delle Sollecitazioni, pages 739-748 , 2000.

R. Pimentel, A. Pavic and P. Waldron, *Vibration Performance of Footbridges Established Via Modal Testing*, In roceedings of IABSE Symposium "Structures for the Future-The Search for Quality", pages 602-609 , 1999.

E. Viola and L. Federici, *Modellazione del comportamento dinamico di un telaio con travi fessurate*, In roceedings of the XXVII Convegno Nazionale dell'Associazione Italiana per l'Analisi delle Sollecitazioni, pages 1155-1164 , 1998.

D. Zonta, *Structural damage detection and localization by using vibrational measurements*, Ph.D. Dissertation, University of Bologna, 2000.

Identification of Modal Parameters with Unknown Input

Maurizio De Angelis [1] , Vincenzo Sepe [2] and Danilo Capecchi [3]

[1] Dipartimento di Ingegneria Strutturale e Geotecnica, University of Rome "La Sapienza", Rome, Italy
[2] Dipartimento di Progettazione, Riabilitazione e Controllo delle Strutture Architettoniche, University of Chieti-Pescara "G. D'Annunzio", Pescara, Italy
[3] Dipartimento di Scienza delle Costruzioni, University of Naples "Federico II", Naples, Italy

Abstract. The techniques developed for structural identification are typically devoted to obtain a structural model on the base of information on the response and on the forcing action. In many situations, however, it can be necessary to refer only to the measured responses. In this paper we describe the theoretical and computational aspects of a technique we have recently developed to identify the modal model in the frequency domain when the input is unknown. We show that this identification problem has an unique solution when at least three responses are known. Anyway, when one of the modal parameters is known, the other ones can be obtained on the base of two responses only. Numerical applications are carried out on plane framed structures ranging from three to eight stories. The modal model may also be incomplete, including only the lowest modes. The results obtained are satisfactory in most cases.

1 Introduction

The structural identification techniques are typically devoted to obtain a structural model on the basis of experimental data. The model can be parametric or non-parametric, but always some assumptions on its structure must be made. Often it is stated that the model suitable to reproduce the structural response is linear elastic. In such a case the use of a parametric modal model is common: the structure is described by a set of modal vectors, natural frequencies and modal dampings (Ewins, 1986; Beck and Jennings, 1980; McVerry, 1980; Goyder, 1980).

The experimental data usually consist of time-histories both from the excitation and from the response. In many situations, however, it can be necessary, or simply useful, to refer only to the measured response, above all in the civil engineering field (Torkamani and Ahmadi, 1988; Brownjohn *et al.*, 1989 and 1992; Wilson and Liu, 1991; Katafygiotis *et al.*, 1998; Katafygiotis and Yuen, 1998; Farrar and James, 1997; Quek *et al.*, 1999). This is the case when the forcing cannot be measured easily, e.g. due to accessibility reasons, or if only a very small number of sensors can be used, in the limit case one sensor only, due to economic reasons. Basically two very different situations can be considered. In the first one the unknown input is assumed as a stationary random process, in the limit a white noise process; this occurs for ambient excitation, e.g. traffic, wind or micro-earthquakes. In the second situation the input can no longer be assumed as stationary; this occurs for most earthquakes, for explosions and also for artificially induced forces.

Identification techniques are very different in the two cases. In the stationary excitation situation, which is more commonly considered, reliable techniques both in time domain and frequency domain are available (Ibrahim, 1977; Luz and Wallaschek, 1992; Wallaschek, 1989). On the other hand, consolidated procedures do not exist for non-stationary unknown excitation. In the time domain the method applied by Wang and Haldar (1994, 1997) to unknown seismic excitation is quite interesting, although its limits are not clear. For the same kind of excitation, the technique in the frequency domain proposed by Benedetti and Gentile (1994) should also be signalled.

This paper describes theoretical and computational aspects of a technique we have recently developed when the input is unknown (Capecchi *et al.*, 2002*a,b*; Sepe *et al.*, 2001). Assuming a linear behaviour of the structure, the objective is to identify a parametric modal model in the frequency domain; the input can be either stationary or non-stationary. The basic idea is that the ratio of the Fourier transform of the time-history responses at different points of a structure is input-independent and then it can be an experimental source for modal model identification.

To simplify, the paper deals only with seismic situation, but many of the conclusions can be extended to different situations. As shown theoretically in the quoted papers, for this kind of input and when at least three time-histories are known, the identification problem has a limited number of isolated mathematical solutions, among which the effective one can easily be chosen, as the only one that is meaningful from the structural point of view.

It is discussed the accuracy that can be obtained when only two responses are known, but one of the modal parameters can be estimated in some other way (e.g. a natural frequency can be get by visual inspection of the Fourier transform of the responses). The possibility of identifying a reduced model, taking into account only the lowest modes, is also discussed.

2 Theoretical Aspects

When considering the frequency domain context, the input-output relation of a linear system under an imposed base motion has the simple linear algebraic form:

$$Y(\omega) = H(\omega)\, A(\omega) \tag{1}$$

where $Y(\omega)$ is the vector of the responses Fourier transforms (FT) of the n degrees of freedom (dofs) of the system, $A(\omega)$ is the FT of the base motion acceleration and $H(\omega)$ is the vector of transfer functions from A to Y. Although devices for measuring structural displacements and velocities are available (e.g. optical transducers described in Sepe and Zingali, 2001), in the experimental practice usually absolute time-history accelerations are measured. Therefore, in the following $Y(\omega)$ will be assumed to be an absolute acceleration and consequently $H(\omega)$ is the absolute acceleration transfer function vector.

Relation (1) can be rewritten for the single i-th component of the response in the form:

$$Y_i(\omega) = H_i(\omega)\, A(\omega) \tag{2}$$

and the i-th transfer function can be given the following expression, as function of modal parameters:

$$H_i(\omega) = 1 - \omega^2 \sum_{k=1}^{n} \frac{v_{ik}}{\omega^2 - 2i\xi_k \omega\omega_k - \omega_k^2} \tag{3}$$

where v_{ik} is the i-th component of the vector $v_k = p_k\, u_k$, with p_k the k-th participation factor and u_k the k-th normal mode normalized to mass matrix, ω_k and ξ_k are the k-th natural frequency and modal damping, respectively. The summation is extended over the n modes of the system, equal in principle to the number n of dofs of the structure.

Consider the response of two points, i and j, of the structure, whose response is represented by $Y_i(\omega)$ and $Y_j(\omega)$. Both of them depend on the input $A(\omega)$ and on the modal parameters of the structure, but their ratio is independent of the input, as it is clear from relation (1),

$$R_{ij}(\omega) = \frac{Y_i(\omega)}{Y_j(\omega)} = \frac{H_i(\omega)A(\omega)}{H_j(\omega)A(\omega)} = \frac{H_i(\omega)}{H_j(\omega)} \tag{4}$$

and depends only on the modal parameters. By collecting them in the vector x, the relation (4) can be rewritten as:

$$R_{ij}(\omega) = \frac{H_i(\omega, x)}{H_j(\omega, x)} \tag{5}$$

or equivalently as:

$$Y_i(\omega)H_j(\omega, x) - Y_j(\omega)H_j(\omega, x) = 0 \tag{6}$$

Relation (6) expresses a non-numerable set of equations for the unknown x, one for each value of ω, and the question rises whether or not it has a solution. Before considering the problem it must however be noticed that not all the unknowns are independent, because the following completeness relations hold for the modal vector components (De Angelis *et al.*, 2001):

$$1 - \sum_{k=1}^{n} v_{\alpha k} = 0, \quad \alpha = 1,2,\ldots,n \tag{7}$$

2.1 Solvability Conditions

It can been shown that relations (5) or (6) are not sufficient to get a unique solution for x, as it follows from the reasoning below, which summarises results obtained by Capecchi *et al.* (2002*a,b*).

Both H_i / H_j and R_{ij} can be given the polynomial rational form:

$$\frac{H_i(\omega)}{H_j(\omega)} = \frac{a_1^i \omega^{2n-1} + a_2^i \omega^{2n-2} + \ldots + a_{2n}^i}{a_1^j \omega^{2n-1} + a_2^j \omega^{2n-2} + \ldots + a_{2n}^j}$$
$$\frac{Y_i(\omega)}{Y_j(\omega)} = \frac{b_1^i \omega^{2n-1} + b_2^i \omega^{2n-2} + \ldots + b_{2n}^i}{b_1^j \omega^{2n-1} + b_2^j \omega^{2n-2} + \ldots + b_{2n}^j} \tag{8}$$

The first expression of (8) derives from the definition of $H_r(\omega)$; the second one can be obtained from the experimental data function $Y_r(\omega)$ with a polynomial fitting procedure, which however, at least in principle, in absence of noise furnishes exact results. By comparing the two ratios (8) and the equation (6), the following relations can be written:

$$\begin{aligned}
& a_1^i(x) = b_1^i; \quad a_1^j(x) = b_1^j \\
& \ldots \\
& \ldots \\
& a_{2n-2}^i(x) = b_{2n-2}^i; \quad a_{2n-2}^j(x) = b_{2n-2}^j \\
& a_{2n-1}^i(x) = a_{2n-1}^j(x) = b_{2n-1}^i = b_{2n-1}^j \\
& a_{2n}^i(x) = a_{2n}^j(x) = b_{2n}^i = b_{2n}^j
\end{aligned} \tag{9}$$

Notice that while writing relations (9) it has been taken in to account that $a^i_{2n-1} = a^j_{2n-1}$ and $a^i_{2n} = a^j_{2n}$ (Capecchi *et al.*, 2002*a,b*), so that coefficients of the polynomial with lowest degree give only two equations instead of four.

The left hand side of the two relations (9) must be considered as function of the modal parameters x, the right hand side are instead known quantities. So relations (9) represent an algebraic system of equations with the modal parameters as unknowns. There are $4n$ equations, $4n - 2$ corresponding to the number of coefficients a_i, a_j and two more equations because of completeness relations (7). There are also $4n$ unknowns, n modal frequencies, n modal damping coefficients and $2n$ modal components (n for each one of the two dofs i and j). Therefore, the system seems to be determined, at least in principle. Unfortunately, the knowledge of the ratio $R_{ij}(\omega)$ does not allow to have all the coefficients b_i. As it is clear from (8), these coefficients can be determined only scaled by a constant c that, being irrelevant in the ratio, can assume any arbitrary value and must therefore be added to the list of unknowns, which so become $4n+1$, against $4n$ equations.

There is then an indeterminacy of order one, in the sense that if only a modal parameter is specified, for instance the first natural frequency, the problem has an unique solution. To make the problem fully determinate, one more response $Y_s(\omega)$ must be known besides the response functions $Y_i(\omega)$ and $Y_j(\omega)$, so that one more ratio, $R_{is}(\omega)$ or $R_{js}(\omega)$, can be evaluated and a new equation like (6) considered. In this case, indeed, the problems becomes over-determined.

3 Objective Function

To solve the over-determined problem expressed by relation (6) and the corresponding relations obtained considering dofs i and s instead of i and j, it is followed the classical approach of system identification by assuming the objective function:

$$l(\mathbf{x}) = \sum_{r=1}^{m} g_r^{ij} \left| Y_i(\omega_r) H_j(\omega_r, x) - Y_j(\omega_r) H_i(\omega_r, x) \right|^2$$
$$+ \sum_{r=1}^{m} g_r^{is} \left| Y_i(\omega_r) H_s(\omega_r, x) - Y_s(\omega_r) H_i(\omega_r, x) \right|^2 + \sum_{\alpha=i,j,s} g_\alpha \left(1 - \sum_{k=1}^{n} v_{\alpha k} \right)^2 \tag{10}$$

where g_r^{ij} and g_r^{is} and g_α are weight coefficients. It is clear that in absence of measurement errors the search for the minimum of this objective function is equivalent to solve the equations (6); the objective function (10) can be easily generalised for the cases when more than two ratios are available.

The expression of the objective function (10) is quadratic with respect to most of the x component, more precisely to the components that represent the modal vectors. As it will be shown later, the occurrence of quadratic terms makes simpler the numerical problem.

Up to now a noiseless situation has been considered; in practice, however, noise is present. One of the main consequence of the presence of noise is that the higher modes contributions on the response cannot be appreciated. In practice, instead of using a complete modal model, an incomplete model must be used, that considers only $n^* < n$ modes. The missing modes can be represented by a unique mode, with the transfer function assuming the following expression:

$$H_i(\omega) = 1 - \omega^2 \sum_{k=1}^{n^*} \frac{v_{ik}}{\omega^2 - 2i\zeta_k \omega \omega_k - \omega_k^2} - \frac{V_i}{\Omega} \omega^2 \tag{11}$$

where V_i and Ω are associated to the missing modes.

Moreover, some a priori information is sometimes available for the modal quantities. For example the first natural frequencies can be estimated from the visual inspection of responses $Y(\omega)$. From this considerations it follow that the most general expression for the objective function to be used is:

$$L(x) = l(x) + \sum_{i=1}^{N} g_i (x_i - x_i^0)^2 \tag{12}$$

where $l(x)$ is defined in Eq.(10), x_i^0 represents an *a priori* estimate for the i-th modal quantities and g_i express our confidence on this estimate.

The minimum of this objective function is found by using a properly modified Gauss-Newton algorithm. Being $l(x)$ quadratic with respect to some of the x components, a two step identification procedure is implemented. Vector x is partitioned in two parts $x \equiv (x_\omega \mid x_v)$, where

x_ω gathers the natural frequencies and modal damping while x_v gathers the modal vector components. To start the iterations, an a priori estimate x_ω^0 for x_ω is assumed and $l(x)$ is considered as a function of x_v only, that is $l(x)= l(x_v)$. Because $l(x_v)$ is a quadratic objective function, its minimum can be found in a single step by solving a linear system of equations, which gives the estimate x_v^1. Assuming $x_v = x_v^1$, $l(x)$ is then considered as a function of x_ω, that is $l(x) = l(x_\omega)$. Objective function $l(x_\omega)$ is now minimised by using the Gauss-Newton method. The procedure is iterated until the estimate vectors x_v^i and x_ω^i stabilise.

The minimum problem may be badly conditioned and this aspect can give problems in the presence of experimental or modelling errors. Some of these problems can be avoided with an appropriate choice of the weights g_r^{hk}. The choice which, in the numerical investigation performed, has been shown to be the most effective one has been to assume:

$$g_r^{hk} = \left| \frac{1}{H_k(\omega,\hat{x})Y_h(\omega)} \right|^2 \tag{13}$$

where $\hat{x}$ is an estimate of x. In the iteration procedure for the i-th step it is assumed for $\hat{x}$ the value of x estimated at the i-1-th step. In presence of errors, it is useful to assume weights also during the linear estimation phase of modal vector components.

Notice that the objective function (10) with the weights as given by relation (13), when it is assumed $\hat{x} = x$ is similar to the objective function

$$l(x) = \sum_{r=1}^{m} \left| R_{ij}(\omega) - \frac{H_i(\omega_r, x)}{H_j(\omega_r, x)} \right|^2 + \sum_{r=1}^{m} \left| R_{is}(\omega) - \frac{H_i(\omega_r, x)}{H_s(\omega_r, x)} \right|^2 + \sum_{\alpha=i,j,s} g_\alpha \left(1 - \sum_{k=1}^{n} v_{\alpha k} \right)^2 \tag{14}$$

(it would be the same if complex values were assumed for the weights g_r^{hk}); $l(x)$ as defined in Eq.14, due to the independence of R_{ij} from the input (Eq.4), would seem the most natural objective function. However, the direct use of the objective function (14) is not advisable, because it is not quadratic in the modal vector components and because the ratio between the experimental responses can be very sensitive to errors.

4 Numerical Applications

The procedure presented in previous sections has been applied to two simple structures. A five stories frame (non shear-type) and a eight stories shear type structure. In both cases the absolute acceleration time histories of the stories are evaluated in the ideal noiseless situation under a base motion represented by the NS component of El Centro earthquake.

After the modal model has been identified, the input has been reconstructed, in the first case, according to the relation:

$$A(f) = \frac{Y_i(f)}{H_i(f)} \tag{15}$$

where f denotes the frequency (Hz), used from now on instead of the angular frequency ω.

4.1 Five stories frame

The frame is the same studied in De Angelis *et al.* (2001) (Figure 1), with proportional damping ξ=0.02 for all the modes and modal frequencies f_1=1.24 Hz, f_2=3.99 Hz, f_3=7.45 Hz, f_4=11.7 Hz, f_5=16.4 Hz.

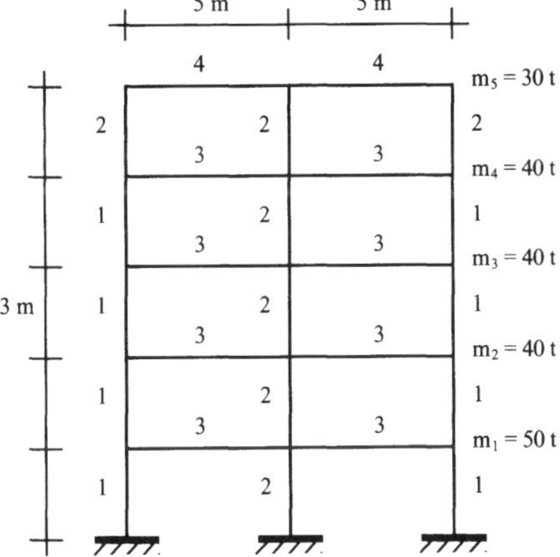

SECTION		EI (Nm^2)
1	HEB 400	1.13 10^8
2	HEB 360	8.45 10^7
3	IPE 400	4.53 10^7
4	IPE 300	3.18 10^7

Figure 1. Geometrical and mechanical characteristics of the frame

Knowledge of three responses. In the first case, ratio R_{13} and R_{15} are considered (Figure 2), with floors ordered from bottom to top, without any further information, and identification is carried out by using objective function (10).

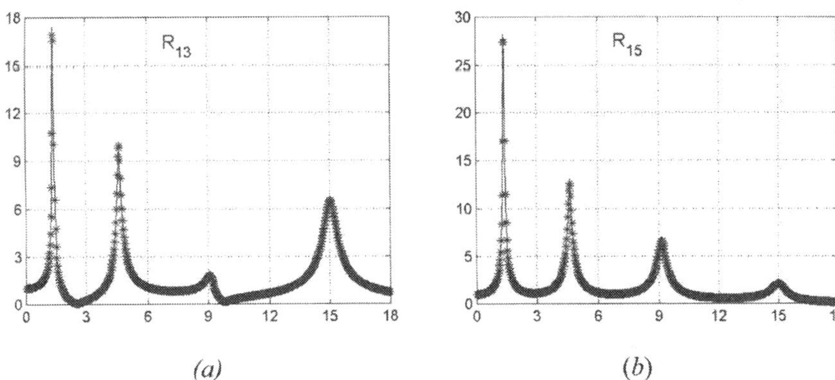

Figure 2. Absolute values of R_{13} (*a*) and R_{15} (*b*) *vs* frequency (Hz); exact values (stars) and reconstructed ones (continuous line)

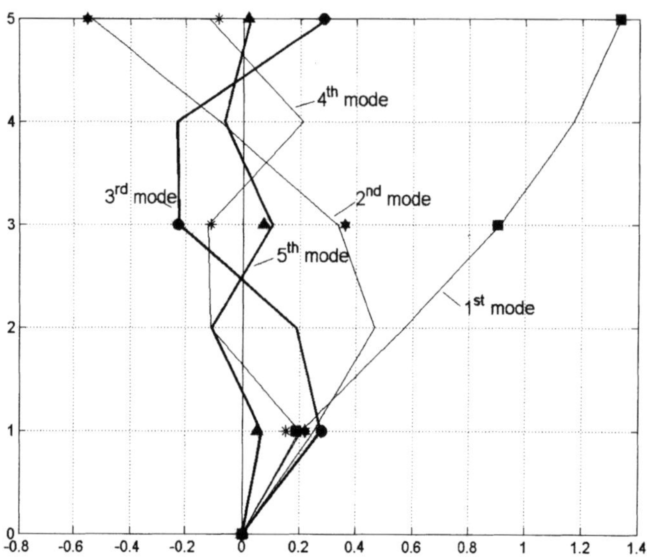

Figure 3. Reconstructed and exact eigenvectors

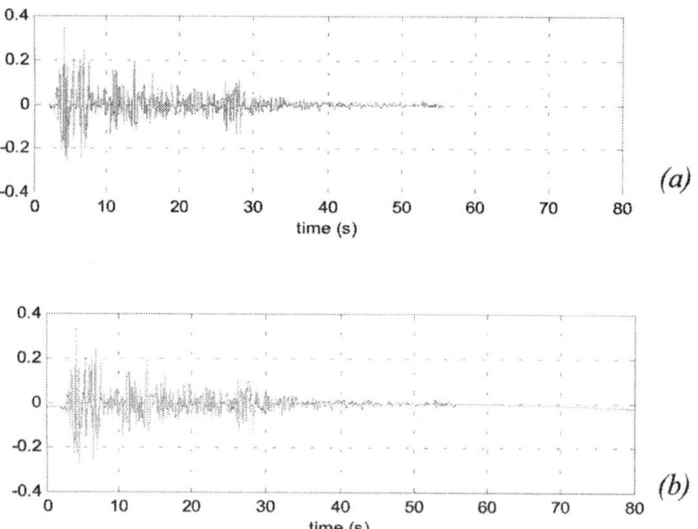

Figure 4. "True" time-history (*a*) and reconstructed time-history (*b*) of the input acceleration (El Centro) normalised with respect to $g = 9.81$ m/s^2

The identification procedure works very well and the identified values of parameters show a small error for all of them, which depends on the rounding-off due to the various numerical computations. Figure 2 compares the reconstructed values (continuous line) and the exact values (stars) for $R_{13}(f)$ and $R_{15}(f)$. Figure 3 compares the reconstructed eigenvectors and the exact ones, and Figure 4 compares the "true" input time-history and the reconstructed input for the same case. As is clear from the figures, the identified results fit very well the exact one, confirming the robustness of the numerical approach.

Knowledge of two responses. In the second case only R_{12} is considered (Figure 5). In this case the problem is undetermined and a unique solution can be obtained only when one more data is known. Here, the first natural frequency f_1 is assumed as a supplementary information; this frequency can be estimated from the FT of the response. The identified values, obtained as an average of values given by $Y_1(\omega)$ and $Y_2(\omega)$ shown in Figure 6, is f_1=1.22 Hz.

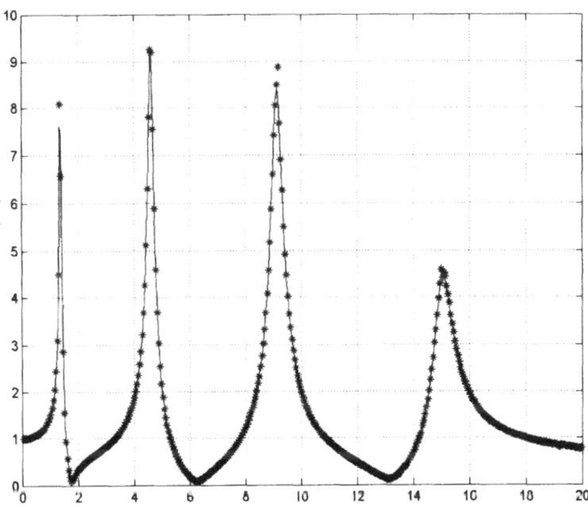

Figure 5. Absolute value of R_{12}

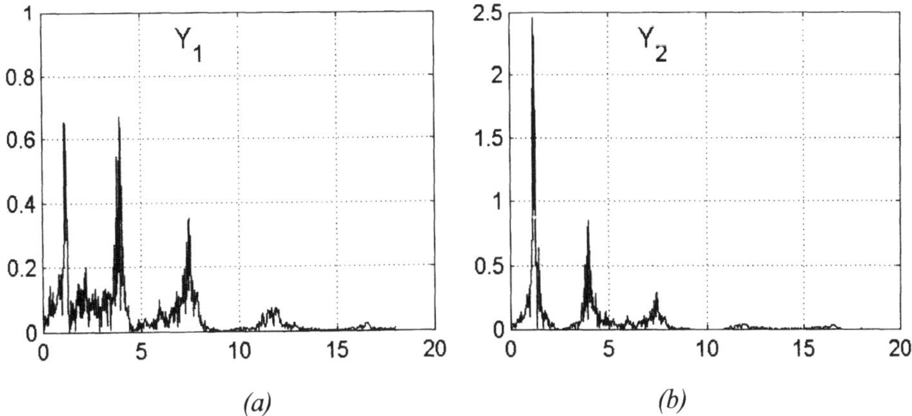

Figure 6. FT transforms (absolute values) of Y_1 and Y_2

Figure 7 shows the errors in modal parameters, defined as:

$$\varepsilon(x_i) = \frac{x_i^{estim} - x_i^{exact}}{x_i^{exact}} \tag{16}$$

as a function of the errors in the estimate of the first natural frequency. An examination of this figure shows that errors are amplified only in a few cases. Similar results were found in Capecchi *et al.* (2002b) for a 3 dofs system, with the use of an analytical-numerical approach.

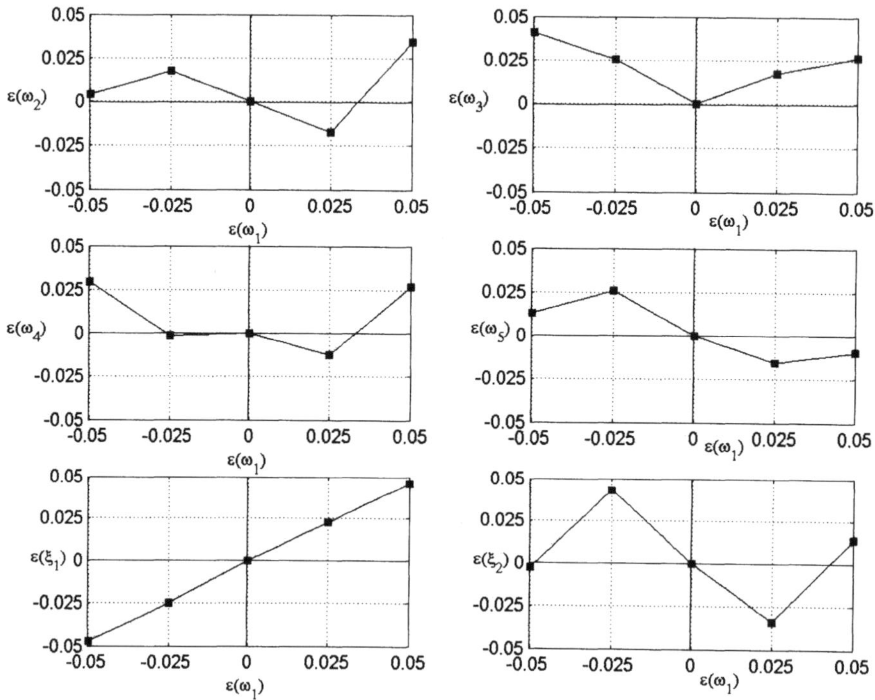

Figure 7. Errors for estimate of some modal parameters as functions of errors on first natural frequency

Use of an incomplete model. As clear form Figure 6, even in simple case the Fourier transform of the output allows to estimate only the lowest frequencies with sufficient accuracy. By increasing the number of dofs it is then necessary to make recourse to an incomplete model. This because the lack on knowledge of an initial estimate for frequencies and damping coefficients makes it difficult the convergence of the proposed numerical algorithm. Moreover, in presence of measurement or modelling errors, the estimate of higher frequencies would be always not reliable independently of the numerical approach used for identification.

Figure 8 shows the comparison between the ratios R_{13} and R_{15} used for the modal parameters identification. They correspond to the data shown in Figure 2 truncated at the frequency of 8 Hz, as whether data for f>8Hz were strongly corrupted by noise. The incomplete modal model contains only thc first three modes.

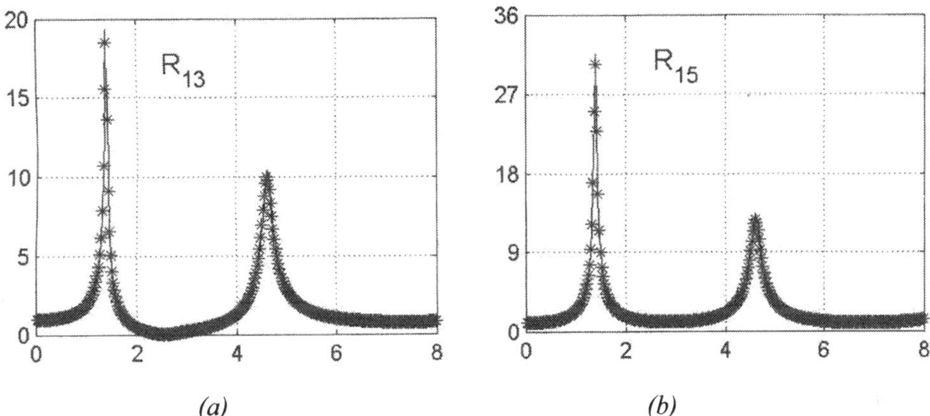

Figure 8. Absolute values of R_{13} *(a)* and R_{15} *(b)* truncated at 8 Hz; exact values (stars) and reconstructed ones (continuous)

Tables 1 compares exact frequency and damping values with those obtained with the incomplete model. As it can be seen, the errors of the identified model is quite low.

Table 1. 3 dofs model for the 5 stories frame. Exact and identified values of natural frequencies and damping coefficients

Mode	1	2	3
f_{exact} [Hz]	1.243	3.993	7.448
f_{estim} [Hz]	1.247	3.868	7.536
ξ_{exact}	0.020	0.020	0.020
ξ_{estim}	0.020	0.020	0.0180

4.2 Eight dofs shear-type frame

Similar conclusions hold for the other example considered (taken from Lus *et al.*, 1999), an eight dof shear type frame, with modal dampings $\xi_1=\xi_2=\xi_3=\xi_4=0.02$, $\xi_5=\xi_6=\xi_7=\xi_8=0.05$ and modal frequencies (Hz) f_1=0.922, f_2=2.73, f_3=4.45, f_4=6.02, f_5=7.38, f_6=8.49, f_7=9.32, f_8=9.82.

For this example, ratios R_{14} and R_{18} have been considered. Data are truncated at 6.2 Hz and the incomplete model contains the first four modes.

Figure 9*a,b* and Table 2, that compare exact and identified values of these ratios, confirm the robustness of the proposed procedure.

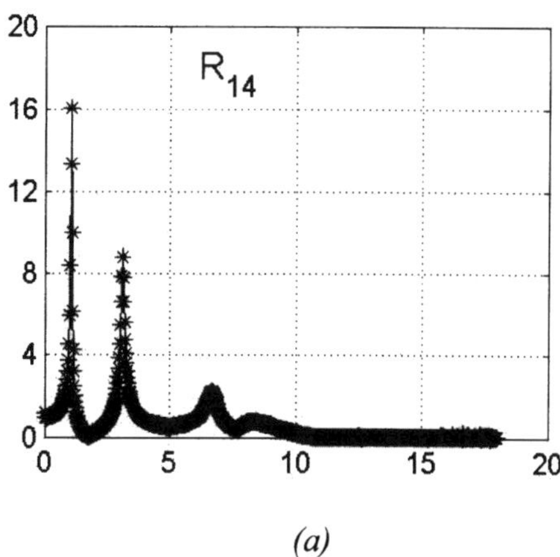

(a)

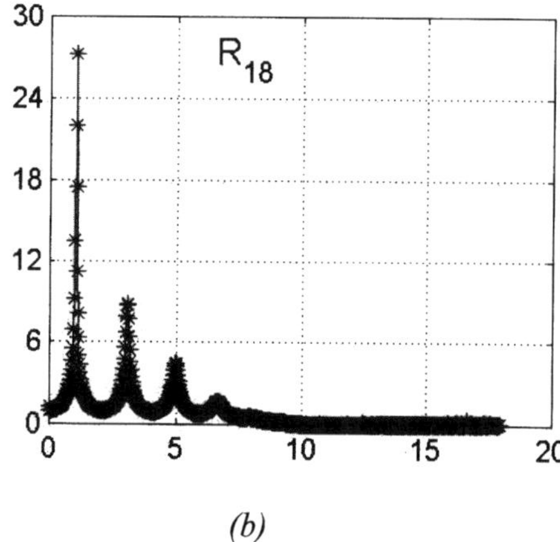

(b)

Figure 9. Eight dof shear type frame. Absolute values of R_{14} (*a*) and R_{18} (*b*) *vs* frequency (Hz); exact values (stars) and reconstructed ones (continuous line).

Table 2. 4 dofs model for the 8 stories frame. Exact and identified values of natural frequencies and damping coefficients.

Mode	1	2	3	4
f_{exact} [Hz]	0.922	2.730	4.450	6.020
f_{estim} [Hz]	0.927	2.738	4.442	6.122
ξ_{exact}	0.020	0.020	0.020	0.020
ξ_{estim}	0.020	0.020	0.0190	0.023

5 Concluding remarks

It is shown that the modal parameters of a linear structure subject to an unknown base-motion can be identified when at least three time-histories of the response are known; in this case, in fact, the identification problem has a limited number of isolated mathematical solutions, among which the effective one can be chosen as the only one that is meaningful from the structural point of view.

However, the search of the effective solution requires an ad hoc computational procedure, that is discussed in this paper. The numerical results confirm the robustness of the proposed algorithm.

Numerical applications are developed for a five and eight degrees-of-freedom systems. Pseudo-experimental response is given by time-history accelerations. Results obtained are satisfactory in most cases, even when the modal model is incomplete, in the sense that only the lowest modes are included. In the case when only two responses are known, modal parameters can be identified if at least one of them can be estimated in some other way. In the paper it is studied how much the reliability of the estimate of the unknown parameters depends on the reliability of a natural frequency, namely the first one, obtained by visual inspection of FT of the responses; it is shown that the errors on the identified modal parameters are of the same order of magnitude of the errors on the assumed frequency.

References

Beck J.L., Jennings P.C., Structural identification using linear models and earthquake records, *Earthquake Engineering and Structural Dynamics*, vol.8, pp. 145-160, 1980

Benedetti D., Gentile C., Identification of modal quantities from two earthquake responses, *Earthquake Engineering and Structural Dynamics*, 23: 447-462, 1994

Brownjohn J.M., Dumanoglu A.A., Severn R.T., Blakeborough A., Ambient vibration survey of the Bosporus suspension bridge, *Earthquake Engineering and Structural Dynamics*, vol.18, pp. 263-283, 1989

Brownjohn J.M., Dumanoglu A.A., Severn R.T., Ambient vibration survey of the Fatih Sultan Mehmet (second Bosporus) suspension bridge, *Earthquake Engineering and Structural Dynamics*, vol.21, pp. 907-924, 1992

Capecchi D., De Angelis M. & Sepe V., 2002, Identification of structures subject to unknown seismic excitation, *3rd Joint Conference GIMC and AACME*, Giulianova (Italy), June 2002*a*

Capecchi D., De Angelis M., Sepe V., Modal model identification with unknown nonstationary base motion, *submitted to Meccanica*, 2002b

De Angelis M., Sepe V., Vestroni F., Identificazione dei parametri modali di una struttura eccitata alla base, *Ingegneria Sismica*, XVIII (3), pp.42-50, 2001 (in Italian)

Ewins D.J., Modal testing: theory and practice, *Research Study Press – J. Wiley & Sons*, 1986

Farrar C.R., James G.H. III, System identification from ambient vibration measurements on a bridge, *Journal of Sound and Vibration*, vol.205, n.1, pp.1-18, 1997

Goyder H.D.G., Methods and application of structural modelling from measured frequency response data, *J. of Sound and Vibration*, 68, 1980, pp. 209-230

Ibrahim S.R., Random Decrement Technique for modal identification of structures, *Journ. Of Spacecraft and Rockets*, vol. 14 n.11, Nov. 1977, pp.696-700

Katafygiotis L.S., Mickelborough N.C., Yuen K.V., Statistical identification of structural modal parameters using ambient response data, Proceed. *3rd Internation Conference on Computational Stochastic Mechanics*, Island of Santorini, Greece, pp.93-102, 1998

Katafygiotis L.S., Yuen K.V., Statistical modal identification using ambient or string wind response data, Proceed. *4th Internation Conference on Stochastic Structural Dynamics*, University of Notre Dame, Indiana (USA), pp.387-394, 1998

Lus H., Betti R., Longman R. W., Identification of Linear Structural Systems Using Earthquake - Induced Vibration Data, *Earthquake Engineering and Structural Dynamics*, Vol.28, pp.1449-1467, 1999

Luz E., Wallaschek J., Experimental modal analysis using ambient vibration, *The International Journal of Analytical and Experimental Modal Analysis*, vol. 7 (1), Jan 1992, pp. 29-39

Mc Verry G.H., Structural identification in the frequency domain from earthquake records, *Earthquake Engineering and Structural Dynamics*, Vol. 8, 1980, pp. 161-180

Quek S.T., Wang W., Koh C.G., System identification of linear mdof structures under ambient excitation, *Earthquake Engineering and Structural Dynamics*, vol.28, pp. 61-77, 1999

Sepe V., De Angelis M, Capecchi D., Identificazione delle strutture sotto forzanti dinamiche non note, *10° Convegno Nazionale "L'ingegneria Sismica in Italia*, Potenza-Matera 9-13 settembre 2001 (in Italian)

Sepe V., Zingali A.E., Wind induced response of an elevated steel water tank, *Wind and Structures*, Vol.4, n.5, pp.383-398, Techno-Press, 2001

Torkamani M.A.M., Ahmadi A.K., Stiffness identification of a tall building during construction period using ambient test, *Earthquake Engineering and Structural Dynamics*, vol.16, pp. 1177-1188, 1988

Wang D., Haldar A., Element-level system identification with unknown input, *ASCE Journ. Eng. Mech.*, vol.120 (1), Jan. 1994, pp.159-176

Wang D., Haldar A., System identification with limited observations and without input, *ASCE Journ. Eng. Mech.*, vol.123 (5), May 1997, pp.504-511

Wallaschek J., An experimental modal analysis technique for large-scale structures, *Ingenieur-Archiv*, vol. 60, 1989, pp.117-123

Wilson J.C., Tao Liu, Ambient vibration measurements on a cable-stayed bridge, *Earthquake Engineering and Structural Dynamics*, vol.20, pp. 723-747,1991

Assessment of Historical Buildings via Ambient Vibration Measurements: Experiences on Bell-Towers

Alessandro De Stefano[1] and Rosario Ceravolo[1]

[1] Department of Structural Engineering and Geotechnics, Politecnico di Torino, Turin, Italy

Abstract. In the recent past, some proposals for new identification methods were formulated to handle current types of non-stationary excitation. A novelty of this approach is that identification is based solely on the properties of slowly modulated harmonics. This paper reports on some recent applications of ambient vibration testing over two ancient towers sited in Piedmont, Italy. The results of a structural identification from the signals constituted a first stage for the definition of a predictive model, which might prove useful for structural rehabilitation.

1 Introduction

Structural diagnostics is often associated to experimental modal analysis techniques, since deviations in modal parameters may reflect changes in local mechanical properties. In civil engineering applications, measurements carried out in environmental excitation conditions assume a special importance, in fact: a) measurements can be performed during the structure's regular servicc; b) they require no special excitation equipment; c) it is not necessary to measure the excitation.

In recent years, time domain techniques have been used rather successfully (e.g. Asmussen, 1997), thanks to the spectral resolution offered by these methods in the analysis of complex systems, and thanks to their modal uncoupling capability. Frequency domain techniques, though affected by the disturbance due to the needing of a edge-smoothing time window on digital samples, are again made interesting by the powerful cleaning effect of Single Value Decomposition (e.g. Maia and Silva, 1997). The common limitation of time and frequency approaches lies in the fact that they have both been conceived for the analysis of response signals that do not deviate much from stationarity, though in actual fact they are widely used for the analysis of signals whose main characteristic is far from being stationarity (bridges excited by vehicle traffic, towers exposed to wind gusts, etc.).

In structural response signals, the main characteristic appears to be the slow variation in modal energy; the response might be viewed as a time-varying combination of modulated harmonic functions. Furthermore, in non-stationary conditions, classical Fourier analysis should be replaced by the more general time-frequency analysis. A method has been recently proposed that works out instantaneous quantities (Bonato et al 2000, Ceravolo et al 2001), such as the phase difference and the amplitude ratio between channels, as a function of frequency. In linear systems, modal components are recognised since they show estimator values that are characterised by stability over time. The estimators are defined on the basis of the time-frequency analysis of vibration response signals, so these techniques might be placed into a new class of time-frequency domain methods.

2 Time-Frequency Based Linear Identification

Let us assume that a structure, subjected to unknown excitation, is instrumented with simultaneous acquisition channels according to some of the N degrees of freedom. In the time-frequency representation of the response signals, the energy appears to be concentrated around the modal frequencies and modulated according to the evolution of the time-frequency transform of the modulating waveform. In other words, though the input is randomly varying without any stationary character, the modal response remains nearly stable on the modal frequency for a time interval that is long, at least, as the free decay after an impulse.

Due to the fact that in the (t,f) plane at each modal frequency the shape of the modulating waveform is similar for all N records, it can be demonstrated that the amplitude ratio and phase difference between two measured signals $s_i(t)$ and $s_j(t)$ can be determined directly from their bilinear time-frequency auto and cross representations $D_{si}(t,f)$, $D_{sj}(t,f)$, $D_{si,sj}(t,f)$ in the following manner (for details see Bonato et al 2000):

$$AR(t,f)=\sqrt{\frac{D_{s_i}(t,f)}{D_{s_j}(t,f)}} \tag{1}$$

$$PH(t,f)=phase\{D_{s_i,s_j}(t,f)\} \tag{2}$$

$AR(t,f)$ represents the time-frequency estimator for the amplitude ratio, and $PH(t,f)$ the time-frequency estimator for the phase difference. In multi-component signals, when interference and auto-terms are superimposed, the use of a kernel capable of attenuating the interference or cross terms leads to an increase of the reliability of the amplitude ratio estimation. After this filtering operation, the only risks for the estimators to be inconsistent regard a possible close coupling among different modal components. In time-frequency plane, close coupling occurs when components interfere in both time and frequency simultaneously, this being an advantage in comparison with classical frequency analysis. By operating on the kernel parameters it is also possible to increase the resolution capacity in narrow coupling.

In frequency intervals where a single modal component is predominant, the estimators tend to become steady in time. As this property increases progressively up to a peak at the modal frequencies, the latter can be identified by searching the minima of the estimators' standard deviation as a function of frequency. The strategy of the method consists in the following procedure: the stability of the phase difference along the time axis between two contemporarily recorded signals is checked step by step for increasing values of frequency. For instance, in classically damped systems, the RMS value of that difference oscillates around $\pi/2$, except for the modal frequencies, revealed by a sharp downward peak, corresponding to a RMS phase difference value converging to zero. At a given identified modal frequency, the AR estimators supply amplitude relationships between channels as a function of time, and, by averaging, an estimate of the modal shapes.

In structural dynamics, usually we expect that the release of dynamic energy from a system will be gradual enough to give rise to a response characterised by a number of modulated waveforms. Under these conditions, the identification of modal frequencies and shapes reduces to separating modal components and does not call for strict model assumptions (in the method described, only the stability over time of modal shapes was assumed). In the following this method is referred to as Time Frequency Instantaneous Estimator (TFIE) method.

At first sight, the extension of the TFIE approach above to the evaluation of damping may appear direct. Actually, in this case new problems arise:

1) Damping affects the modulation of the signal rather than its instantaneous frequency;
2) In practical applications damping evaluation is critical, and any available data about the system's behaviour should not be neglected by the identification procedure;

Based on the foregoing considerations, a time-frequency damping identification approach may include the following stages (Ceravolo et al 2001):

1) Modal frequency and shape identification, based on the respective time-frequency estimators.
2) Time-frequency transformation of the response signals and regularisation. The latter point is achieved by imposing, at any time t, the linear model formulation in the frequency domain, or, in other words, by searching the Frequency Response Functions (FRF) that best fit the instantaneous response spectra.
3) The final value of the damping is obtained by averaging over time the instantaneous damping diagrams. Results coming from different channels may be averaged, or combined by resorting to multi-criteria techniques.

Given a time-frequency representation of a displacement signal, instantaneous parameters associated with the Frequency Response Function (FRF), H, can be estimated by minimising the following functional at each time t:

$$D(f,t)-\left|H(f,t)\right|^2 \tag{3}$$

In this minimisation there is an explicit assumption that the instantaneous energy spectrum $D(t,\omega)$ associated to a general time-frequency transform, approaches a scaled version of the squared modulus of the FRF (the theoretical bases of the method are in Ceravolo 2001). In the case of a linear and proportionally damped system one may write:

$$H(f)=\sum_k AR_k \Big/ \left(f_k{}^2-f^2+2j\zeta_k f_k f\right) \tag{4}$$

and instantaneous parameters for optimisation may be chosen among modal amplitudes, AR_k, damping ζ_k, or modal frequency, f_k. A similar relationship could be written for velocity or acceleration signals.

It has been observed that estimation accuracy depends on the relative energetic importance of the modes. In other words, the instantaneous estimator is more accurate in the temporal segments where the mode to be identified is predominant and is not affected by the residuals of the other modes. The definition of an instantaneous damping makes it possible to work out temporal weighted averages or to select temporal segments to improve the estimation of weaker modes. A possible solution can be the introduction of functions allowing for the relative energetic importance of the k-th mode as a function of time. In this connection, a proposal consists of using the following function, $w_k(t)$, either in the computation of weighted averages, or, having selected a threshold value, in the selection of time intervals where the estimate is deemed reliable (Ceravolo et al 2001):

$$w_k(t) = \frac{AR(t, f_k)/(2f_k{}^2\bar{\zeta}_k)}{\sum_l AR(t, f_l)/\left(f_l{}^2 - f_k{}^2 + j2\pi\bar{\zeta}_l f_l f_k\right)} \tag{5}$$

where k identifies the mode in which damping is estimated and the sum is extended to all the modal components contained in the signal. The damping parameters that appear in the second member of Equation (5) should be construed as tentative values (e.g., by averaging the instantaneous estimator over the entire temporal support considered).

3 Applications to SS. Annunziata Bell-Tower in Roccaverano

The first case study reported is the SS. Annunziata church bell-tower in Roccaverano. A vibration test campaign was carried out on a XVI century church bell-tower rising in Roccaverano (Asti-Italy) whose style is inspired to the school of Bramante (Figure 1). In the past the church was exposed to a strong earthquake which caused serious damage both on the facade and on the bells-tower and subsequently some interventions of restoration were made. The tower has been subjected to an extensive experimental investigation both under ambient vibrations and actions induced by the bells.

Vibration measurements were performed on the bell tower only, by placing the accelerometers on the landings, arranged in the horizontal direction. Each set-up is made up of the signals relating to four acquisition channels, of which two were fixed as reference channels and two were moved to the levels of the different landings. The measurements were made separately in the E-W and N-S directions, but one of the tests was conducted with two accelerometers arranged in the orthogonal direction, according to the two main axes of the bell tower, in a central position. This made it possible to correlate the modal shapes observed in the two main directions and to build up space modal shapes, of special importance in connection with torsional modes.

Different types of excitation were used, and namely, the one generated by bell tolling in two different directions, the one produced by pulses applied to the bells and finally the one arising from environmental noise. An aspect to be noted is the absolute absence on Fourier spectra of important components in the 2.5-10 Hz range, namely a frequency range within which the second flexural mode and the first two torsional modes are typically located in structures of this type.

Let us now examine the results obtained with the TFIE method. For the sake of brevity, in lieu of the two-dimensional estimator, we give the charts of the standard deviation of the estimator as a function of frequency, as this makes it possible to identify each modal frequency as a relative minimum in scatter due to the constancy of modal amplitude ratios or phase differences. In the following we shall refer to the phase estimators, which admits a much clearer interpretation than the amplitude ones. Figure 2 shows a sample phase difference estimator; this kind of diagrams displays very clear minimum points and it also identify frequencies which did not appear at all in the PSD's. The results are listed in Table 1.

It was clearly observed the existence of a first predominant mode in the N-S direction, i.e. at 1.66 Hz according to the less stiff direction, and another prevalent mode in the E-W at 2.26 Hz. These modes, which were also quite evident in Fourier spectra, were then immediately associated with the first two flexural modes according to the two main axes. The mode at 4.67 Hz (Figure 3), observed in either direction, was associated with a first torsional mode.

All in all, the results of the identification process based on the time-frequency estimators offer information which is totally new with respect to the peaks observed in energy spectra. This prompted the need to construct a simplified and yet reliable FE model to work out a correct classification of the modes (details of this application are in Bonato et al 2000a).

Figure 1. View of SS. Annunziata church and bell-tower in Roccaverano (Asti), Italy.

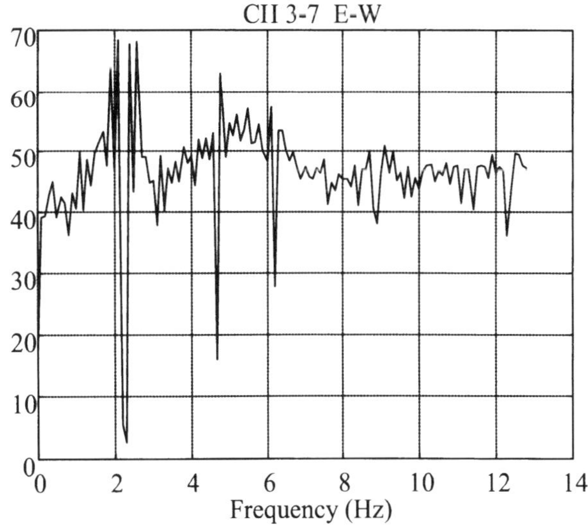

Figure 2. SS. Annunziata bell-tower: standard deviation of sample phase difference estimator as a function of frequency.

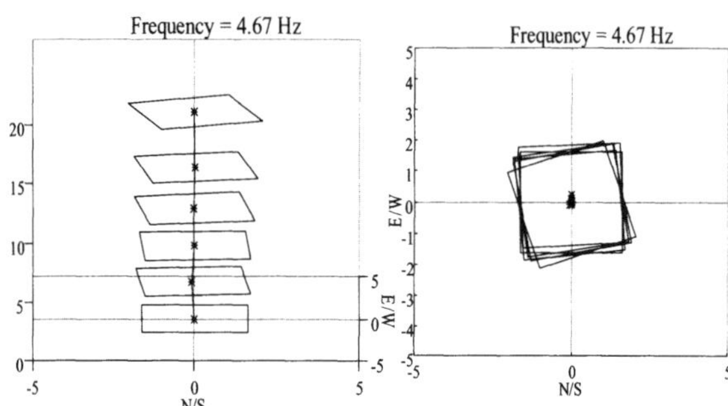

Figure 3. SS. Annunziata bell tower: a sample mode identified in the range 1-5 Hz (1st torsional).

Table 1. SS. Annunziata bell tower: first six modal frequencies as identified through TFIE method

Mode	Identified frequencies (Hz)
1	1.66
2	2.26
3	4.67
4	6.18
5	6.40
6	8.90

The following data were used to build up a simplified truss model of the bell-tower:

1) The geometric data of the structure;

2) Frequencies for which solid identification and physical evidence had been collected: 1.66 Hz, first flexural mode in the N-S and 2.26 Hz direction, first flexural mode in the E-W direction.

The chosen model was a space truss (Figure 4), whose horizontal members represent the flexural stiffness of the bell tower and the diagonals simulate the shear stiffness of the masonry panels. Albeit approximate, this model makes it possible to exploit the geometric data available in the determination of stiffness values, thereby reducing the number of parameters to be calibrated to the elastic properties of the material.

The geometric data were exploited according to the following criteria:

1) The plan dimensions of the model reflect the distances between the outer walls;
2) The uprights of the lattice, which are assigned a very high stiffness value, are situated at the landings, where the accelerometers were placed;
3) From a careful examination of the vertical sections of the bell tower and the preliminary modal shapes observed, it can be seen that over a certain height, of ca 3.5 m, the bell tower constitutes a single body with the church building;
4) Additional constraints were introduced in the form of 4 horizontal struts placed at the level of the first landing, to represent the influence of the church;

5) The areas of the vertical members of the lattice were determined so that, at each storey, the moments of inertia of the lattice along the two main axes would equate the moment of inertia of the structure. The elasticity modulus (E) was assumed to bc uniform over the entire tower.
6) The stiffness of the diagonals was determined so as to equate the shear stiffness of the panels. The area of the panels was known, whilst the tangential elastic modulus (G) remained unknown. For the two storeys having no openings, a fictitious modulus (G^*), which cannot be directly correlated with the material properties, was considered.

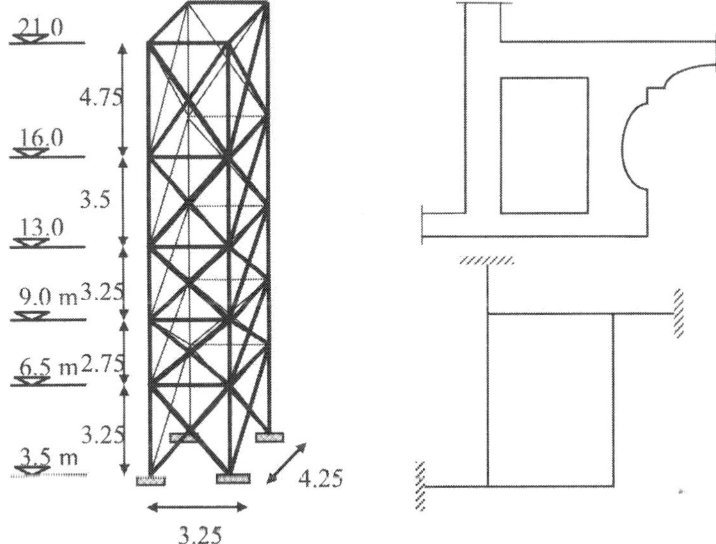

Figure 4. SS. Annunziata bell tower: simplified truss model.

The unknowns of the calibration process boiled down to five: the E and G moduli, the fictitious modulus G^* and the stiffness parameters, k_1 and k_2, of the struts simulating the restraint.

The calibration process, conducted by means of an iterative procedure, consisted of minimising a cost function: $E = \sum_{i=1}^{n} \left(\frac{f_{mi} - f_{ci}}{f_{ci}} \right)^2$

where f_{ci} are the frequencies calculated through the calibration process, and f_{mi} are the measured frequencies. It should be noted that, in addition to making the calibration process more significant, the use of a physical model related to the geometric data actually measured also reduces the risk of dealing with an ill-conditioned problem.

A first test was conducted by including in the cost function only the first two frequencies, namely the first flexural frequencies along the two principal axes, which are the most accurate and reliable one. After checking the reliability of the modes identified this far, a new calibration process was performed to include the fourth frequency, i.e. the second flexural frequency in the N-S direction. The quality of the results was significantly improved. Table 2 provides an overview of the frequencies of interest supplied by the calibrated model.

The values taken on by the parameters after the calibration process are given in Table 3. In particular, the elastic moduli have values comparable with those obtained on masonry of similar structural types. It should also be noted that torsional modal frequencies are underestimated by the model which, being of a lower order, is not perfectly verified. In fact, a calibration process extended to include the torsional mode, of reliable attribution, was not successful.

With the aid of the updated model, it was possible to complete the direct identification process on the higher frequency (Table 2). Furthermore, the model made it possible to classify the identified frequencies. Figure 3 shows one of the 13 modes identified by the proposed technique. At this second stage, the low coefficient of modal participation explained the total absence of some of these modal components in PSD diagrams. In a further sensitivity study, it was seen that the restraint parameters have a negligible effect on lower modes (Ceravolo et al, 1999), but in the field of frequencies of over 10 Hz, the bell tower loses its predominant influence on the modes, as the church building it is connected to prevails, so that mode classification calls for a global model reflecting the interaction between these two structural elements. A damage assessment stage would require the introduction of a further local stiffness parameters in the calibration process.

The repair consisted of mortar injections and prestressed cables at each landing level. The results of new test campaign on SS. Annunziata tower (Ceravolo et al, 1999), after a repair intervention, are summarized and compared with the original ones (Tables 2 and 3). One may notice the splitting effect on the two fundamental frequency, already reported on repaired structures and due probably to non-linearity.

The as a 20-30% increase in the flexural frequencies (Young modulus) and a slight decrease in the torsional one (shear modulus). Thus, we can draw the following conclusions:

1) A slight increase in *G*, probably a fictitious effect of the mass increase, show that the tower's shear behaviour has not significantly changed.
2) A considerable increase in the Young modulus and bending stiffness demonstrates the efficacy of the intervention.

Table 2. SS. Annunziata: identified and updated model frequencies, before and after the repair.

Mode	Before repair		After repair	
	Measured frequencies (Hz)	Model frequencies (Hz)	Measured frequencies (Hz)	Model frequencies (Hz)
1	1.66	1.67	1.97 2.12	1.97
2	2.26	2.25	2.34 2.54	2.54
3	4.67	3.86	4.30	4.00
4	6.18	5.96		6.70
5	6.40	6.55		7.33
6	8.90	7.45		7.98

Table 3. SS. Annunziata: parameters of the updated model.

Updated parameters	Before repair	After repair
E	$12.3 \cdot 10^8$ N/m^2	$18.8 \cdot 10^8$ N/m^2
G	$2.5 \cdot 10^8$ N/m^2	$2.0 \cdot 10^8$ N/m^2
G^*	$0.2 \cdot 10^8$ N/m^2	$0.3 \cdot 10^8$ N/m^2

4 Applications to the Bell-Tower of Alba's Cathedral

A second test campaign regarded the bell tower of Alba's Cathedral (Figure 5). The tower has been subjected to an experimental investigation under three types of excitation: ambient vibrations, bell tolling and actions induced by a corer. Acceleration measurements were carried out on the bell tower only, by placing three accelerometers per landing, arranged along the two horizontal axes. Two set-up were used, each one made up of the signals relating to 12 acquisition channels, of which three were fixed at the top level (43 m height) as reference channels and ten were moved to different landings. The measurements were made contemporarily in the E-W and N-S directions and the three fixed channels allowed to correlate different set-up and build spatial modal shapes. During the signal pre-processing stage it was observed that the signals picked up under corer excitation were noisy and disturbed by strong harmonic components, hence they were not used in the elaborations.

Figure 5. S. Lorenzo Cathedral bell-tower in Alba

The signals were analyzed applying two different methods. The first one, in the time domain, is based on the ERA algorithm (Maia and Silva, 1997) applied on the estimated Random Decrement functions (Asmussen, 1997). The second one, referred to as TFIE method, was applied by choosing Choi-Williams transforms with kernel parameter σ=5. ERA and TFIE methods were applied to different measurements, this resulting in occurrence diagrams for modal frequencies as depicted in Figure 6.

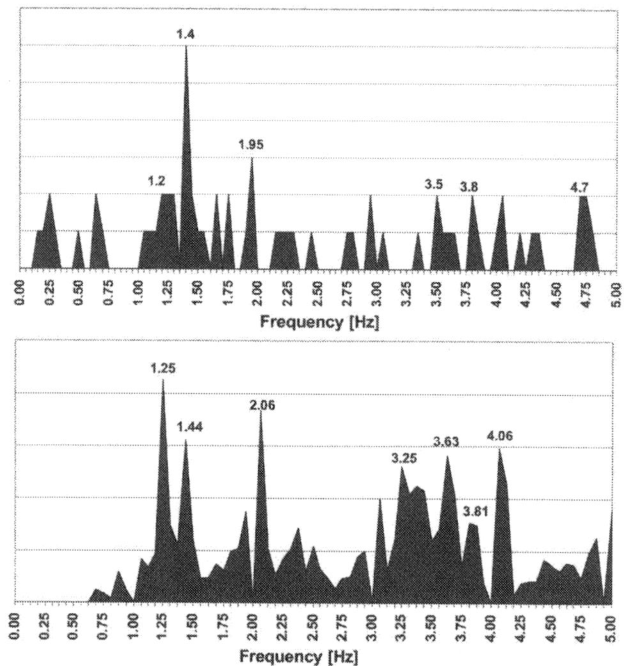

Figure 6. S. Lorenzo Cathedral bell-tower: occurrence of modal frequency values resulting from different identification sessions

For brevity's sake, only a part of the identified modes is presented in Table 4, where the results coming from the two techniques may be compared. The first evident advantage in using TFIE was the possibility to work on several parameters (kernel selectivity, type of transform, etc.) that may improve the estimators and filter spurious components, as well as check modal energy changes in time. If compared to the the previous application, the connection to the church in this case seems to play a more important role on the torsional components of the tower's modal shapes. A further proof of this influence is that the TFIE method has identified additional modes, here omitted for brevity's sake, that have been attributed to the church. On the other hand, the subspace and RD based algorithms showed to be more suitable when one has to perform a fast identification or when the system's dynamic behaviour is very clear.

Table 4. S. Lorenzo Cathedral bell-tower. Modal frequencies as identified through the TFIE method

TFIE frequencies [Hz]				
1st	2nd	3rd	4th	5th
1.25	1.44	3.63	3.81	4.12

ERA frequencies [Hz]				
1st	2nd	3rd	4th	5th
1.22	1.44	3.58	3.80	4.05

5 Conclusions

The first conclusion results from the ambient vibration tests presented, which showed that experimental modal analysis techniques provide important information about the actual structural behaviour. Although no information can be directly deduced about material strength and ultimate bearing capacity, a validated and updated reference FE model is paramount, at least in order to discard models intrinsically too weak (non upgradable), or more often to assess the basic properties of the materials and construction details.

The direct use of the experimental modal identification for diagnostic purpose, in particular for such complex structures as ancient masonry buildings, is still under investigation, and no definitive methods and techniques are already enough consolidated. Nevertheless, it is widely recognized that higher modes can be effective in local damage detection and that they can be extracted only if the sensitivity and accuracy of the measurement chain is high and the noise-signal ratio is low. In this connection, time-frequency identification may prove useful, especially in case of non-stationary input, because it does nor require the stationarity assumption. In fact short segments of the signal can be analysed and this circumstance is favorable for a correct structural identification, since it may be assumed that modal parameters remain constant in time. Recent results obtained by the authors confirm the effectiveness of the TFIE method also in damping evaluation.

Time domain approaches are recognized to be reliable in assessing modal frequencies and shapes, and in fact, on the bell-tower in Alba, both ERA and TFIE approaches have been used and results fitted quite well. Damping evaluation via time-domain methods is generally reliable only if signals are not too far from stationarity, whilst the TFIE method does not retain this limitation. On the other hand, the TFIE method requires for each identified mode two sensors in a sensitive location, hence a test design is needed, based on an initial model.

References

Asmussen JC. (1997). Modal Analysis based on Random Decrement Technique. Ph.D. thesis Aalborg University, Denmark. 1997.

Bonato P, Ceravolo R, De Stefano A, Molinari F. (2000). Cross-Time-Frequency Techniques for the Identification of Masonry Buildings. *Mechanical Systems and Signal Processing*, 14(1): 91-109.

Bonato P, Ceravolo R, De Stefano A, Molinari F. (2000). Use of cross Time-Frequency Estimators for the Structural Identification in Non-Stationary Conditions and under Unknown Excitation. *Journal of Sound and Vibration*. 237(5): 775-791.

R. Ceravolo, C. Genovese, A. Pavese, G. Pistone, D. Zorgniotti (1999). Finalizzazione all'Intervento di Restauro dell'Identificazione Strutturale di una Torre Campanaria. *Atti del IX Convegno Nazionale "L'Ingegneria Sismica in Italia"*, CD-ROM, Torino 20-23 settembre.

Ceravolo R, De Stefano A, Molinari F. (2001). Developments and Comparisons on the Definition of an Instantaneous Damping Estimator for Structures under Unknown and Non-Stationary Excitation. *Key Engineering Materials*. 204-205: 231-240.

Maia NMM., Silva JMM eds (1997). *Theoretical and Experimental Modal Analysis*. Research Studies Press.

On Damage Identification in Vibrating Beams from Changes in Node Positions

Michele Dilena

Department of Civil Engineering, University of Udine, Udine, Italy.

Abstract This paper summarizes some recent results about the effect of damage on the nodes of the mode shapes of a thin beam in axial or bending vibrations. The damage, a crack, is simulated by a massless spring. It can be shown that the direction by which nodal points move may be useful for predicting damage location. Analytical results agree well with experimental tests performed on cracked steel beams.

1 Introduction

In this paper we seek to detect a single crack from the knowledge of the damage-induced shifts in the nodes of the mode shapes of a beam in axial or bending vibration. Despite a very extensive literature on damage identification through frequency measurements, see for instance Adams et al. (1978), Rizos et al. (1990), Hearn and Testa (1991), Liang et al. (1992), Narkis (1994), Davini et al. (1995), Vestroni and Capecchi (1996), Ruotolo and Surace (1997), Morassi (2001), very little work has been done on using changes in the nodes of the modal shapes to identify localized damage. Natke and Cempel (1991) observed that mode shapes are more sensitive near nodes (or nodal lines) that at maximum amplitudes, and showed how the presence of a hole in a plate may dramatically change its nodal lines. Chen and Garba (1988) investigated, both analytically and experimentally, the modal pattern change of a large space structure due mainly to damage of one of its members.

One of the first works in which the effect of a crack on the nodes of mode shapes is considered was carried out by Gladwell and Morassi (1999) for longitudinally vibrating beams. It was shown that *nodes move toward the crack*, that is, every node located on the left of the crack in the undamaged configuration moves to the right, and every node on the right moves to the left. Then, for every principal mode, which has at least two nodes, there is exactly one neighboring pair of nodes which move toward each other, and the crack is located between them. However, that paper left an important question unsolved, and namely the capability to extend this result to cracked beams in bending vibration. Although desirable for the application of diagnostic methods, the existence of such a result is not obvious. In fact, the monotonicity property found in (Gladwell and Morassi, 1999) is a consequence of the *oscillation and separation theorems* for solutions of the Sturm-Liouville operator governing the axial vibrations of a thin rod. An analogue property for the fourth-order differential operator governing the bending vibrations of a beam involves *conjugate points* and crack location, that is *conjugate points move toward the crack.* Conjugate points are points of the beam axis given by the intersection of two special solutions of the fourth-order differential operator, see, for instance, the classical paper by Leighton and Nehari (1958). Unfortunately, unlike nodal points, conjugate points lack a clear physical interpretation; therefore, since their variations cannot be measured, damage location cannot be deduced. Considering the effect of a damage on the nodal points of flexural modes, the presence of a crack in some areas of the longitudinal axis of the beam causes some nodes to shift away from damaged area. Despite no monotonicity property being found to exist between nodes and crack location, an exact analysis of the direct problem proves useful to characterize the areas of the beam axis where a crack causes nodes always to shift along the same direction, for different levels of damage severity.

Until now, only ideal undamped systems have been considered. In the last part of the paper we shall briefly consider an axially vibrating beam under dissipative boundary conditions. The damped case is deeply different from the undamped one. In particular, the eigenvalues and the corresponding eigenfunctions are complex-value functions and this fact has a strong influence on the damage detection procedure. Anyway, it is always possible to factorize the eigenfunctions in their real and imaginary parts and to assess the damaged-induced shifts of the nodes (zeroes), see also Geist and McLaughlin (1994) for a study on the effect of structural damping on nodes of an Euler–Bernoulli beam.

From an experimental point of view, node positions are easier to measure than modal shapes; they only require the detection of modal component sign, rather than a set of measurements of magnitudes. Dynamic tests and numerical applications performed on damaged steel beams support the proposed methods for the localization of a crack.

2 Rod in Axial Vibration

In order to illustrate the main idea of the diagnostic technique via mode shift measurements, let us consider an uniform elastic beam of length L with constant cross-section of area A, Young's modulus E and linear mass density ρ. We choose this simple case because it allows for a close form expression of the mode shapes, both undamaged and damaged configuration. The infinitesimal free longitudinal vibrations of the rod under free–free boundary conditions are governed by the following equation

$$EA\frac{\partial^2 w}{\partial x^2} - \rho\frac{\partial^2 w}{\partial t^2} = 0, \qquad x \in (0, L),\ t > 0, \tag{2.1}$$

with

$$\frac{\partial w}{\partial x}(0, t) = 0, \qquad \frac{\partial w}{\partial x}(L, t) = 0, \qquad t > 0, \tag{2.2}$$

where $w = w(x, t)$ is the longitudinal displacement of the bar at the cross section of abscissa x and at a moment of time t. The dynamic problem is completed by assigning some initial conditions, at the moment of time $t = 0$, on w and $\frac{\partial w}{\partial t}$ in the whole interval $[0, L]$.

The dimensionless eigenvalue problem related to the problem (2.1) – (2.2) is given by

$$u_U''(\xi) + {\lambda_U}^2 u_U(\xi) = 0, \qquad \xi \in (0, 1), \tag{2.3}$$

$$u_U{}'(0) = u_U{}'(1) = 0, \tag{2.4}$$

where $u_U = u_U(\xi)$ is the spatial variation of the free longitudinal vibration of the undamaged rod and $u_U{}' \equiv \frac{du_U}{d\xi}$.

The dimensionless quantities ξ and ${\lambda_U}^2$ are defined as follows

$$\xi = \frac{x}{L} \in (0,1), \qquad {\lambda_U}^2 = \frac{\rho L^2}{EA}\omega^2, \tag{2.5}$$

where ω is the natural frequency of vibration. Suppose now that a crack appears at the cross section of abscissa $s \in (0, L)$. Assuming that the crack remains always open during the longitudinal vibration, by modelling it as a massless translational spring of stiffness K_A, as suggested in Freund and Herrmann (1976), the dimensionless eigenvalue problem for the damaged rod is

$$u''(\xi) + \lambda^2 u(\xi) = 0, \qquad \xi \in (0,\sigma) \cup (\sigma, 1), \tag{2.6}$$

$$u'(0) = u'(1) = 0, \tag{2.7}$$

where the jump conditions

$$[u'(\sigma)] = 0, \qquad [u(\sigma)] = \mu u'(\sigma), \tag{2.8}$$

hold at the cross-section where the crack occurs. In the equations above $[\phi(\sigma)] \equiv (\phi(\sigma^+) - \phi(\sigma^-))$ denotes the jump of the function $\phi(\xi)$ at $\xi = \sigma$. The dimensionless quantities σ and μ are defined as follows

$$\sigma = \frac{s}{L}, \qquad \mu = \frac{EA}{K_A L}. \tag{2.9}$$

By using standard techniques to solve the boundary value problem (2.6)-(2.8), the eigenvalues λ^2 are the infinite sequence $0 = \lambda_0^2 < \lambda_1^2 < \ldots < \lambda_m^2 < \ldots$ of zeroes of the *characteristic equation*

$$\sin\lambda + \lambda\mu \sin\lambda\sigma \sin\lambda(1-\sigma) = 0. \tag{2.10}$$

Only strictly positive eigenvalues will be considered hereafter, that is the rigid vibrating mode associated to λ_0^2 will be omitted. The vibrating mode corresponding to the eigenvalue λ^2 is given by

$$u(\xi) = B_1 \cos\lambda\xi, \qquad \xi \in (0,\sigma), \tag{2.11}$$

$$u(\xi) = -B_2 \cos\lambda(1-\xi), \qquad \xi \in (\sigma, 1), \tag{2.12}$$

where coefficients B_1 and B_2 may be written as

$$B_1 = 1, \qquad B_2 = \frac{\sin\lambda\sigma}{\sin\lambda(1-\sigma)} \tag{2.13}$$

The undamaged rod corresponds to $K_A \to \infty$, or in an equivalent form to $\mu \to 0$. Let $(u_U^{(m)}(\xi), \lambda_{Um}^2)$ be the mth eigenpair, where $u_U^{(m)}(\xi) = \cos m\pi\xi$, $\lambda_{Um}^2 = m^2\pi^2$, $m =$

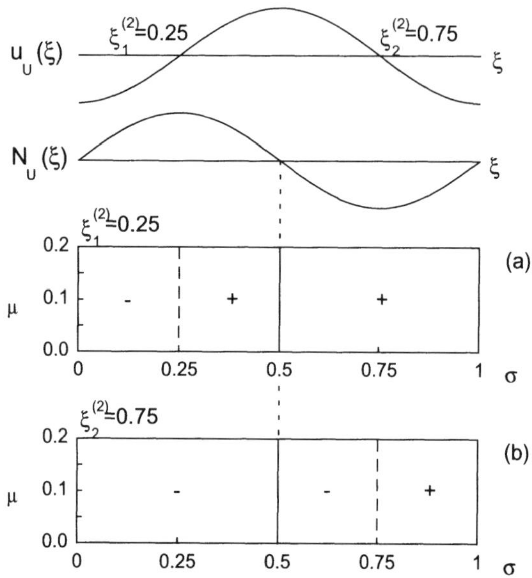

Figure 1. Positive (+) and negative (-) nodal displacement domain of the first (a) and the second (b) node of the second axial mode.

$1, 2, \ldots$. It is well known that $u_U^{(m)}(\xi)$ has m simple internal zeros (*nodes*) $\{\xi_i^{(m)}\}_{i=1}^m$ in the interval $(0, 1)$. The axial force $N_U^{(m)}(\xi) \equiv EAu_U^{(m)'}(\xi)$ vanishes at $\xi = 0$ and $\xi = 1$, and it has $m - 1$ simple zeros $\{\xi_{Ni}^{(m)}\}_{i=1}^{m-1}$ in the same interval $(0, 1)$, for $m \geq 2$.

We consider now how nodes of the eigenfunctions change with damage. If the crack is located at a cross-section of abscissa σ for which $N_U^{(m)}(\sigma) = 0$, then the related eigenpair is insensitive to damage, that is $(u_U^{(m)}(\xi), \lambda_{Um}^2) \equiv (u^{(m)}(\xi), \lambda_m^2)$. In fact, it can be easily seen that the undamaged vibrating mode identically satisfies the differential equation (2.6), boundary conditions (2.7) and jump conditions (2.8). On the contrary, if $N_U^{(m)}(\sigma) \neq 0$, then one can prove that nodes move toward the crack location. In fact, the variational formulation of the eigenvalue problem shows that the eigenvalues of the rod are decreasing functions of the damage parameter μ, that is, $\lambda_m < \lambda_{Um}$. Therefore, taking into account the expression (2.11) of the mode shapes, it is easy to see that every node located in the interval $(0, \sigma)$ moves to the right, that is $\xi_i^{(m)} > \xi_{Ui}^{(m)}$. Similarly, every node located in $(\sigma, 1)$ moves to the left, $\xi_i^{(m)} < \xi_{Ui}^{(m)}$. It is worth noticing that the property above holds true for axially vibrating rods with different boundary conditions and variable stiffness and inertia coefficients, see Gladwell and Morassi (1999).

Figure 1 represents the signs of the shift affecting the nodes of the second vibrating mode of a beam under free-free boundary conditions. The plane domain (σ, μ) is divided into two regions corresponding to positive signs (*positive nodal displacement domain*, PNDD) and negative signs (*negative nodal displacement domain*, NNDD, complementary to the former) of the shift. Negative domains, NNDD, (and positive PNDD, respectively) correspond to the values of position/damage severity pairs (σ, μ) which, for the node under exam, produce shifts to the left (and to the right, respectively) in the coordinate system chosen to represent the points of the beam axis. The boundary line separating the two sections is independent by the damage severity μ. Note that, if the crack is located at the mid–point of the beam, for which $N_U^{(2)}(\sigma = 0.50) = 0$, then the corresponding mode is insensitive to damage and the node shift is equal to zero.

3 Beam in Bending Vibration

The effect of a crack on the nodes of mode shapes of a free–free uniform beam in bending vibration was recently investigated by Dilena and Morassi (2002). In this case the problem is more complicated because a monotonicity property between crack location and nodes shifts does not hold. For the sake of simplicity, let us consider now an uniform Euler-Bernoulli beam with supported boundary conditions. Assuming that the crack remains always open during the bending vibration, it is represented by a massless rotational elastic spring of stiffness K_B at the cross-section of abscissa s. The eigenpair $(v(\xi), \lambda^4)$ of the undamped bending vibrations of the cracked beam satisfies the following dimensionless boundary value problem

$$v^{IV}(\xi) = \lambda^4 v(\xi) \qquad \xi \in (0,\sigma) \cup (\sigma, 1), \tag{3.1}$$

$$v(\xi) = v''(\xi) = 0 \qquad \text{at } \xi = 0 \text{ and } \xi = 1, \tag{3.2}$$

where the jump conditions

$$[v(\sigma)] = [v''(\sigma)] = [v'''(\sigma)] = 0, \tag{3.3}$$

$$\mu v''(\sigma) = [v'(\sigma)], \tag{3.4}$$

hold at the cross-section where the crack occurs. The dimensionless quantities ξ, σ, λ^4 and μ are defined as follows

$$\xi = \frac{x}{L} \in (0,1), \qquad \sigma = \frac{s}{L}, \qquad \lambda^4 = \frac{\rho\omega^2}{EI}L^4, \qquad \mu = \frac{EI}{K_B L}, \tag{3.5}$$

where L is the beam length, I represents the moment of inertia of the cross-section, E is the Young's modulus of the material, ρ is the linear mass density and the natural frequency of vibration is denoted by ω. The eigenvalues λ^4 are the infinite sequence $0 < \lambda_1^4 < \ldots < \lambda_m^4 < \ldots$ of zeroes of the *characteristic equation*

$$\begin{aligned} -4\sin\lambda\sinh\lambda + \mu\lambda(-\cosh\sigma\sin\sigma + \cosh\lambda(1-2\sigma) \\ -\cos\lambda\sinh\lambda + \cos\lambda(1-2\sigma)\sinh\lambda) = 0. \end{aligned} \tag{3.6}$$

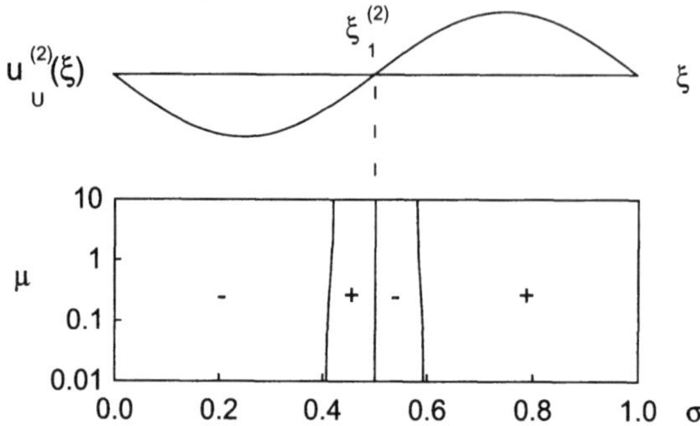

Figure 2. Positive (PNDD) and negative (NNDD) nodal displacement domain of the node of the second bending mode.

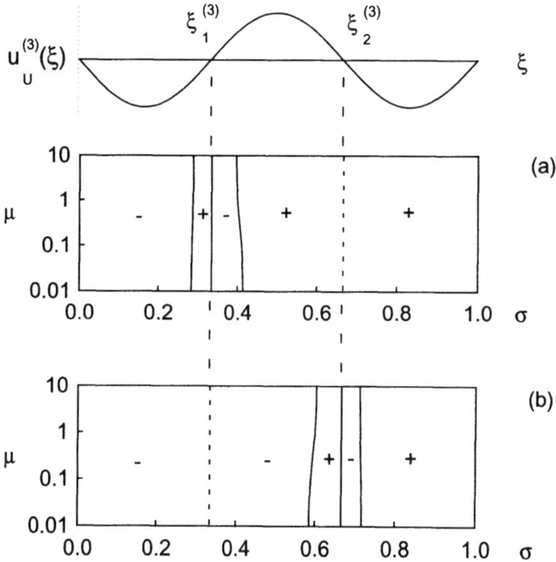

Figure 3. Positive (PNDD) and negative (NNDD) nodal displacement domain of the first (a) and the second (b) node of the third bending mode.

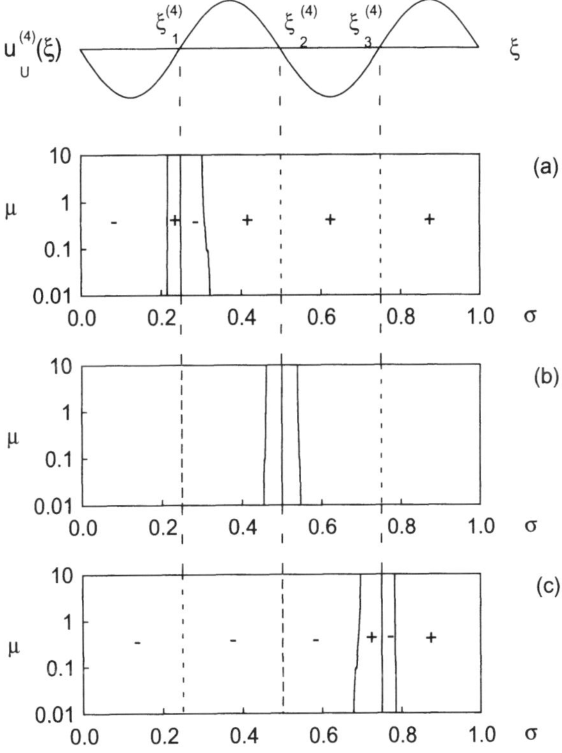

Figure 4. Positive (+) and negative (-) nodal displacement domain of the first (a), the second (b) and the third (c) node of the fourth bending mode.

The vibrating mode corresponding to the eigenvalue λ^4 is given by

$$v(\xi) = B_1 \sin \lambda\xi + B_2 \sinh \lambda\xi, \qquad \xi \in (0, \sigma), \tag{3.7}$$

$$v(\xi) = B_3 \sin \lambda(1-\xi) + B_4 \sinh \lambda(1-\xi), \qquad \xi \in (\sigma, 1), \tag{3.8}$$

where coefficients B_i, $i = 1, \ldots, 4$, may be written as

$$\begin{array}{ll} B_1 = 1, & B_2 = \frac{\sin\lambda \sinh\lambda(1-\sigma)}{\sin\lambda(1-\sigma)\sinh\lambda}, \\ B_3 = \frac{\sin\lambda\sigma}{\sin\lambda(1-\sigma)}, & B_4 = \frac{\sin\lambda \sin\lambda\sigma}{\sin\lambda(1-\sigma)\sinh\lambda}. \end{array} \tag{3.9}$$

The undamaged beam corresponds to $K_B \to \infty$, or to $\mu \to 0$. Let $(v_U^{(m)}(\xi), \lambda_{Um}^4)$ be its mth eigenpair. It is well known that $v_U^{(m)}(\xi)$, $m \geq 2$, has $m-1$ simple internal zeros (*nodes*) $\{\xi_i\}_{i=1}^{m-1}$ in the interval $(0,1)$, *e.g.* $0 < \xi_1^{(m)} < \ldots < \xi_{m-1}^{(m)} < 1$, where $\xi_i^{(m)} = \frac{i}{m}$, $i = 1, \ldots, m-1$. Moreover, the bending moment $M_U^{(m)}(\xi) \equiv -EIv_U^{(m)''}$ is proportional to the eigenfunction $v_U^{(m)}$. Then, if the crack is located at a cross-section of abscissa

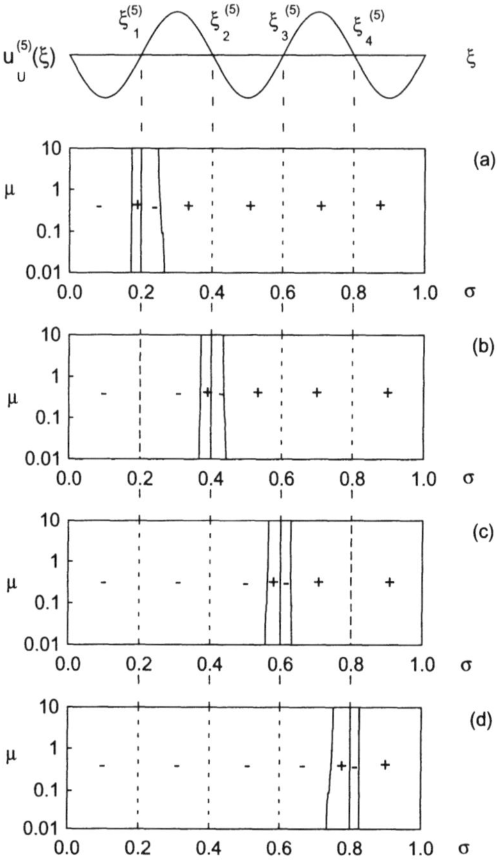

Figure 5. Positive (+) and negative (-) nodal displacement domain of the first (a), the second (b), the third (c) and the fourth (d) node of the fifth bending mode.

σ for which $M_U^{(m)}(\sigma) = 0$, then the related eigenpair is insensitive to damage, that is $(v_U^{(m)}(\xi), \lambda_{Um}^4) \equiv (v^{(m)}(\xi), \lambda_m^4)$.

From the exact expression of the damaged vibration modes one can parametrically assess the effect of a crack on the nodes of bending modes when the normalized values of position, σ, and damage severity, μ, vary. In particular, Figures 2–5 illustrate the plane domains (σ, μ) corresponding to positive signs (PNDD) and negative signs (NNDD) of the shift affecting a generic node in the first five vibration modes. We recall that negative domains, NNDD, (and positive PNDD, respectively) correspond to the values of position/damage severity pairs (σ, μ) which, for the node under exam, produce shifts to the left (and to the right, respectively) in the coordinate system of the beam axis. The variance range for the dimensionless parameter μ is selected so as to include the most significant cases arising in applications, that is, low and medium damage. Generally speaking, for a generic node, the plane (σ, μ) is divided into several subsections, where the boundaries separating the subsections belonging to either NNDD or to PNDD are little affected by large variations of μ and they are therefore approximately parallel to the μ-axis. In particular, let us consider the ith nodal point of the mth vibration mode, where $i = 1, 2, \ldots, m-1$ and $m \geq 2$. Negativc domains, NNDD, are the union of

i) a rectangular band $\xi < \xi_i^{(m)}$, to which a left band adjacent to the line $\xi = \xi_i^{(m)}$ and the bands located alternately to the right and to the left of the lines $\xi < \xi_k^{(m)}$, where $1 \leq k \leq i-1$, must be subtracted. The latter bands are very narrow along ξ direction (only few thousandths of beam length) and are represented in Figures 2–5 by dashed lines.;

ii) a right band adjacent to the line $\xi = \xi_i^{(m)}$;

iii) narrow bands along ξ direction (only few thousandths of beam length), alternately to the left and to the right of the lines $\xi = \xi_k^{(m)}$, where $i+1 \leq k \leq m-1$.

The positive domains, PNDD, are complementary to the negative ones.

4 Axially vibrating rod with dissipative boundary conditions

Let us consider the thin straight elastic rod of Section 2. Let us assume that the rod left end is free and that the right end is constrained by a lumped linear damper of viscous damping coefficient $c > 0$. The dimensionless eigenvalue problem for the cracked rod in axial vibration is governed by

$$u''(\xi) - \lambda^2 u(\xi) = 0, \qquad \xi \in (0, \sigma) \cup (\sigma, 1), \tag{4.1}$$

with boundary conditions

$$u'(0) = 0, \qquad u'(1) + \gamma\lambda u(1) = 0, \tag{4.2}$$

and jump conditions

$$[u'(\sigma)] = 0, \qquad [u(\sigma)] = \mu u'(\sigma), \tag{4.3}$$

The dimensionless parameters ξ, σ and μ are defined as in expressions (2.5) and (2.9); moreover

$$\lambda^2 = -\frac{\rho L^2 \omega^2}{EA}, \qquad \gamma = c(\rho EA)^{\frac{1}{2}} \neq 1. \tag{4.4}$$

where $\omega \in \mathbb{C}$. It can be shown that for the eigenvalue problem (4.1)–(4.3) there is an infinite sequence of simple eigenvalues corresponding to the complex-value roots of the characteristic equation

$$0 = \sinh\lambda + \gamma\cosh\lambda + \lambda\mu\sinh\lambda\sigma(\sinh\lambda(1-\sigma) + \gamma\cosh\lambda(1-\sigma)) = 0. \tag{4.5}$$

The vibrating mode corresponding to the eigenvalue λ is given by

$$\begin{aligned} u_1(\xi) &= C_1\cosh\lambda\xi, & \xi \in (0,\sigma), \\ u_2(\xi) &= C_2\cosh\lambda\xi + C_3\sinh\lambda\xi, & \xi \in (\sigma,1), \end{aligned} \tag{4.6}$$

where the coefficients C_2 and C_3 may be written in terms of C_1,

$$C_2 = C_1(1 + \lambda\mu\sinh\lambda\sigma\cosh\lambda\sigma), \qquad C_3 = C_1(-\lambda\mu\sinh^2\lambda\sigma), \tag{4.7}$$

and C_1 is such that the normalizing condition

$$\int_0^1 u^2(x)\,dx + \frac{\gamma(u(1))^2}{2\lambda} = 1 \tag{4.8}$$

is satisfied. Unlike the undamped case, here the eigenvalues and the corresponding eigenfunctions are complex-value functions. In particular, by representing each eigenfunction in its real and imaginary part, we can parametrically study the effect of a crack on the nodes of the real and the imaginary part of the eigenfunctions, when the location σ and the damage severity μ vary. In order to take into account the effect of the damping coefficient γ on identification results, five levels of damping ($\gamma = 0.05, 0.15, 0.25, 0.35, 0.50$) were studied and considered in detail in (De Tuoni, 2002).

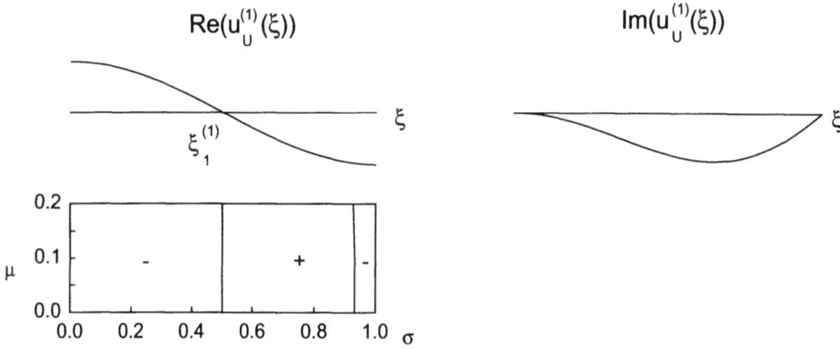

Figure 6. Positive (+) and negative (-) nodal displacement domain of the node of the real part of the first axial mode.

Figures 6–8 illustrate the positive (PNDD) and negative (NNDD) nodal displacement domains, (σ, μ), for $\gamma = 0.25$, relative to the real part (first column) and the imaginary part (second column) of the low order eigenfunctions. Concerning the real part, we can observe that, differently from the ideal undamped case, there is no monotonicity property between nodes shift and crack location. In fact, the plane domain (σ, μ) is divided into several subsections whose boundaries are approximately parallel to the μ–axis. The extension of the narrow bands depends on the damping coefficient. These regions are very thin for small values of γ, and become wider when γ increases.

Considering now the effect of a crack on the nodes of the imaginary part (second column of Figure 7 and 8), we can see that the plane domain (σ, μ) is the union of PNDD and NNDD regions alternately located and well separated on the whole beam length. The boundaries separating two consecutive subsections are rather sensitive to variations of μ, while it can be shown that they are little affected by variations of the damping coefficient γ.

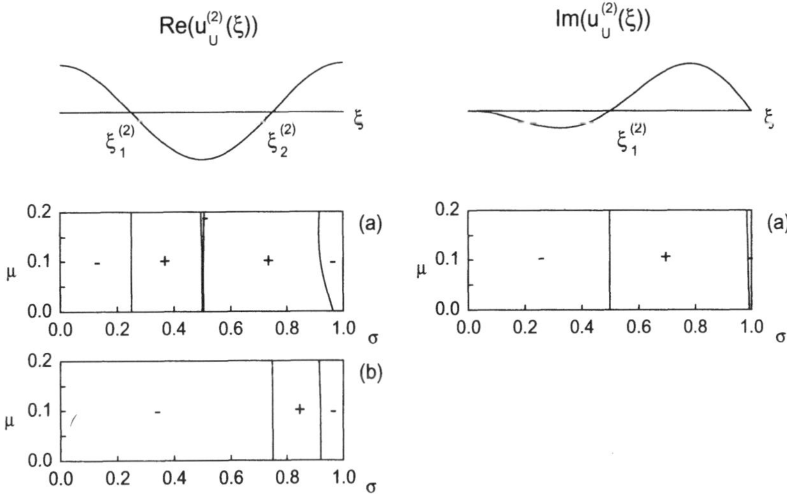

Figure 7. Positive (+) and negative (-) nodal displacement domain of the first (a) and second (b) node of the second axial mode. First column: real part of the eigenfunction. Second column: imaginary part of the eigenfunction.

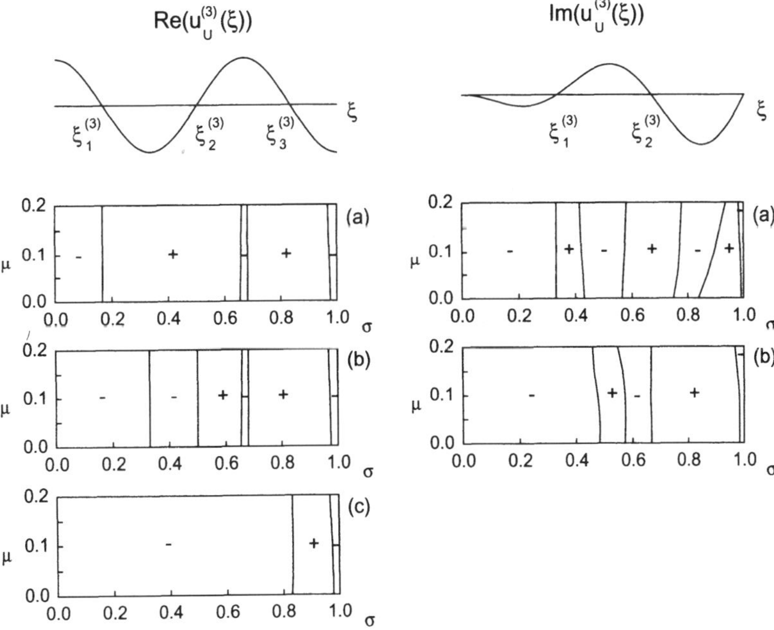

Figure 8. Positive (+) and negative (-) nodal displacement domain of the first (a), second (b) and third (c) node of the third axial mode. First column: real part of the eigenfunction. Second column: imaginary part of the eigenfunction.

5 Applications

Previous sections dealt with the description of NNDD and PNDD domains of the nodes of the first modes of an uniform beam in axial or bending vibration under different boundary conditions. Their practical utility as diagnostic tool is obvious; tests show the signs of node shifts, which are then entered in the (σ, μ) plots to identify the areas where damage is likely to be located. A cross-analysis of results of each node for the mode under consideration and the intersection of the areas where damage is likely to be located can lead to greater accuracy in detection of the crack.

Aiming to account for the prospective practical use of the proposed identification technique within the analsys of real cases, the present section is devoted to outlining some applications of experimental and analytical character.

In (Gladwell and Morassi, 1999) a double T steel beam in axial vibration was tested in undamaged and two levels of damaged configurations; the damage was made by saw-cutting the transversal cross-section of the beam. The technique used to estimate the position of a node is based on detecting the phase angle change of the inertance response function between two points of the beam axis including the node. The key notion is that, although the amplitude of each inertance term is very small near each node, the phase angle is clearly measurable, and should be equal to $+90^o$ or -90° in the ideal undamped case. The node is located by finding two neighboring points between which the phase angle changes sign.

Measurements on undamaged and damaged configurations were interpreted by using the mathematical model discussed in Section 2. In both undamaged and damaged cases, the agreement between experimental and analytical frequencies is very good with percentage errors less then 1.5%, see Table 1. Taking into account the results of Section 2, Table 2 can be used for localizing the crack. With few exceptions, there is a good agreement between the experimental and theoretical values of node location: there are only two anomalies concerning the first and the forth eigenmode; these are due to the inadequacy of the one dimensional mechanical model of the cracked rod to describe the region close to the damaged cross-section. The severity of the damage and the features of the specimen were probably decisive for the success of the identification; with less severe levels of damage, the sensitivity of the problem to the accuracy of the data may become important.

Table 1. Experimental and analytical frequencies of the undamaged and damaged rod. Frequency values in Hz. Data from (Gladwell and Morassi, 1999). $\Delta f_n\% = (f_n(\text{model}) - f_n(\text{exp.}))/f_n(\text{exp.}) \times 100$.

f_n	Undamaged rod			Damaged rod		
[Hz]	model	exper.	$\Delta f_n\%$	model	exper.	$\Delta f_n\%$
f_1	1173.750	1173.750	0.00	958.375	958.375	0.00
f_2	2347.500	2348.625	-0.05	2140.488	2141.375	-0.04
f_3	3521.250	3512.500	0.25	3387.078	3378.875	0.24
f_4	4695.000	4716.000	-0.45	4032.904	3975.875	1.43
f_5	5868.750	5870.000	-0.02	5739.806	5741.250	-0.03

Table 2. Node locations of an undamped axially vibrating beam in undamaged configuration and sign of the node shifts due to the crack. Abscissa of the cracked cross-section: $\sigma = 0.374$. Data from (Gladwell and Morassi, 1999).

Mode [-]	First node [m]	Second node [m]	Third node [m]	Forth node [m]	Fifth node [m]
1	0.500 $\longleftarrow$				
2	0.250 $\longrightarrow$	0.750 $\longleftarrow$			
3	0.167 $\longrightarrow$	0.500 $\longleftarrow$	0.833 $\longleftarrow$		
4	0.125 $\longrightarrow$	0.375 $\longleftarrow$	0.625 $\longleftarrow$	0.875 $\longleftarrow$	
5	0.100 $\longrightarrow$	0.300 $\longrightarrow$	0.500 $\longleftarrow$	0.700 $\longleftarrow$	0.900 $\longleftarrow$

About the bending vibration case, some numerical applications of the technique proposed in Section 3 will be presented hereafter. The inverse problem of damage detection is solved for different cases, using pseudo-experimental data, that is, the eigenvalues are obtained from the direct problem in undamaged and damaged conditions. Identification results are shown in detail for a damage location corresponding to $\sigma = 0.45$. Two different levels of damage are considered; the first case, $\mu = 0.20$ is characterized by moderate damage, that is, the value of stiffness K_B is such that the variations of the first few eigenvalues are about 8% of the initial values, while the second case (severe damage, $\mu = 0.50$) corresponds to variations of the same quantities of about 16%.

Table 3. Node locations of a bending vibrating beam in undamaged configuration and sign of the node shifts due to a crack. Abscissa of the cracked cross-section: $\sigma = 0.45$. Configurations: (1) Damage D1, $\mu = 0.20$. (2) Damage D2, $\mu = 0.50$.

State	Mode	1st node	2nd node	3rd node	4th node
Damage D1	2	0.500 $\longrightarrow$			
	3	0.333 $\longrightarrow$	0.667 $\longleftarrow$		
	4	0.250 $\longrightarrow$	0.500 $\longleftarrow$	0.750 $\longleftarrow$	
	5	0.200 $\longrightarrow$	0.400 $\longrightarrow$	0.600 $\longleftarrow$	0.800 $\longleftarrow$
Damage D2	2	0.500 $\longrightarrow$			
	3	0.333 $\longrightarrow$	0.667 $\longleftarrow$		
	4	0.250 $\longrightarrow$	0.500 $\longleftarrow$	0.750 $\longleftarrow$	
	5	0.200 $\longrightarrow$	0.400 $\longrightarrow$	0.600 $\longleftarrow$	0.800 $\longleftarrow$

The measurements of the nodes location and of the nodes shift are summarized in Table 3. The graphs of Figures 2-5 allow the identification of the areas where damage is likely to be located. The knowledge of the sign changes of the second and the third modes only allows locating the crack within the interval $(0.41, 0.50)$. Considering higher nodes, the interval of possible damage position is reduced to $(0.43, 0.46)$. It should be noted that the insertion of the crack caused the node of the first vibration mode to shift away from the damage section. More exhaustive results and some experimental applications can be found in (Dilena and Morassi, 2002), where the identification technique was employed to localize a crack in two double T free–free steel beams of the series HE100B in bending

Table 4. Node locations of the real and imaginary part (in brackets) of the first four eigenfunctions for a damped vibrating beam in undamaged configuration and sign of the node shifts due to a crack. Abscissa of the cracked cross-section: $\sigma = 0.45$. Damage severity: $\mu = 0.05$. Damping coefficient: $\gamma = 0.25$

Mode	1st node	2nd node	3rd node	4th node
1	0.500 $\longleftarrow$			
2	0.250 $\longrightarrow$	0.750 $\longleftarrow$		
	(0.500) $(\longleftarrow)$			
3	0.167 $\longrightarrow$	0.500 $\longleftarrow$	0.667 $\longleftarrow$	
	(0.333) $(\longleftarrow)$	(0.667) $(\longleftarrow)$		
4	0.125 $\longrightarrow$	0.375 $\longrightarrow$	0.625 $\longleftarrow$	0.875 $\longleftarrow$
	(0.250) $(\longrightarrow)$	(0.500) $(\longleftarrow)$	(0.667) $(\longleftarrow)$	

vibration. Generally speaking, a good agreement was observed between experimental behaviour and analytical expectations.

Finally, we consider briefly the case of an axially vibrating beam with a damper located at the right end. To fix ideas, we suppose that a crack of damage severity $\mu = 0.05$ appears in a cross section located at the $\sigma = 0.45$. It can be shown that the knowledge of the nodes shift of the real part of the first five eigenfunctions allows locating the crack within the interval $(0.38, 0.50)$ for different values of damping coefficient γ. Taking into account the imaginary part of the eigenfunctions, the interval of possible damage positions is reduced to $(0.42, 0.48)$, see Table 4.

Bibliography

R. D. Adams and P. Cawley and C. J. Pye and B. J. Stone, A vibration technique for non-destructively assessing the integrity of structures. *Journal of Mechanical Engineering Science*, 20:93–100, 1978.

J. C. Chen and J. A. Garba, Structural Safety Evaluation Based on System Identification Approaches. In *Structural damage assessment using a system identification tecnique*, H. G. Natke and C. Cempel editors. Wiesbaden: Vieweg Braunschweig, 1988.

C. Davini and A. Morassi and N. Rovere, Modal analysis of notched bars: tests and comments on the sensitivity of an identification technique. *Journal of Sound and Vibration*, 179:513–527, 1995.

M. De Tuoni, Identificazione del danno in travi vibranti con condizioni al contorno di tipo dissipativo. *Thesis* (in Italian), University of Udine, Udine, Italy, 2002.

M. Dilena and A. Morassi, Identification of crack location in vibrating beams from changes in node positions. *Journal of Sound and Vibration*, 255: 915–930, 2002.

L. B. Freund and G. Herrmann, Dynamic fracture of a beam or plate in plane bending. *Journal of Applied Mechanics*, 76-APM-15: 112–116, 1976.

B. Geist and J. R. McLaughlin, The effect of structural damping on nodes for the Euler–Bernoulli beam: a specific case study. *Appl. Math. Lett.*, 3(7): 51–55, 1994.

G. M. L. Gladwell and A. Morassi, Estimating damage in a rod from changes in node positions. *Inverse Problems in Engineering*, 7: 215–233, 1999.

G. Hearn and R. B. Testa, Modal analysis for damage detection in structures. *ASCE Journal of Structural Engineering*, 117: 3042–3063, 1991.

W. Leighton and Z. Nehari, On the oscillation of solutions of self-adjoint linear differential equations of the fourth order. *Transactions of the American Mathematical Society*, 89: 325–377, 1958.

R. Y. Liang and J. Hu and F. Choy, Theoretical study of crack-induced eigenfrequency changes on beam structures. *ASCE Journal of Engineering Mechanics*, 118: 384–396, 1992.

A. Morassi, Identification of a crack in a rod based on changes in a pair of natural frequencies. *Journal of Sound and Vibration*, 242: 577–596, 2001.

Y. Narkis, Identification of crack location in vibrating simply supported beams. *Journal of Sound and Vibration*, 172: 549–558, 1994.

H. G. Natke and C. Cempel, Fault detection and localisation in structures: a discussion. *Journal of Mechanical Systems and Signal Processing*, 5: 345–356, 1991.

P. F. Rizos and N. Aspragathos and A. D. Dimarogonas, Identification of crack location and magnitude in a cantilever beam from the vibration modes. *Journal of Sound and Vibration*, 138: 381–388, 1990.

R. Ruotolo and C. Surace, Damage assessment of multiple cracked beams: numerical results and experimental validation. *Journal of Sound and Vibration*, 206: 567–588, 1997.

F. Vestroni and D. Capecchi, Damage evaluation in cracked vibrating beams using experimental frequencies and finite element models. *Journal of Vibration and Control*, 2: 69–86, 1996.

Damage Identification of Beams Using Static Test Data

Mario Di Paola[‡] and Cristiano Bilello[‡]

[‡] Dipartimento di Ingegneria Strutturale e Geotecnica, Universitá degli Studi di Palermo, Viale delle Scienze, 90128, Palermo, Italy.

Abstract A damage identification procedure for beams under static loads is presented. Damage is modelled through a damage distribution function which determines a variation of the beam stiffness with respect to a reference condition. Using the concept of the equivalent superimposed deformation, the equations governing the static problem are recast in a *Fredholm's integral equation* of the second kind in terms of bending moments. The solution of this equation is obtained through an iterative procedure as well as in closed form. The latter is explicitly dependent from the damage parameters, thus, it can be conveniently used to set-up a damage identification procedure. Some numerical results are presented both to prove the validity of the proposed solution procedure, and to show its preformance in damage identification in presence of measurement noise.

1 Introduction

Damage can occur in structures during their service life due to extreme events and/or aggressive environmental conditions. It causes a reduction of structural stiffness and it manifests itself as a change of the static as well as the dynamic structural behavior.

Most of non-destructive damage identification techniques are based on the comparison between the measured response from the damaged structure and that obtained from an analytical model of the undamaged one. Depending on the nature of the experimental data, such methods can be clarified into two categories: static and dynamic damage identification procedures. It is generally easier to obtain dynamic data and the literature corresponding to the latter case is quite extensive and includes the use of modal data (Adams et al., 1991; Davini et al., 1992; Gounaris et al., 1996; Cerri and Vestroni, 2000), curvature measures (Pandey et al., 1991; Luo and Hanagud, 1997; Ratcliffe, 1999), frequency response functions (Wang et al., 1997; Thyagarajan et al., 1998; Sampaio et al., 1999; Lee and Shin, 2002), the strain energy (Hearn and Testa, 1993; Shi et al., 1998; Cornwell et al., 1999).

However, for certain type of structures, like beam-type, performing static tests can be as easy as performing dynamic tests; in this case, it becomes unjustifiable to complicate the damage identification problem by introducing other uncertainties coming from mass and damping factors. In spite of the large number of works presented in the field of dynamic identification, relative few papers have considered the possibility of using static data for damage identification. Static and modal data have been used for damage identification using finite element models (Hajela and Soeiro, 1989, 1990; Hjelmstad et al., 1991; Wang et al., 2001). Incomplete measures of the structure degree-of-freedoms have been taken into account through condensation of system matrices (Sanayei and Onipede, 1991; Sanayei and Scampoli, 1991). The influence of noisy measurements, number and spatial distribution of measured data, and number and character of load cases, on the performance of the identification procedure has been investigated (Banan et al., 1994a,b) and probabilistic indices of identified parameters have been proposed to distinguish real damage from noise (Hjelmstad and Shin, 1997).

In this work a damage identification procedure for beams under static loads is presented. In section 2, on the basis of the observation outlined in Di Paola (2002) for the static analysis of linear elastic truss structures with uncertainties, the static problem for a damaged beam is reformulated using the concept of the equivalent superimposed deformation. It is shown that the given problem is governed by a *Fredholm's integral equation* of the second kind in terms of bending moments. The solution of this equation is obtained through an iterative procedure as well as in an closed form. The latter is explicitly dependent from the damage parameters, thus, it can be conveniently used to set-up a damage identification procedure based on the comparison between theoretical and measured structural response from the damaged and undamaged beam (section 3). Finally in section 4, through numerical tests, the proposed solution is first validated, and, then, used for two damage identification problems with pseudo-experimental data affected by noise.

2 Formulation of the Static Problem

The problem is formulated here for an Euler-Bernoulli beam but it can be readily extended to higher order structures including Timoshenko beams, plate structures and, in the most general case, three-dimensional elastic continua. It will be assumed that changes in the beam response due to damage are represented through a variation of its bending stiffness (Luo and Hanagud, 1997; Wang et al., 1997; Sato, 1983; Yuen, 1985; Bilello, 2001) given as

$$EI\,(x) = EI\,[1 + d\,(x)] \tag{2.1}$$

where EI is a reference stiffness, say the undamaged one and $d\,(x)$ is a damage distribution function depending on some damage parameters that must be defined according to a given damage model (as shown later). The upper bound of the damage distribution function $d\,(x) = 0$ corresponds to the undamaged case while the lower bound $d\,(x) = -1$ indicates a local failure due to damage.

2.1 Non-Redundant Beams

For clarity sake the case of non-redundant beams is first presented. The static equations of an Euler-Bernoulli beam subjected to the external load $f\,(x)$ are easily obtained in the form

$$Q'(x) = -f(x), \; Q(x) = M'(x) \tag{2.2}$$

$$\chi(x) = -\frac{M(x)}{EI(x)} \tag{2.3}$$

$$w'(x) = \vartheta(x), \; \vartheta'(x) = \chi(x) \tag{2.4}$$

where $Q(x)$ is the shear force, $M(x)$ is the bending moment, $w(x)$, $\vartheta(x)$ and$\chi(x)$ are, respectively, the deflection, the slope and the curvature, and the apex prime denotes derivatives with respect to the spatial variable x. Combining Eqs. (2.2-2.4) yields a fourth order differential equation in the unknown function $w(x)$

$$[EI(x)\,w''(x)]'' = f(x) \tag{2.5}$$

that, associated with the given boundary conditions, governs the static problem.

If the beam is non-redundant integration of Eq. (2.5) is not strictly required. Indeed, the shear force and the bending moment distributions are known from equilibrium equations. Thus, one can substitute the known bending moment in (2.3) and integrate Eqs. (2.4) directly. However, in order to have an explicit dependence of the beam response on damage parameters, it may be observed that the variation of bending stiffness due to the presence of damage may be modelled as a superimposed curvature depending on the bending moment and on the damage distribution $d(x)$; indeed, taking into account Eq. (2.1), one may rewrite Eq. (2.3) in the form

$$\chi(x) = \chi_e(x) + \chi_d(x) \tag{2.6}$$

where

$$\chi_e(x) = -\frac{M(x)}{EI} \tag{2.7}$$

$$\chi_d(x) = \frac{M(x)}{EI} g_d(x); \quad g_d(x) = \frac{d(x)}{[1 - d(x)]} \tag{2.8}$$

From Eqs. (2.6-2.8) it is easily shown that the given damaged beam (*original beam*) can be sought as two undamaged beams (*reference beam*) under different load cases: the first one (*principal beam*) is subjected to the external load $f(x)$ while the second one (*auxiliary beam*) is subjected to the superimposed curvature $\chi_d(x)$. For a simply supported beam this is represented in Figure 1. Hereinafter, the response of the principal beam and of the auxiliary beam will be denoted with $w^0(x)$ and $\tilde{w}(x)$, respectively.

To obtain the response of the given beam, the superposition principle is applied, so that the following equations hold

$$Q(x) = Q^0(x) + \tilde{Q}(x), \; M(x) = M^0(x) + \tilde{M}(x) \tag{2.9}$$

$$\chi(x) = \chi^0(x) + \tilde{\chi}(x) \tag{2.10}$$

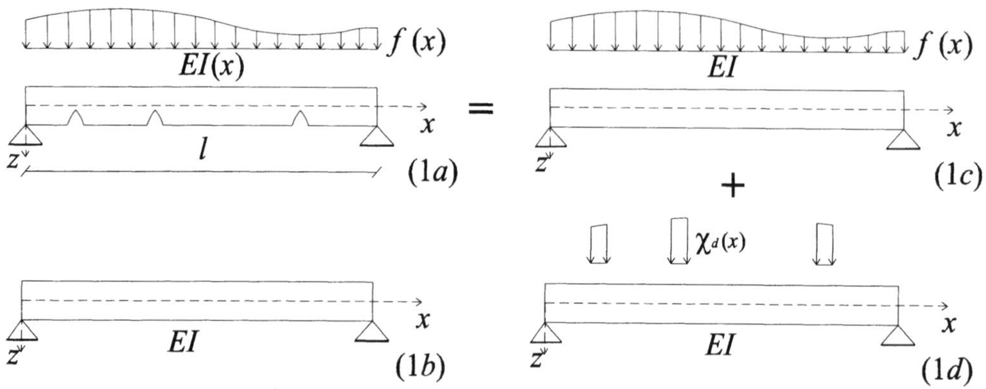

Figure 1. Simply-supported damaged beam. (1a) Given (original) beam, (1b) reference beam ,(1c) principal beam and (1d) auxiliary beam.

$$w(x) = w^0(x) + \tilde{w}(x), \ \vartheta(x) = \vartheta^0(x) + \tilde{\vartheta}(x). \tag{2.11}$$

It is worth noticing that in the case of non-redundant beams the curvature $\tilde{\chi}(x)$ in (2.10) coincides with $\chi_d(x)$, since $\tilde{Q}(x) = \tilde{M}(x) = 0$. The given problem is now broken into two sub-problems: the first one, for the principal beam subjected to the external load $f(x)$,

$$Q'^o(x) = -f(x), \ Q^0(x) = M'^0(x) \tag{2.12}$$

$$\chi^0(x) = -\frac{M^0(x)}{EI} \tag{2.13}$$

$$w'^0(x) = \vartheta^0(x), \ \vartheta'^0(x) = \chi^0(x) \tag{2.14}$$

and the second one, for the auxiliary beam subjected to the superimposed curvature $\chi_d(x)$,

$$\tilde{\chi}(x) = \frac{M^0(x)}{EI} g_d(x) \tag{2.15}$$

$$\tilde{w}'(x) = \tilde{\vartheta}'(x), \tilde{\vartheta}'(x) = \tilde{\chi}(x). \tag{2.16}$$

In writing Eq.(2.15) it has been taken into account the fact that, in this case, the bending moments in the given beam coincide with those calculated from the principal beam.

The solution for the principal beam can be easily obtained for any load distribution $f(x)$, then, the solution for the auxiliary beam is obtained from Eqs.(2.15) and (2.16). The solution for the given beam is finally reconstructed in closed form from Eqs.(2.9-2.11).

2.2 Redundant Beams

In this case the shear force and the bending moment distributions are unknown and Eq. (2.5) should be solved in order to obtain the unknown deflection $w(x)$. However this would not provide, in general, an explicit dependence of the beam response on the variation of stiffness parameters due to damage.

Based on the observations outlined in the previous section, an alternative solution procedure is presented herein. The given redundant beam is decomposed in two reference beams: the principal one subjected to the applied external load $f(x)$ and the auxiliary one subjected to the superimposed curvature. The response of the principal beam is easily obtained for a generic load distribution $f(x)$ and it is assumed known. Unlike the case of non-redundant beams, the response of the auxiliary beam cannot be calculated directly because the superimposed curvature depends now on the actual bending moment distribution (Eq. 2.8) that is still unknown.

To solve this problem, let us express the bending moment distribution in the auxiliary beam in terms of the superimposed curvature $\chi_d(x)$ as

$$\tilde{M}(x) = \int_0^l G_\chi^M(x,y)\, \chi_d(y)\, dy \tag{2.17}$$

in which the function $G_\chi^M(x,y)$ represents the bending moment distribution in the auxiliary beam due to a superimposed concentrated curvature $\chi_d(x,y) = \delta(x-y)$ where $\delta(\cdot)$ is the Dirac's delta function. In a similar way, we define the functions $G_\chi^w(x,y)$, $G_\chi^\vartheta(x,y)$ and $G_\chi^Q(x,y)$ as the corresponding deflection, slope and the shear force distributions. These functions, hereinafter, will be called *curvature Green's functions.* It can be shown that they satisfy the following set of equations

$$G_\chi^Q(x,y) = c_1 \tag{2.18}$$

$$G_\chi^M(x,y) = \int G_\chi^Q(x,y)\, dx + c_2 \tag{2.19}$$

$$G_\chi^\vartheta(x,y) = -\int \frac{G_\chi^M(x,y)}{EI} dx + U(x-y) + c_3 \tag{2.20}$$

$$G_\chi^w(x,y) = \int G_\chi^\vartheta(x,y)\, dx + c_4 \tag{2.21}$$

in which $U(\cdot)$ is the Unit step function and the coefficients c_1, c_2, c_3 and c_4 are calculated from the given boundary conditions. The solution of Eqs. (2.18-2.21) can be easily obtained for different boundary cases in order to provide a full set of curvature Green's functions (Di Paola and Bilello, 2002). Moreover, it may be easily shown that $G_{\chi}^{M}(x,y)$ is always in the form

$$G_{\chi}^{M}(x,y) = EI\sum_{k=1}^{n} X_k(x)X_k(y) \tag{2.22}$$

where $X_1, X_2, \ldots X_n$ is a set of n linearly independent polynomial functions belonging to the class L_2 of *square summable* functions in $[0\ l]$. Eq. (2.22) shows also the symmetry of this function, i.e., $G_{\chi}^{M}(x,y) = G_{\chi}^{M}(y,x)$.
Substituting Eq. (2.8) in (2.17) and using Eq. (2.10) one obtains the integral equation

$$M(x) = M^0(x) + \int_0^l K(x,y)\, M(y)\, dy \tag{2.23}$$

whose kernel $K(x,y)$ has the expression

$$K(x,y) = \frac{G_{\chi}^{M}(x,y)\, g_d(y)}{EI} \tag{2.24}$$

Eq. (2.23) is a *Fredholm's integral equation* of the *second kind* the solution of which can be always written in compact form as

$$M(x) = M^0(x) + \int_0^l H(x,y)\, M^0(y)\, dy \tag{2.25}$$

where $H(x,y)$ is the "*resolving kernel*". At this stage two possible procedures may be pursued to evaluate $H(x,y)$: the first one is by invoking the *Neumann's series* expansion, the second one using directly an analytical expression. Both are presented herein.

Following the first approach, the *resolving kernel* is obtained through the series expansion

$$H(x,y) = \sum_{j=1}^{\infty} K_j(x,y) \tag{2.26}$$

The *iterative kernels* $K_j(x,y)$ in Eq. (2.26) are calculated from the recursive equation

$$K_j(x,y) = \int_0^l K(x,z)\, K_{j-1}(z,y) dz \tag{2.27}$$

and $K_1(x,y) = K(x,y)$. It can be shown that the series (2.26) is *uniformly* and *absolutely convergent* if $\|K\| < 1$; in this case Eq. (2.25) is the only solution of the integral equation (2.23) (Tricomi, 1958). It is worth noting that, if the beam is non-redundant $G_{\chi}^{M}(x,y)$ is identically zero and Eq. (2.25) yields $M(x) = M^0(x)$ which is still the exact solution of the given problem. The solution obtained so far represents the extension to a continuous beam of what presented in Di Paola (2002) for the case of linear elastic truss structures with uncertainties.

Following the second approach, the properties of the curvature Green's function $G_{\chi}^{M}(x,y)$ allow one to obtain an analytical expression for $H(x,y)$. Indeed, using Eq. (2.22) in (2.24) yields

$$K(x,y)=\sum_{k=1}^{n} X_k(x)X_k(y)\,g_d(y) \tag{2.28}$$

In virtue of Eq. (2.28) and of the properties of the functions X_k, $K(x,y)$ is said to be a *Pincherle-Goursat's* kernel. For this case the resolving kernel $H(x,y)$ is evaluated in the form (Tricomi 1958)

$$H(x,y)=a^{-1}\mathbf{X}^{T}(x)\,\mathbf{B}\mathbf{X}(y)\,g_d(y) \tag{2.29}$$

In Eq. (2.29) $\mathbf{X}$ is an $n-$dimensional vector collecting the functions X_k, a is the determinant of the n x n symmetric matrix

$$\mathbf{A}=\mathbf{I}_n-\int_0^l \mathbf{X}(z)\,\mathbf{X}^{T}(z)\,g_d(z)dz, \tag{2.30}$$

$\mathbf{I}_n$ is the n-dimensional identity matrix and finally $\mathbf{B}$ is an n x n matrix whose elements B_{ij} are the *cofactors* of the elements A_{ij} in $\mathbf{A}$. Substituting Eq. (2.29) in (2.25) yields

$$M(x)=M^{0}(x)+a^{-1}\mathbf{X}^{T}(x)\,\mathbf{B}\int_0^l \mathbf{X}(y)\,g_d(y)\,M^{0}(y)\,dy \tag{2.31}$$

This equation provides an analytical relation between the bending moment distribution and damage parameters implicitly included in the expression for a, $\mathbf{B}$ and g_d. Then, using Eqs. (2.9-2.11), the deflection, the slope and the shear force for the given beam can be calculated as

$$w(x)=w^{0}(x)+\int_0^l G_{\chi}^{w}(x,y)\,g_d(y)\,M(y)\,dy \tag{2.32}$$

$$\vartheta(x)=\vartheta^{0}(x)+\int_0^l G_{\chi}^{\vartheta}(x,y)\,g_d(y)\,M(y)\,dy \tag{2.33}$$

$$Q(x)=Q^{0}(x)+\int_0^l G_{\chi}^{Q}(x,y)\,g_d(y)\,M(y)\,dy \tag{2.34}$$

It is worth noticing that the procedure just described is valid in any case including the case of a damaged beam with a varying cross section along x (Figure 2a); however, from a computational point of view, the better way to handle with this problem is in considering always the reference beam as an uniform beam (Figure 2b) and including the known geometrical stiffness variation in the superimposed curvature (Figure 2d). Following the same approach, one can treat partially restrained beams (see Di Paola and Bilello, 2002 for more details), thus, the curvature Green's functions remain the same whatever the variation of structural parameters are and independently from the presence of partially restrained boundaries.

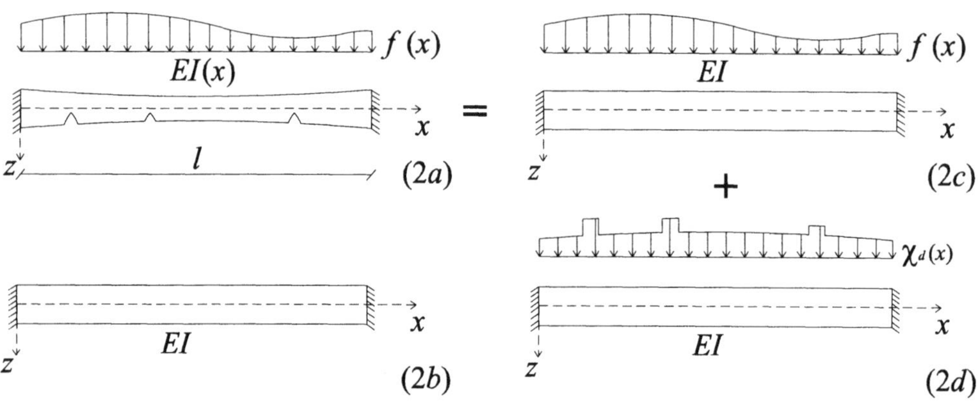

Figure 2. Redundant damaged beam with varying cross section along x. (2a) Given (original) beam, (2b) reference beam, (2c) principal beam and (2d) auxiliary beam.

Eqs. (2.31) and (2.32-2.34) provide the exact solution for the static problem of Euler-Bernoulli beams with arbitrary bending stiffness variation. They give an explicit dependence from the parameters that govern the stiffness variation, thus, they may be conveniently used for damage identification, that is $d(x)$ depends on unknown parameters to be identified, or, as another example, for probabilistic analysis of beam with uncertainties, when $d(x)$ is a given random field.

3 Damage Identification

3.1 Damage Model

There are no restriction in the expression for $d(x)$ and in the most general case it could be given as a linear combination of given functions by unknown coefficients to be identified. In this way one may be able to identify an arbitrary stiffness distribution.

However, in this study damage is physically sought as saw-cuts located at different positions $x = x_r$ along the beam length l. The cuts' depths to beam height ratios are denoted by β_r $(0 < \beta_r < 1)$ and represent measures of damage extents. The rth damage is modelled through a uniformly distributed reduction of the bending stiffness in a beam segment $2\Delta x_r$-wide with its midpoint at $x = x_r$ (Lee and Shin, 2002). The stiffness reduction coefficient is denoted by D_r. The damage distribution function can be written as

$$d(x, \mathbf{x}_d, \boldsymbol{\beta}_d) = \sum_{r=1}^{q} D_r(\beta_r)\left\{U\left[x - x_r - \Delta x_r(\beta_r)\right] - U\left[x - x_r + \Delta x_r(\beta_r)\right]\right\} \tag{3.1}$$

where q is the number of damaged sections while $\mathbf{x}_d$ and $\boldsymbol{\beta}_d$ are q-dimensional vectors of damage locations and extents, respectively. For a rectangular cross-section beam, the functions $D(\beta_r)$ and $\Delta x_r(\beta_r)$ have been experimentally evaluated and are given by (Bilello, 2001)

$$D(\beta_r) = -\frac{\beta_r(3 - 2\beta_r)}{2 - \beta_r} \tag{3.2}$$

$$\Delta x_r(\beta_r) = \frac{\beta_r h}{\mathrm{Tan}\theta} \tag{3.3}$$

where $\theta = 0.7328$.

3.2 Identification Procedure

The damaged identification procedure proposed in this paper is based on the minimization of the objective function defined as

$$J(\mathbf{x}_d, \boldsymbol{\beta}_d) = \frac{1}{2}\sum_{i=1}^{nlc}\sum_{j=1}^{nm} \alpha_i\left[\Delta\eta_i^t(x_j, \mathbf{x}_d, \boldsymbol{\beta}_d) - \Delta\eta_i^e(x_j)\right]^2 \tag{3.4}$$

in which α_i is the weight associated to the ith load case, nlc is the number of load cases, nm is the number of measurements, $\Delta\eta_i^t(x_j, \mathbf{x}_d, \beta_d)$ and $\Delta\eta_i^e(x_j)$ are, respectively, the theoretical and measured variations of a structural response characteristic due to damage. The identification algorithm is then formulated as a constrained nonlinear least-squared minimization problem, i.e.,

$$\underset{\mathbf{x}_d, \boldsymbol{\beta}_d}{\text{minimize}}\ J(\mathbf{x}_d, \boldsymbol{\beta}_d) \text{ subject to } \mathbf{0} < \frac{\mathbf{x}_d}{l} < \mathbf{1} \text{ and } \mathbf{0} < \boldsymbol{\beta}_d < \mathbf{1} \tag{3.5}$$

where $\mathbf{0}$ and $\mathbf{1}$ are q-dimensional zero and unit vectors, respectively.

For application purposes, use is made of a simple algorithm recently proposed for damage identification through frequency data (Vestroni and Capecchi, 1996, 2000; Cerri and Vestroni, 2000) and here recast for the use of static data. The optimization process is broken in two phases. For each scenario of damage locations the following reduced objective function is defined

$$\breve{J}(\mathbf{x}_d) = \underset{\boldsymbol{\beta}_d}{\text{minimize}}\ J(\mathbf{x}_d, \boldsymbol{\beta}_d) \text{ subject to } \mathbf{0} < \boldsymbol{\beta}_d < \mathbf{1} \tag{3.6}$$

that, for each $\mathbf{x}_d$ gives the best value of $\boldsymbol{\beta}_d$ to minimize $J(\mathbf{x}_d, \boldsymbol{\beta}_d)$. The final estimate is obtained from the minimization of $\breve{J}(\mathbf{x}_d)$. When a limited numbers of unknown damage locations is expected this method provides a simple and effective optimization tool.

It is worth noticing that, under the assumption of determinism of damage parameters, the minimum number of data required to solve the nonlinear optimization problem (3.5) (or (3.6)) is obtained from the relation (Banan et al., 1994a; Hjelmstad and Shin, 1997; Vestroni and Capecchi, 2000)

$$nlc \text{ x } nm \geq 2q \tag{3.7}$$

However, this represents only a necessary condition for damage identifiability, i.e, if not satisfied then damage parameter estimates are certainly not reliable while the converse is generally not true.

3.3 Simulation of experimental data

In this work experimental data are simulated through a perturbed analytical response, i.e., the kth data sample in (3.4) will be given as

$$\Delta\eta_i^{e(k)}(x_j) = \left[1 + \mathrm{N}_{ij}^{(k)}\right] \Delta\eta_i^t\left(x_j, \hat{\mathbf{x}}_d, \hat{\boldsymbol{\beta}}_d\right) \quad i = 1, 2, ...nlc \; j = 1, 2, ...nm \; k = 1, 2.....NS \tag{3.8}$$

in which $\hat{\mathbf{x}}_d$ and $\hat{\boldsymbol{\beta}}_d$ are the actual damage parameter vectors (assigned), $\mathrm{N}_{ij}^{(k)}$ is a uniformly distributed random variable with zero mean and given amplitude range and NS is the number of simulated tests. The solution of the optimization problem corresponding to the set of data $\Delta\eta_i^{e(k)}(x_j)$ is denoted with $\breve{\mathbf{x}}_d^{(k)}$ and $\breve{\boldsymbol{\beta}}_d^{(k)}$.

The performance of the identification procedure is, then, analyzed using two global indices (Banan et al., 1994b) the normalized Average Mean Error (AME),

$$\text{AME} = \frac{1}{2q} \sum_{r=1}^{q} \frac{\left|\mathrm{E}\left[\breve{x}_r\right] - \hat{x}_r\right|}{\hat{x}_r} + \frac{\left|\mathrm{E}\left[\breve{\beta}_r\right] - \hat{\beta}_r\right|}{\hat{\beta}_r} \tag{3.9}$$

and the normalized Average Standard Deviation (ASD),

$$\text{ASD} = \frac{1}{2q} \sum_{r=1}^{q} \frac{\sigma_{\breve{x}_r}}{\hat{x}_r} + \frac{\sigma_{\breve{\beta}_r}}{\hat{\beta}_r} \tag{3.10}$$

where $E[\cdot]$ and σ denote, respectively, mean and standard deviation of the output variables. The lower AME and ASD are the better the estimation is while the worst condition corresponds to a large AME with a low ASD since in this case the identification procedure provides poor parameter estimations with high level of confidence. The identification procedure is consistent if both indices converge to zero as the amount of available data increases (Banan et al., 1994b).

4 Numerical examples

In this section some numerical results are presented first, to prove the validity of the solution procedure presented in section 2, and then, to show its efficacy for damage identification problems.

A clamped-supported rectangular cross-section beam is considered. In the first example, damage is located at $\hat{x} = 0.6l$ and $\hat{\beta} = 0.5$ and the beam is subjected to a uniform distributed load p (Figure 3). The damage reduction coefficient $D(\beta_r)$ and the width of the damaged segment $2\Delta x_r(\beta_r)$ are calculated according to Eqs. (3.2, 3.3).

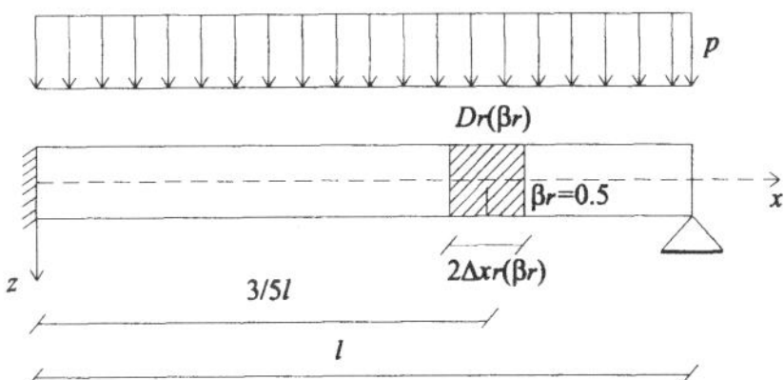

Figure 3. Clamped-supported damaged beam.

In Figure 4 the results in terms of non-dimensional deflection $\bar{w} = w/l$ and non-dimensional bending moment $\bar{M} = Ml/EI$ are reported versus the non-dimensional coordinate $\bar{x} = x/l$. The iterative solution (Eq. (2.26)) is compared with the exact one (Eq. (2.29)). The response of the undamaged beam is also reported. Besides the effects of damage on the beam response, it is remarkable that the iterative solution converges to the exact one using only two terms of the series expansion (2.26).

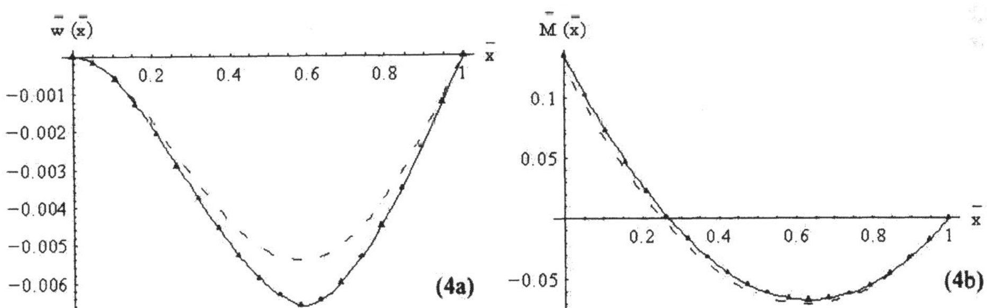

Figure 4. Comparison between iterative and exact solution. (4a) Non-dimensional deflection and (4b) non-dimensional bending moment: undamaged beam (dashed), damaged beam exact solution (solid), damaged beam iterative solution $j = 2$ (▲).

In the second example an identification problem is presented. A single damage configuration is analyzed. The actual damage location and extends are, respectively, $\hat{x} = 0.75l$ and $\hat{\beta} = 0.5$. A single ($nm = 1$) measurement station at $x = 0.5l$ and six load cases (nlc

= 6) are taken into account. Each load case is obtained by applying, alternatively from the left- and the right- end of the beam, a force P at different locations equally spaced $\Delta x = l/(nlc + 1)$. The maximum amplitude of noise is varied up to 10%. For each noise level and load conditions two hundreds data samples are generated. Bending moment measurements have been used to define the objective function in (3.4).

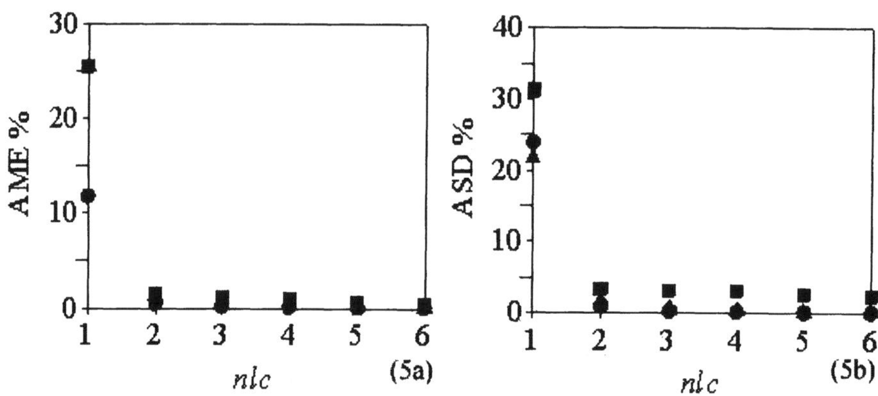

Figure 5. Influence of the amount of available data and of noise on the performance of the identification procedure. (5a) AME and (5b) ASD: maximum amplitude of noise 1% (•), 5% (▲) and 10% (■).

Figure 5 shows the variation of AME and ASD with respect to number of load cases for different maximum amplitude of noise. It is worth noticing that, independently from the noise level, a larger amount of information (load cases) will always provide a better estimation. However for large amplitude of noise (10%) the ASD does not change with *nlc*. This means that the precision of the estimation cannot be further improved by increasing the amount of information. The effects of different boundary conditions and load configurations have also been investigate (see Di Paola and Bilello, 2002) but are not reported here for space reasons.

A double-damage case has also been investigated. Since the objective of this application is to show the influence of different damage extents on the identification process, damage locations ($\hat{x}_1 = 0.5l$ and $\hat{x}_2 = 0.75l$) are assumed known and only damage extents must be identified. In Figure 6 the AMD and ASD are reported versus the actual damage extents $\hat{\beta}_1$ and $\hat{\beta}_2$ for $nlc = 4$, and maximum amplitude of noise equal to 5%. Loads are applied in the same way as described before. It can be pointed out that better results are obtained when damage extents are comparable. This is a consequence of the physical fact that a larger damage extent determines, generally, larger structural response perturbation; thus its effect becomes predominant with respect to the others. In this case the larger damage extent is identified with high accuracy and precision whereas this is not the case for the smaller ones.

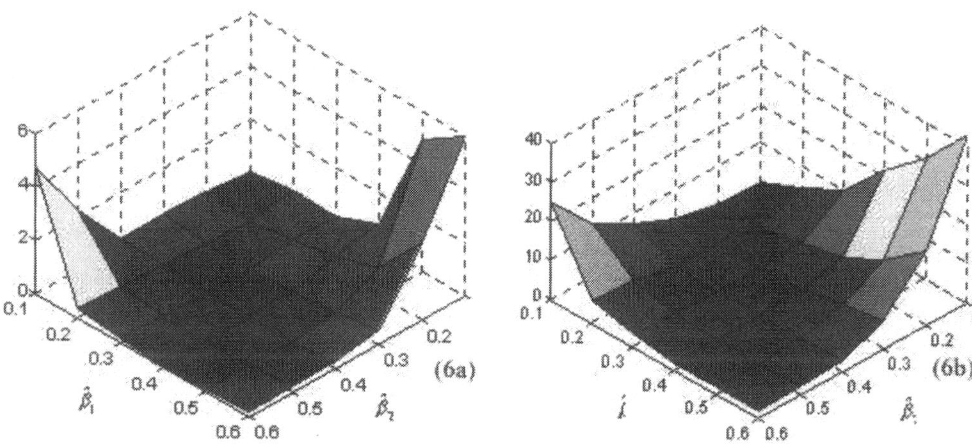

Figure 6. Influence of different damage extents on the performance of the identification procedure. (6a) AME and (6b) ASD.

5 Conclusions

A new approach for the static analysis of linear elastic damaged Euler-Bernoulli beams has been presented. Damage is modelled through a generic variation in the beam stiffness with respect to a reference condition, say the undamaged one. Using the concept of the equivalent superimposed deformation the associated static problem is reduced to the solution of a Fredholm's integral equation of the second kind characterized by a Pincherle-Goursat's kernel. This solution has been obtained through an iterative approach as well as in analytical form for any arbitrary stiffness variation. Numerical examples have shown the validity of the proposed solution procedures.

Moreover, the analytical solution provides an explicit dependence of the structural response on damage parameters, thus, it can be conveniently used to set-up a damage identification procedure based on a least square constrained nonlinear minimization problem. A proper objective function is defined as a function of theoretical and measured variations of a structural response characteristic in presence of damage. The effects of noisy measurements and of the amount of available data on the performance of the identification procedure have been investigated through numerical simulations and the introduction of statistical indices measuring the accurateness and precision of the estimation.

The results have shown that, for adequately small amplitude of noise, both mean errors and standard deviations of estimated parameters reduce with increasing amount of information so proving the consistency of the estimation. In a double-damaged beam, accurate estimations have been obtained when damage extents are comparable while in the general case, the identification procedure provides more accurate estimations of larger damage extents than smaller ones.

6 Acknowledgements

This work is part of a larger national research project coordinated by Prof. C. Davini and supported by M.U.R.S.T through grant 2000-2002 number MM08342598-003. The financial support is gratefully acknowledged.

Bibliography

R. D. Adams, P. Cawley, C. J. Pye, and B. J. Stone. A vibration technique for non-destructively assessing the integrity of structures. *Journal of Mechanical Engineering Science*, 20:93–100, 1991.

M. R. Banan, M. R. Banan, and K. D. Hjelmstad. Parameter estimation of structures from static response. I computational aspects. *Journal of Structural Engineering*, 120: 3243–3258, 1994a.

M. R. Banan, M. R. Banan, and K. D. Hjelmstad. Parameter estimation of structures from static response. II numerical simulation studies. *Journal of Structural Engineering*, 120:3259–3283, 1994b.

C. Bilello. *Theoretical and experimental Investigation on Damaged Beams under Moving Systems.* PhD. Thesis, Dip. Ing. Strut. e Geotec., Università degli Studi di Palermo, 2001.

M. N. Cerri and F. Vestroni. Detection of damage in beams subjected to diffused cracking. *Journal of Sound and Vibration*, 234:259–276, 2000.

P. Cornwell, S. W. Doebling, and C. R. Farrar. Application of the strain energy damage detection method to plate-like structures. *Journal of Sound and Vibration*, 224:359–374, 1999.

C. Davini, F. Gatti, and A. Morassi. A damage analysis of steelbeams. *Meccanica*, 22: 321–332, 1992.

M. Di Paola. A new approach for the static analysis of linear elastic redundant trusses with uncertainties. *Probabilistic Engineering Mechanics*, under review, 2002.

M. Di Paola and C. Bilello. An integral equation for damage identification of Euler-Bernoulli beams under static loads. *Journal of Engineering Mechanics*, under review, 2002.

G. D. Gounaris, C. A. Papadopoulus, and A. D. Dimarogonas. Crack identification in beams by coupled response measurements. *Computer & Structures*, 58:299–302, 1996.

P. Hajela and F. J. Soeiro. Recent developments in damage detection based on system identification methods. *Structural Optimization*, 2:1–10, 1989.

P. Hajela and F. J. Soeiro. Structural damage detection based on static and modal analysis. *AIAA Journal*, 28:1110–1115, 1990.

G. Hearn and R. B. Testa. Modal analysis for damage detection in structures. *Journal of Structural Engineering*, 117:1798–1803, 1993.

K. D. Hjelmstad and S. Shin. Damage detection and assessment of structures from static response. *Journal of Engineering Mechanics*, 123:568–576, 1997.

K. D. Hjelmstad, S. L. Wood, and S. J. Clark. Mutual residual energy method for parameter estimation in structures. *Journal of Structural Engineering*, 118:223–242, 1991.

U. Lee and J. Shin. A frequency response function-based structural damage identification method. *Computer & Structures*, 80:117–132, 2002.

H. Luo and S. Hanagud. An integral equation for changes in the structural dynamics characteristics of damaged structures. *Int. Journal of Solids and Structures*, 34:4557–4579, 1997.

A. K. Pandey, M. Biswas, and M. M. Samman. Damage detection from changes in curvature mode shapes. *Journal of Sound and Vibration*, 145:321–332, 1991.

C. P Ratcliffe. A frequency and curvature based experimental method for locating damage in structure. *Journal of Vibration and Acoustic*, 226:1029–1042, 1999.

R. P. C. Sampaio, N. M. M. Maia, and J. M. M. Silva. Damage detection using frequency-response-function curvature method. *Journal of Sound and Vibration*, 226:1029–1042, 1999.

M. Sanayei and O. Onipede. Assessment of structures using static test data. *AIAA Journal*, 29:1156–1179, 1991.

M. Sanayei and S. F. Scampoli. Structural element stiffness identification from static test data. *Journal of Engineering Mechanics*, 117:1021–1036, 1991.

H. Sato. Free vibration of beams with abrupt changes of cross-section. *Journal of Sound and Vibration*, 89:59–64, 1983.

Z. Y. Shi, S. S. Law, and L. M. Zhang. Structural damage localization from modal strain energy changes. *Journal of Sound and Vibration*, 218:825–844, 1998.

S. K. Thyagarajan, M. J. Schulz, and P. F. Pai. Detecting structural damage using frequency response functions. *Journal of Sound and Vibration*, 210:162–170, 1998.

F. G. Tricomi. *Lezioni sulle Equazioni Integrali.* Editore Gheroni, 1958.

F. Vestroni and D. Capecchi. Damage evaluation in cracked vibrating beams using experimental frequencies and finite elements models. *Journal of Vibration and Control*, 2:69–86, 1996.

F. Vestroni and D. Capecchi. Damage detection in beam structures based on frequency measurements. *Journal of Engineering Mechanics*, 126:761–768, 2000.

X. Wang, N. Hu, H. Fukunga, and Z. H. Yao. Damage detection in beam structures based on frequency measurements. *Engineering Structures*, 23:610–621, 2001.

Z. Wang, R. M. Lin, and M. K. Lim. Structural damage detection using measured FRF data. *Computer Methods Applied Mechanics*, 147:187–197, 1997.

M. M. F. Yuen. A numerical study of the eigenparameters of damaged cantilever beam. *Journal of Sound and Vibration*, 103:301–310, 1985.

A Novel Approach for the Identification of Material Elastic Constants

Elena Ferretti, Antonio Di Leo and Erasmo Viola

DISTART – Department of Structures Transportations Waters Survey and Territory Engineering, University of Bologna, Bologna, Italy

Abstract The present study is part of an identifying programme for constitutive parameters in damaged materials, termed the "effective parameters". The programme starting point is that the experimental response depends not only on constitutive parameters, but also on structural mechanics and interaction with the test-machine. It is showed how the load-displacement diagram of compressed concrete cylinders is affected by crack propagation, through the resistant structure modification. Moreover, it is analytically demonstrated that the effective stress (σ_{eff})-effective strain (ϵ_{eff}) curve exhibits a strictly positive derivative at the point corresponding to the average stress ($\overline{\sigma}$)-average strain ($\overline{\varepsilon}$) curve peak. Finally, it is proposed a new identification procedure which provided satisfactory results, giving monotone strictly non-decreasing, size-effect insensitive and failure mechanism insensitive $\sigma_{eff} - \varepsilon_{eff}$ curves.

1 Introduction

In order to derive a constitutive law in uniaxial compression from experimental data, it is common practice to define the average stress $\overline{\sigma}$ and the average strain $\overline{\varepsilon}$ as shown in Figure 1. The $\overline{\sigma} - \overline{\varepsilon}$ relationship in Figure 1 is known as uniaxial constitutive law for monotone strain processes. The term "constitutive" is associated with the $\overline{\sigma} - \overline{\varepsilon}$ relationship since this relationship is considered as representative of the mechanic behaviour of the material. However, one can make the following remarks concerning the choice of this term:-

1. The $\overline{\sigma} - \overline{\varepsilon}$ law in Figure 1 is size–effect sensitive, while a constitutive law should not exhibit a size effect.
2. The identification procedure in Figure 1 consists of a mere change of scale. Thus, experimental and identified curves are homothetic (Figure 1). In particular, they both exhibit a softening behaviour. Nevertheless, it is not possible to associate a physical meaning with the softening behaviour of a material response, as the concept of instability loses its sense in the infinitesimal neighbourhood of a point (Ferretti, 2001). Strain–softening has been widely regarded as inadmissible by several authors (Bazant et al., 1997), from the beginning of the 20th century forth.

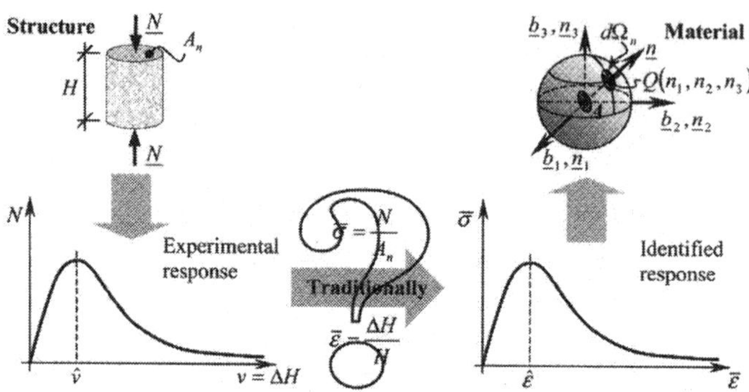

Figure 1. Traditional identification of mono-axial constitutive law by experimental tests.

These inconsistencies come from the impossibility of performing mechanical tests on the material directly: the object in testing is never the material, but a structure interacting with the test–machine (Figure 1, (Ferretti et al., 1999)). Thus, experimental results univocally characterise the behaviour of the specimen–test machine system, while they are not at all representative of the behaviour of the material. In particular, the softening branch has a meaning that is only linked to the structural instability. This branch cannot provide information on the constitutive behaviour, but through an identifying model. To this aim, the evaluation of all factors influencing a test result is needed. Indeed, since the specimen is a structure, experimental results (R) depend not only on constitutive properties (C), but also on structural mechanics (S), interactions between test–machine and specimen (I), and test–machine metrological characteristics (M):

$$R = C + S + I + M. \tag{1.1}$$

It is then necessary to define an identifying procedure from experimental data to material behaviour (inverse problem), which is not affected by the remarks concerning the approach in Figure 1. On the base of partially analogous considerations, Rosati and Natali Sora (2001) have recently proposed a complete response for tensioned concrete.

2 Identification Approach of the Mono-Axial Effective Behaviour

2.1 Identification of the Effective Stress

Name K_C, K_S, K_I, and K_M the weighed contributions assumed by C, S, I, and M, respectively, in the definition of R (Eq. 1.1):

$$C = K_C R, \quad S = K_S R, \quad I = K_I R, \quad M = K_M R. \tag{2.1}$$

From Eqs. 1.1 and 2.1, it follows that $K_C + K_S + K_I + K_M = 1$. All the contributions but the constitutive behaviour can be grouped in one factor K:

$$K = K_S + K_I + K_M. \tag{2.2}$$

From the identifying procedure in Figure 1 it follows that:

$$C \equiv R. \tag{2.3}$$

With the positions in Eqs. 1.1, 2.1 and 2.2, the Eq. 2.3 is replaced by the relationship:

$$C = (1 - K)\,R. \tag{2.4}$$

Eq. 2.4 allows evaluation of the constitutive properties, taking into account the behaviour of the specimen–test machine system. This approach is formally more correct than the approach in Eq. 2.3. Nevertheless, it is not of immediate use, since $K_C = K_C\,(R)$, $K_S = K_S\,(R)$, $K_I = K_I\,(R)$, and $K_M = K_M\,(R)$ are load–step functions. Then,

$$K = K\,(R) \tag{2.5}$$

is a load–step function, and not a constant of the performing test. In conclusion, it is not possible to establish a homothetic correspondence between the experimental load–displacement and the uniaxial stress–strain relationship. Moreover, since Eq. 2.5 is not of objective determination, K can only be estimated, with regard to the material scale. This involves the identification of an effective and not constitutive response.

The loss of homothetic behaviour allows us to obviate to the second remark in §1. In other words, since it is impossible to associate a physical meaning with the strain–softening behaviour, we can aspect that the effective laws must be monotone nondecreasing for any material. An analysis of the reciprocal ratios between K_C, K_S, K_I and K_M for compressed concrete cylinders (Ferretti, 2001) showed that $K \cong K_S$: any specimen can be regarded as composed by a resistant structure (Figure 2), in which crack propagation never occurred, and a volume of incoherent material. To identify the scale factor of the σ axis with respect to the N axis (Figure 1), it is then fundamental to introduce a parameter whose dimensions are those of an area and whose incremental law is linked to the structural scheme variation. In the following, this parameter will be termed the resistant area A_{res}. It was here[1] proposed to estimate the resistant area A_{res} in accordance with the Fracture Mechanics with Damage, by means of the damage parameter D (here D has a scalar value):

$$A_{res} = A_n\,(1 - D)\,. \tag{2.6}$$

The effective stress has here been defined as the average stress acting on A_{res}:

$$\sigma_{eff} = \frac{N}{A_{res}} = \overline{\sigma}\frac{A_n}{A_{res}}. \tag{2.7}$$

The analogy with the manner of operation of the Fracture Mechanics with Damage is limited to Eq. 2.6. In the Fracture Mechanics with Damage, D is analytically formulated and considered as uniformly distributed on A_n. In this study, $D = D\,(R)$ is experimentally evaluated and considered as localised in the volume of incoherent material.

[1] We thank the Italian Ministry for Universities and Scientific and Technological Research (MURST) for its financial support.

Figure 2. Resistant structure at the end of the test.

Algebraic considerations about the formulation of the effective stress. In Eq. 2.7, make explicit the dependence of σ_{eff}, A_{res}, N and D on the displacement v:

$$\sigma_{eff}(v) = \frac{N(v)}{A_{res}(v)}. \tag{2.8}$$

Now, find the derivative of Eq. 2.8 with respect to the variable $\bar{\varepsilon}$:

$$\frac{d\sigma_{eff}}{d\bar{\varepsilon}} = \sigma'_{eff}\frac{dv}{d\bar{\varepsilon}} = H\frac{N' A_{res} - N A'_{res}}{A^2_{res}}. \tag{2.9}$$

The superscript indicates derivation with respect to the variable v, and H is the gage length of $\bar{\varepsilon}$. For the conventions in Figure 1, it follows that:

$$N(v)|_{v=\hat{v}} = N_{\max}, \tag{2.10}$$

where $\hat{v}$ is the value of impressed displacement corresponding to the maximal load. As to the discussion of the sign of Eq. 2.9, it can be stated that:

- N is a monotone nondecreasing function until the peak ($N' \geq 0$, $0 \leq v \leq \hat{v}$), and a monotone strictly nonincreasing function beyond the peak ($N' < 0$, $v > \hat{v}$);
- A_{res} is a monotone nonincreasing function on all the domain ($A'_{res} \leq 0$, $\forall v$), and it can assume a zero tangent only in a neighbourhood of the origin, corresponding to the linear elastic state of the material.

For Eq. 2.6, the assumption of monotonicity for A_{res} involves the monotonicity of the damage law. The experimental results agree with the condition of non zero tangent of A_{res} and D for $v = \hat{v}$, since the crack propagation rate near $v = \hat{v}$ is always very fast:

$$A'_{res}|_{v=\hat{v}} \neq 0, \qquad D'|_{v=\hat{v}} \neq 0. \tag{2.11}$$

It follows immediately that the sign of $d\sigma_{eff}/d\bar{\varepsilon}$ is positive for $0 \leq v \leq \hat{v}$:

$$\frac{d\sigma_{eff}}{d\bar{\varepsilon}} > 0 \qquad 0 \leq v \leq \hat{v}. \tag{2.12}$$

In particular, for $v = \hat{v}$ the Eq. 2.9 assumes the value of:

$$\left.\frac{d\sigma_{eff}}{d\bar{\varepsilon}}\right|_{v=\hat{v}} = -NH\frac{A'_{res}}{A^2_{res}} > 0, \tag{2.13}$$

in which the strict inequality comes from Eq. 2.11. From Eq. 2.13, the first fundamental result follows: a point with strictly positive tangent in the $\sigma_{eff} - \bar{\varepsilon}$ curve corresponds to the point with zero tangent in the $N - v$ curve. This is a notable result, since it has been obtained without having introduced any other assumptions on the shape of the damage law except the condition of non zero tangent in correspondence of the maximal load. The same result for the sign of the tangent can be transposed to the $\sigma_{eff} - \varepsilon_{eff}$ curve, in the point corresponding to the $v = \hat{v}$ point of the $N - v$ curve. The sign of Eq. 2.9 for $v > \hat{v}$ depends on the value of ρ, the ratio between the two terms in the numerator of Eq. 2.9:

$$\rho = \frac{N' A_{res}}{N A'_{res}}; \tag{2.14}$$

$$\frac{d\sigma_{eff}}{d\bar{\varepsilon}} \geq 0 \qquad \forall v > \hat{v},\ 0 \leq \rho \leq 1; \tag{2.15}$$

$$\frac{d\sigma_{eff}}{d\bar{\varepsilon}} < 0 \qquad \forall v > \hat{v},\ \rho > 1. \tag{2.16}$$

In alternative, one can study the sign of the derivative of q, defined as follows:

$$q = \frac{\overline{\sigma}_{\max}}{\sigma_{eff}} \overset{(2.6)}{=} \frac{A_{res}}{A_n} \bigg/ \frac{N}{N_{\max}}; \tag{2.17}$$

$$q' = -\overline{\sigma}_{\max}\frac{\sigma'_{eff}}{\sigma^2_{eff}} = \overline{\sigma}_{\max}\frac{N' A_{res} - N A'_{res}}{N^2}. \tag{2.18}$$

From Eq. 2.18 it can be observed that the sign of q' too is determined by the ratio ρ:

$$q' > 0 \qquad \forall v > \hat{v},\ \rho > 1; \tag{2.19}$$

$$q' \leq 0 \qquad \forall v > \hat{v},\ 0 \leq \rho \leq 1. \tag{2.20}$$

On the other hand, the signs of q' and σ'_{eff} are unconformable $\forall v$ for Eqs. 2.15, 2.16 and 2.18. In conclusion, the sign of $d\sigma_{eff}/d\bar{\varepsilon}$ is surely positive for $0 \leq v \leq \hat{v}$, whereas it is only known when the damage law is known for $v > \hat{v}$. Also this result can easily be transposed to the sign of the derivative of the $\sigma_{eff} - \varepsilon_{eff}$ curve.

2.2 Identification of the Effective Strain

As regards the scale factor of the ε axis with respect to the v axis (Figure 1), the effective strain ε_{eff} has been identified considering that only the conservative forces act in a generic unloading–reloading cycle. In other words, these cycles should be characterised by constant values of resistant area. For this assumption, the instantaneous secant stiffness of the $\sigma_{eff} - \varepsilon_{eff}$ law, $E_s = \tan\alpha$ (Figure 3), is taken equal to the average slope of the unloading–reloading cycle at the current point. Thus, the generic point $\sigma_{eff} - \varepsilon_{eff}$

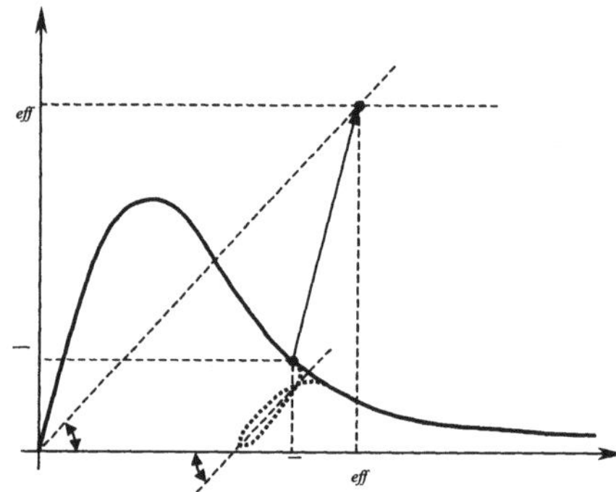

Figure 3. Identification of ε_{eff} starting from the known value of σ_{eff}.

results from the intersection of the two lines $\sigma = \sigma_{eff}$ and $\sigma = E_s\varepsilon$. Figure 3 shows the identification of the effective strain ε_{eff}, starting from the value of effective stress σ_{eff} (Eqs. 2.7 and 2.6) and the knowledge of the damage law D.

3 Sensitivity to the Specimen Slenderness

To operate the transformation in Figure 3 from the $\overline{\sigma} - \overline{\varepsilon}$ diagram to the $\sigma_{eff} - \varepsilon_{eff}$ diagram, the damage law D and the unloading law must be known. In the following, a proposal to experimentally evaluate these laws is presented.

The results on six geometries of cylindrical concrete specimens, with the height–radius ratio variable between three and eight (Ferretti, 2001), will be here presented. Three specimens have been made for each of the six geometries. All the eighteen specimens have been tested in monotone uniaxial load, in the same thermo–hygrometric and curing conditions.

3.1 Unloading Law

The Figure 4 shows how the unloading law is sensibly independent on the slenderness of the specimens. This result supports the assumption for which all the parameters characterising the unloading–reloading cycles, included their average slope, are linked to proprieties of the material and do not depend on the structural mechanics. This happens since the resistant area does not change in the unloading–reloading cycles.

3.2 Experimental Damage Law

To evaluate $D = D(R)$, two experimental damage laws were employed. The first damage law, D_1 (Daponte and Olivito, 1989), relates the damage to the percentage variation of the microseismic signal velocity V at the current point (set–up of the microseismic test in Figure 5.a):

$$D_1 = 1 - V/V_0, \tag{3.1}$$

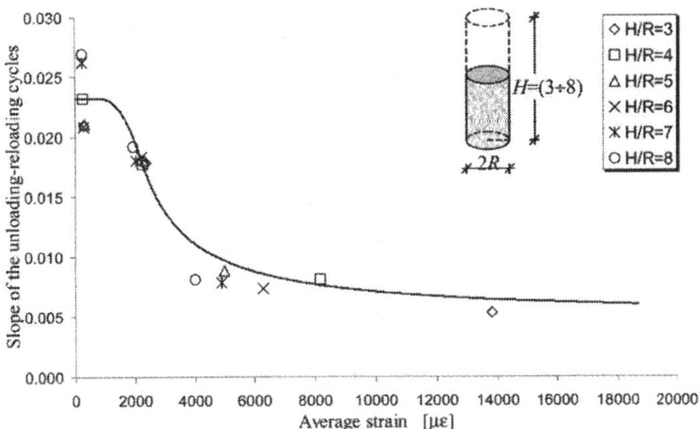

Figure 4. Interpolating law of the unloading-reloading cycles average slope variation for variable slenderness.

where V_0 is the initial microseismic signal velocity. The second damage law, D_2 (Ferretti et al., 1999), relates the damage to the dissipated energy W_d at the current point (Figure 5.b):

$$D_2 = W_d / W_{d,t}, \tag{3.2}$$

where $W_{d,t}$ is the total dissipated energy. D_1 and D_2 turned out to be very close to each other (Ferretti, 2001), until the acceptability threshold of the added noise. This threshold corresponds to the value of deformation beyond which the noise of the crack propagation disturbs the microseismic survey so much that the variations of the microseismic signal cannot be appreciated any longer. For Eq. 2.6, the D_1 and D_2 damage values can be seen in Figure 6 as the one's complement of the percentage resistant area. Only the D_2 damage law has been used in the following, since it has no limitations in the survey field.

The assumption of only conservative forces acting on the unloading–reloading cycles allows evaluation of the specimen defects at the natural state, through the initial damage D_0 (Ferretti and Carli, 1999). In this assumption, indeed, D_0 comes from the ratio

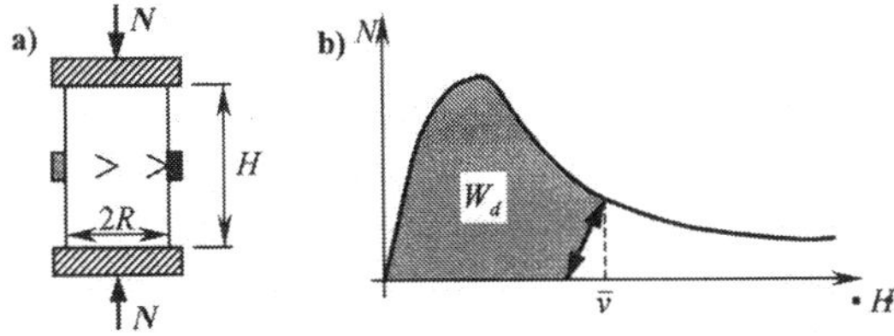

Figure 5. a) Test set-up for the acquisition of D_1; b) Evaluation of W_d.

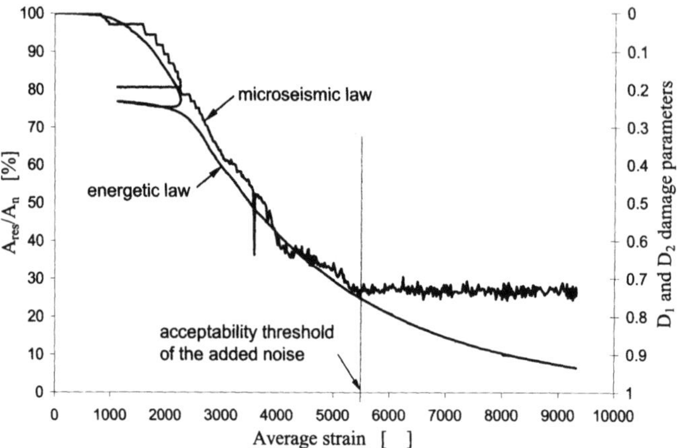

Figure 6. Comparison between the energetic and microseismic laws.

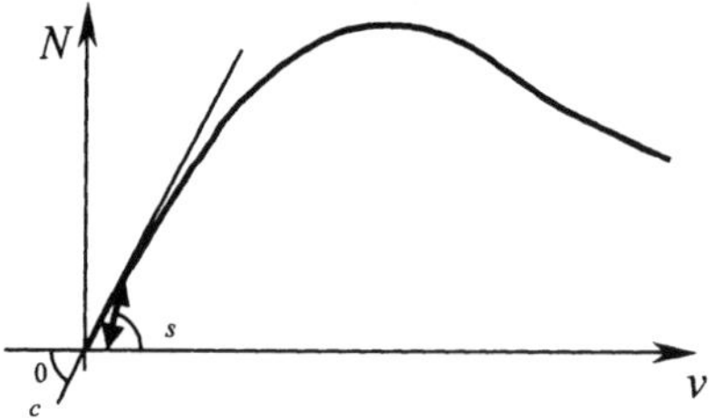

Figure 7. Parameters used to evaluate initial damage D_0 .

between the slope of the stabilising cycle and the tangent to the origin of the $N - v$ diagram (Figure 7):

$$D_0 = 1 - \frac{E^0(\varepsilon)}{E^0_{eff_1}(\varepsilon)} \cong 1 - \frac{\tan\phi^0_c}{\tan\phi^0_s}. \tag{3.3}$$

It must be incidentally recalled that a stabilising cycle is an unloading–reloading cycle that is effectuated for a preloading equal to about the 10% of the maximal presumed load. The stabilising cycle is done in order to limit the influence of the specimen–test machine interaction and test–machine metrological properties on the result. The moderate value of the preloading and the mechanical meaning of the unloading–reloading slopes allow one to associate the difference between the loading and unloading slopes with a damage that is load history independent, characterising the specimen at the natural state.

In accordance with Eq. 3.3, Eq. 2.6 has been modified as follows:

$$A_{res} = A_n (1 - D_{eq}). \tag{3.4}$$

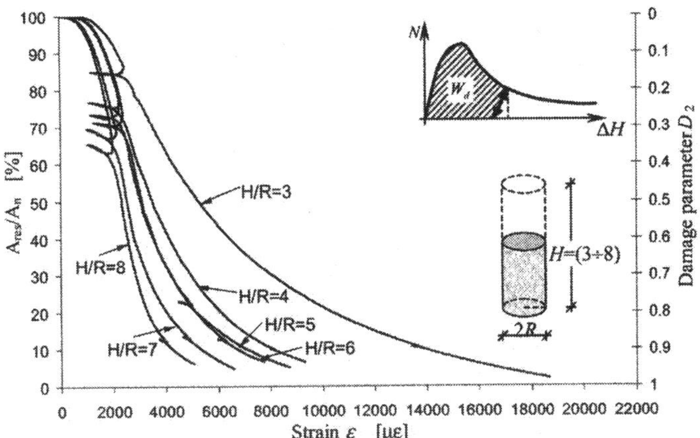

Figure 8. Evolution of resistant area and D_2 damage law for variable slenderness.

D_{eq} is the equivalent damage, comprehensive both of the initial damage D_0 and the damage D_2 due to the monotone loading. D_{eq} can be expressed as:

$$D_{eq} = 1 - (1 - D_0)(1 - D_2) = D_0 + D_2 - D_0 D_2. \tag{3.5}$$

The nominal area deprived of the defects at the natural state, termed the reduced nominal area A'_n, results from:

$$A'_n = A_n (1 - D_0). \tag{3.6}$$

Damage laws were experimentally derived for variable specimen slenderness. The Figure 8 shows the D_2 damage laws obtained for H/R ratios varying from 3 and 8. As can be seen in Figure 8, damage laws are size–effect sensitive. That is, the highest is the H/R ratio, the highest is D for every load–step.

3.3 Sign of the Effective Stress Derivative

As previously stated, the damage can be seen in Figure 9 as the one's complement of the percentage resistant area. The derivative of the experimental damage law in Figure 9, D', turned out to be very close to assume a maximum for $v = \hat{v}$. This result validates the assumption in Eq. 2.11, and corresponds to a fast crack propagation situation.

As regards the sign of the effective stress derivative after the peak, the experimental $q(v)$ function turned out to be a positive–valued, monotone strictly nonincreasing function (Figure 10). The discussion on D' is independent on the single test, and the behaviour in Figure 9 for $v = \hat{v}$ can be generalised. Indeed, the adopted damage law (Eq. 3.2) has a shape that is widely determined by the integral of the $N - v$ curve. Hence, since the $N - v$ curve exhibits a maximum for $v = \hat{v}$, the damage law assumes its maximum derivative in the neighbourhood of $\hat{v}$:

$$q'(v) < 0 \qquad \forall v. \tag{3.7}$$

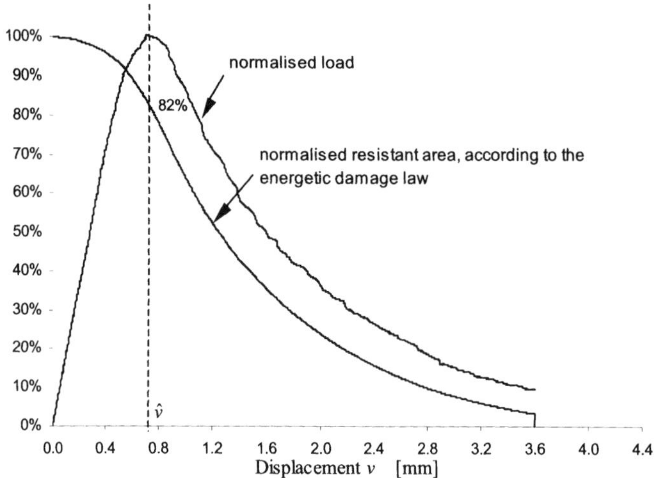

Figure 9. Comparison between the load and resistant area normalised laws.

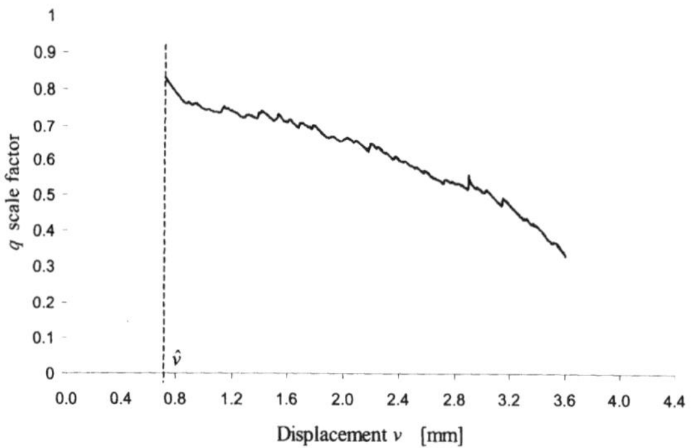

Figure 10. Scale factor between the percentage resistant area and normalised load laws.

This implies that the resistant area decreasing rate was faster than the load decreasing rate. This circumstance and the negative sign between q' and σ'_{eff} (Eq. 2.18) ensure the effective stress derivative to be positive also for $v > \hat{v}$.

In conclusion, for the performed experimental programme and the damage law in Eq. 3.2, the effective stress derivative *always* assumes a finite positive value:

$$\frac{d\sigma_{eff}}{d\bar{\varepsilon}} > 0 \qquad \forall v. \tag{3.8}$$

3.4 Effective Curve

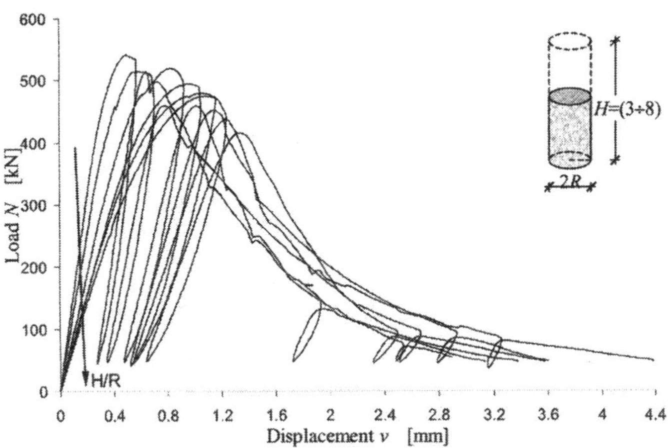

Figure 11. Size effect for the load-displacement diagrams.

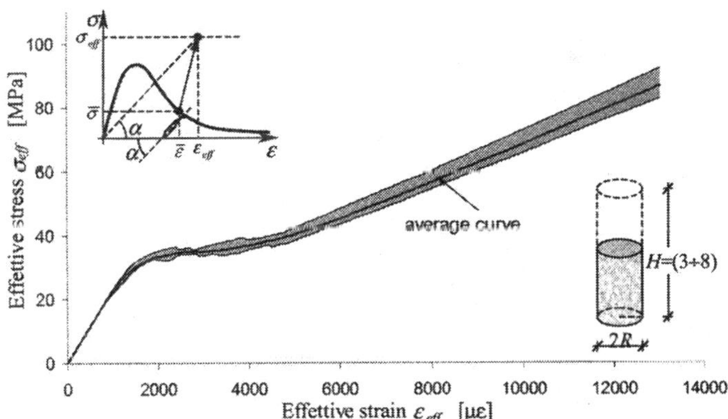

Figure 12. $\sigma_{eff}/\varepsilon_{eff}$ dispersion range for variable slenderness and average curve.

The $N - v$ diagrams for the six specimens with varying slenderness (Ferretti, 2001) are shown in Figure 11. In this Figure, the size–effect in the $N - v$ plane involves both a decrement of the tangent to the origin with the increasing of the H/R ratio, and a decrement of the maximum load with the increasing of the H/R ratio.

The $\sigma_{eff} - \varepsilon_{eff}$ relationships obtained for the six tested geometries fall within the grey region in Figure 12 (dispersion range). As previously stated, a strictly positive derivative for the effective properties curves in the $\sigma_{eff} - \varepsilon_{eff}$ plane directly follows from the independence of the behaviour in Figure 9 on the single test. Moreover, it can be stated that the effective properties curves are size–effect insensitive, since the dispersion range is very little. The average curve shape in the $\sigma_{eff} - \varepsilon_{eff}$ plane is representative of the meso–scale material behaviour.

4 Sensitivity to the Failure Mechanism

The results of a second experimental programme are presented, as to provide a further validation to the proposed identification procedure. Also this experimental programme was performed in the intent to derive a qualitative and not quantitative information. For this reason, the number of the specimens was limited to two. This second time, the geometry of the two specimens is the same, but they are forced to fail in different ways.

As well known, a concrete cylinder in uniaxial compression fails with a dominant bi–cone shaped crack which enucleates in correspondence of the press plates and propagates towards the middle cross–section (Figure 13). To modify the failure mechanism, one of the two specimens was cut along its middle cross–section. Before restoring the specimen, three grommets were fixed on the middle cross–section (Figure 13). The restoration of the specimen has been carried out with mortar. Since the stiffness of the grommets is much higher than the stiffness of the concrete, once loaded the specimen the three grommets have the function to concentrate the load in correspondence of their fixing points. This results in a failure mechanism with the dominant crack enucleating on the middle cross–section, starting from the fixing points of the grommets (Figure 13). The final failure surfaces were the same in the two cylinders (Figure 13).

In Figure 14 the load–displacement experimental curves for the uncut and the cut cylinder are shown. Theese curves are almost superimposed as long as the dominant crack does not start to propagate. That is, the initial stiffness of the restored specimen is almost the same as the initial stiffness of the uncut specimen, stating that the cut and the subsequent restoring have not significantly modified the cylinder stiffness properties.

When the dominant crack starts to propagate, the two curves bifurcate, assuming a different shape (Figure 14). The presence of a bifurcation point confirms that the failure mechanisms of the two specimens were actually different, as it was in the Authors aims.

In Figure 15, the percentage difference from the mean value of load is quoted for intervals of displacement of 0.08 mm. The two local maximums are quoted on the plot.

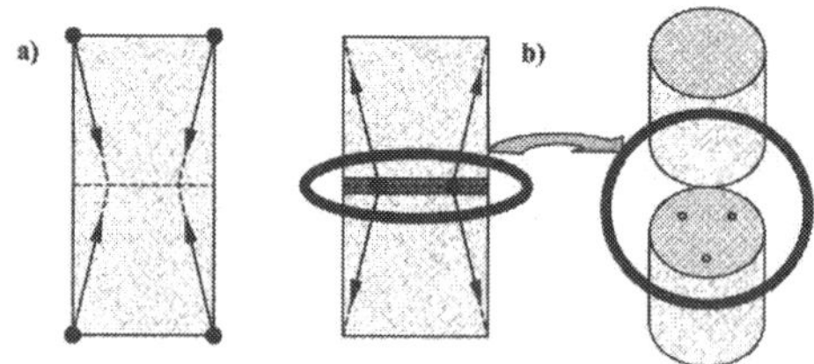

Figure 13. Crack path for the uncut (a) and the cut (b) cylinder.

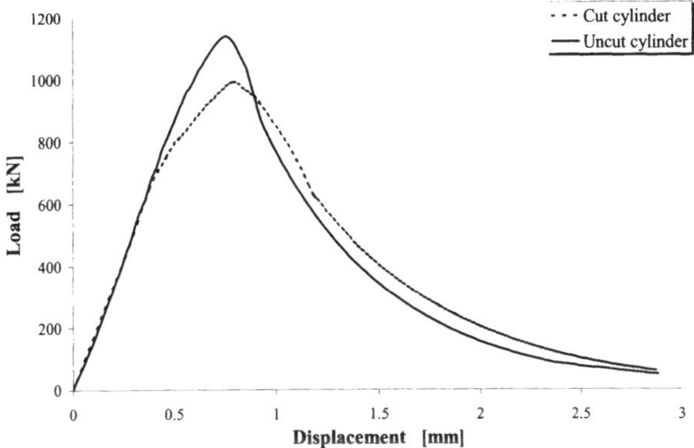

Figure 14. Load-displacement curves for the cut and the uncut cylinder.

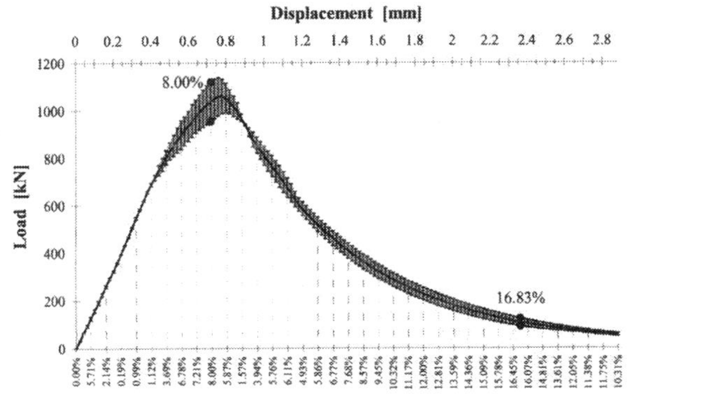

Figure 15. Percentage difference from the mean value of load, at intervals of displacement of 0.08 mm.

The maximum percentage difference from the mean value is reached in advanced phase of softening (Figure 15) and is equal to about 17%. The maximum percentage difference in the neighborhood of the peak is equal to about 8% (Figure 15).

Evaluation of the damage laws D_1 and D_2 has been carried out in accordance with Eqs. 3.1 and 3.2, giving very close results until the acceptability threshold of the added noise this second time too. The initial damage D_0 and the equivalent damage D_{eq} have been identified, respectively, in accordance to Eqs. 3.3 and 3.5, with the reduced nominal area A'_n given by Eq. 3.6. The experimental $q(v)$ function turned out to be a positive–valued, monotone strictly nonincreasing function and the resistant area decreasing rate was faster than the load decreasing rate, as it was in the first experimental programme.

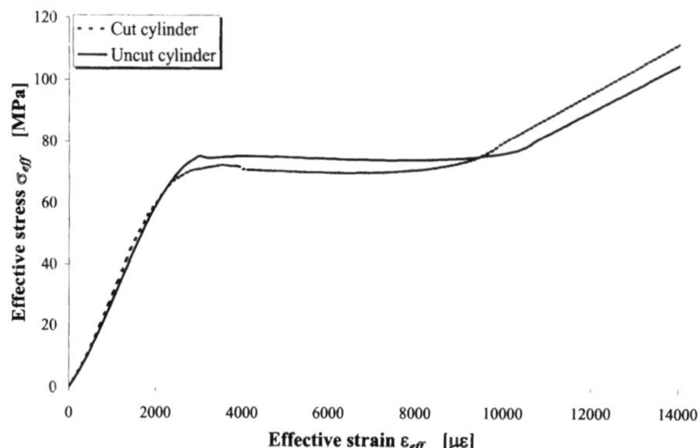

Figure 16. Effective stress-effective strain curves for the cut and the uncut cylinder.

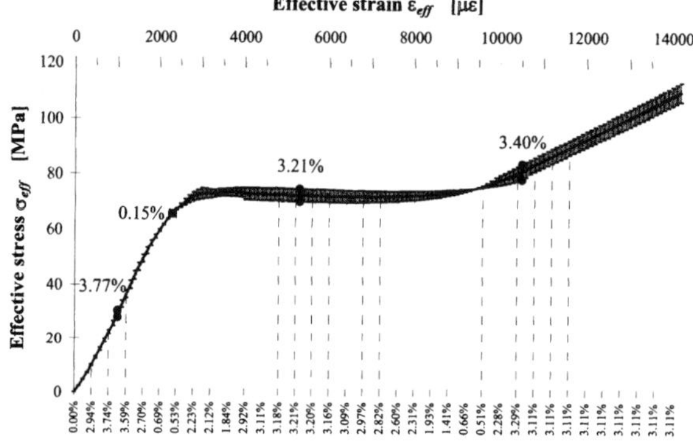

Figure 17. Percentage difference from the mean value of effective stress, at intervals of effective strain of 400 $\mu\varepsilon$.

In Figure 16 the $\sigma_{eff} - \varepsilon_{eff}$ relationships obtained for the two tested cylinders are shown.

In Figure 17, the percentage difference from the mean value of effective stress is quoted for intervals of effective strain of 400 $\mu\varepsilon$. The three local maximums are quoted on the plot. The point of the $\sigma_{eff} - \varepsilon_{eff}$ plane corresponding to the first local maximum in Figure 15 (the point near the peak) is indicated with a square marker. Its percentage difference is quoted on the plot.

As it can be appreciated in Figure 17, the percentage difference from the mean value computed in the $\sigma_{eff} - \varepsilon_{eff}$ plane is noticeably lower than the percentage difference from the mean value computed in the $N - v$ plane.

5 Conclusions

Considering the specimen as a structure interacting with the test–machine, it was demonstrated that the $\sigma_{eff} - \varepsilon_{eff}$ curve of damaging materials exhibits a strictly positive derivative at the point corresponding to the peak of the $\overline{\sigma} - \overline{\varepsilon}$ curve. This result is independent on the adopted damage law.

A procedure to experimentally evaluate the damage law on cylindrical concrete specimens in uniaxial compression has been proposed (first experimental programme). The adopted damage law led to monotone strictly nondecreasing $\sigma_{eff} - \varepsilon_{eff}$ curves for all concrete specimens. Moreover, this curve turns out to be size–effect insensitive.

The two identified $\sigma_{eff} - \varepsilon_{eff}$ curves of the second experimental programme do not exhibit bifurcation points. This means that the identified curve is not sensitive to the change of failure mechanism too. The low value of percentage difference from the mean value in correspondence of the square marker in Figure 17 confirms this conclusion, since for this point the effect of the different failure mechanism reaches a maximum (Figure 15). This result is very important to give validity to the proposed identification procedure, since the independence from the failure mechanism is one of the most important requisites for an identifying procedure of material parameters. Moreover, this result is as much important as the traditional identifying procedure for constitutive laws in uniaxial loading is not insensitive to the change of failure mechanism.

The proposed approach allows identification of a $\sigma_{eff} - \varepsilon_{eff}$ relationship for the description of the meso–scale material behaviour. This relationship, together with adequate failure criteria, leads to structural analysis.

Bibliography

ZP. Bazant, TB. Belytschko and T. Chang. Continuum Theory for Strain-Softening. *Journal of Engineering Mechanics*, 110:1666–1692, 1984.

P. Daponte and RS. Olivito. Crack Detection Measurements in Concrete. In *Proceedings of the ISMM International Conference Microcomputers Applications*, pages 123–127, 1989.

A. Di Leo, A. Di Tommaso and R. Merlari. Danneggiamento per Microfessurazione di Malte di Cemento e Calcestruzzi Sottoposti a Carichi Ripetuti. *Technical note 46, DISTART - University of Bologna - Italy*, 1979.

E. Ferretti. Modellazione del Comportamento del Cilindro Fasciato in Compressione, Ph.D. Thesis: University of Lecce - Italy, 2001.

E. Ferretti and R. Carli. Programma Sperimentale sul Comportamento in Compressione Monoassiale del Calcestruzzo; Parte II: Elaborazione dei Risultati Sperimentali. *Technical note 33, DISTART - University of Bologna - Italy*, 1999.

E. Ferretti, E. Viola, A. Di Leo and G. Pascale. Propagazione della Frattura e Comportamento Macroscopico in Compressione del Calcestruzzo. In *Proceedings of the XIV AIMETA*, on CD-ROM.

G. Rosati and MP. Natali Sora. Tensile Tests on Concretelike Materials: Structural and Constitutive Behaviors. *Journal of Engineering Mechanics*, 127:364–371, 2001.

Computational Aspects and Numerical Simulations in the Elastic Constants Identification

Elena Ferretti, Antonio Di Leo and Erasmo Viola

DISTART – Department of Structures Transportations Waters Survey and Territory Engineering, University of Bologna, Bologna, Italy

Abstract A numerical code for modeling crack propagation using the Cell Method is proposed. The Leon failure surface is used to compute the direction of crack propagation, and the new crack geometry is realized by an intra-element propagation technique. Automatic remeshing is then activated. Applications in Mode I, Mode II and Mixed Mode are presented to illustrate the robustness of the implementation.

1 Comparison between Discrete and Variational Approach

The use of a discrete formulation instead of a variational one is advantageous, since several problems of the variational formulation are avoided by means of the discrete formulation. With the variational formulation:-

- Field variables $f = f(x, y, z, t)$ are needed, which are point and instant dependent. If not directly in possession of point functions, the variational approach obtains them from the global quantities by means of the density notion and limit process. A subsequent phase of equation discretisation and integration is needed to reach the solution.
- The solution is not obtained for the mesh nodes directly, but extrapolated to them.
- It is not possible to attain convergence greater than the second order.
- Heterogeneities represent an obstacle.
- Singularities of the domain contour represent an obstacle.
- Punctual forces represent an obstacle.
- The definition of a model for treating the zone ahead of the crack edge is needed.

On the contrary, with the discrete formulation:-

- Only global variables $g = g(x, y, z, L, A, V, t)$ are needed, which are point and instant, but also line, area and volume dependent. In a word, global functions are domain and not point functions. Balance equations are expressed using global variables, so they do not contain derivatives and do not therefore require integration.
- The solution is obtained for the mesh nodes directly.
- The convergence order is equal to the FEM one, at worst. One can reache convergences of the fourth order.

- The constitutive matrix can vary from one cell to another, with heterogeneities which can be of the same order than the mesh sides dimension.
- Singularities of the domain contour are no longer an obstacle.
- Punctual forces are no longer an obstacle.
- The definition of a model for treating the zone ahead of the crack edge is no longer needed.

2 Crack Extension Criterion

For finding a crack trajectory, the variational principle of the most common crack theories has been used over the past three decades (Parton and Morozov, 1978). Criteria for the initiation of crack propagation can be obtained on the basis of both energy and force considerations. Historically, at first an energy fracture criterion was proposed by Griffith (1920), then Irwin (1957) formulated a force criterion, while the same time demonstrating the equivalence of the two criteria. The Irwin force criterion for crack extension and the equivalent Griffith energy criterion completely solve the question of the limiting equilibrium state of a cracked continuous elastic body. Nevertheless, there exists a number of other formulations also establishing the limiting equilibrium state of a cracked body. Among these, the best known models are those of Lenov and Panasyuk (1959), Dugdale (1960), Wells (1961), Novozhilov (1969), and McClintock (1958).

Usually, the variational problem of finding the limiting load and the crack propagation direction is reduced to that of finding extreme points of a function of several variables (Parton and Morozov, 1978). In the present paper, the variational approach has been abandoned in favour of a discrete formulation of the crack propagation problem. The discrete formulation used in the following is the Cell Method, said the CM (Tonti, 2001).

When studying crack problems for an elastic-perfectly plastic body with the energy equilibrium criterion, the solution is usually given in the case when the plastic deformation is concentrated in a narrow zone ahead of the crack edge. The thickness of this zone is of the order of elastic displacements. Moreover, when the plastic zone ahead of the edge is thin, the problem is reduced to the solution of an elastic problem instead of an elastic-plastic one. This reduction is based on the fact that, in the linearized formulation, a thin plastic zone may be schematically replaced by an additional cut along the face of which are applied forces replacing the action of the plastically deforming material. Attention is then drawn to the fact that the region of plastic non-linear effects in the model under consideration varies with the external load and represents a plastically deforming material in which the state of stress and strain must be determined from the solution of an elastic-plastic problem. With the discrete formulation, on the contrary, no hypothesis on the shape and dimensions of the plastic zone is needed, and the calculation is performed directly, without having to reduce the problem to an equivalent elastic one. The code for crack trajectory analysis with the CM has been developed by Ferretti (2003)[1].

The crack extension direction and the minimum load required to propagate a crack (limiting load) can be determined by using a variety of criteria:-

[1]The Italian Ministry for Universities and Scientific and Technological Research (MURST) is gratefully acknowledged for its financial support.

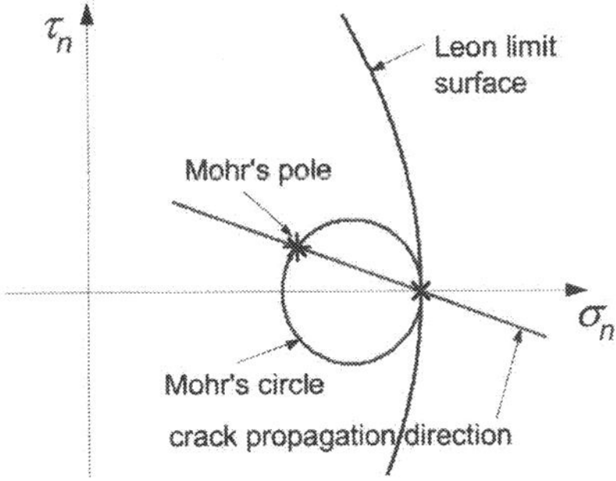

Figure 1. Leon limit surface in the Mohr-Coulomb plane.

- maximal normal stress criterion;
- maximal strain criterion;
- minimum strain energy density fracture criterion;
- maximal strain energy release rate criterion;
- damage law criteria.

In the present paper, the crack extension condition is studied in the Mohr-Coulomb plane. The limiting load is computed as the load satisfying the condition of tangency between the Mohr's circle and the Leon limit surface (Figure 1).

Said c the cohesion, f_c the compressive strength, f_{tb} the tensile strength, τ_n and σ_n, respectively, the shear and normal stress on the attitude of external normal n, the Leon criterion in the Mohr-Coulomb plane is expressed as:

$$\tau_n^2 = \frac{c}{f_c}\left(\frac{f_{tb}}{f_c} + \sigma_n\right). \tag{2.1}$$

The Mohr's circle for the tip neighbourhood is identified by means of the physical significance associated with the CM domain discretization. The CM divides the domain by means of two cell complexes, in such a way that every cell of the first cell complex, which is a simplicial complex, contains one, and one only, node of the other cell complex. In this study, a Delaunay/Voronoi mesh generator is used to generate the two meshes in two-dimensional domains. The primal mesh (the Delaunay mesh) is obtained by subdividing the domain into triangles, so that for each triangle of the triangulation the circumcircle of that triangle is empty of all other sites (Figure 2). The dual mesh (the Voronoi mesh) is formed by the polygons whose vertexes are at the circumcenters of the primal mesh (Figure 2). For each Voronoi site, every point in the region around that site is closer to that site than to any of the other Voronoi sites. The conservation law is enforced on the dual polygon of every primal vertex. To identify the Mohr's circle for the tip neighbourhood, an hexagonal element was inserted at the tip (Ferretti (2003), Figure 2). When the mesh generator is activated, the hexagonal element is divided into equilateral Delaunay triangles and a quasi-regular tip Voronoi cell is generated (the cell

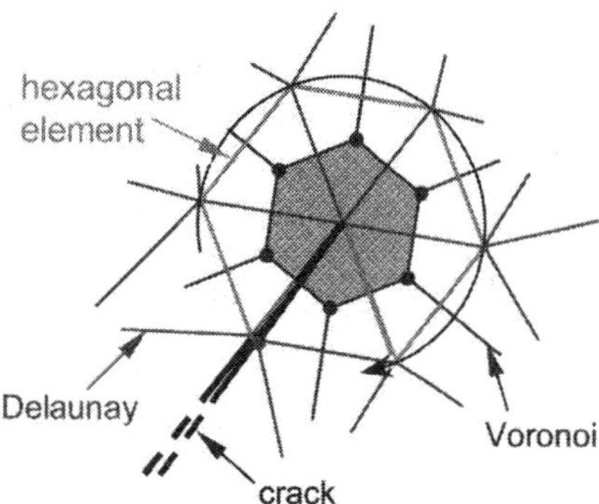

Figure 2. Hexagonal element for analysis in the Mohr-Coulomb plane.

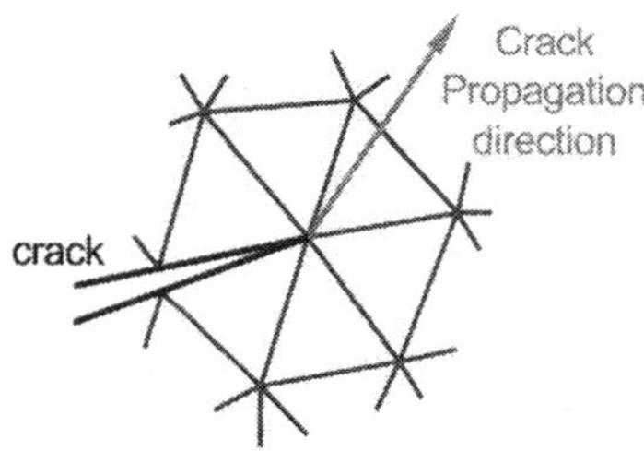

Figure 3. Crack propagation technique.

filled in gray in Figure 2). This allow us to establish a correspondence between the tip stress field and the attitudes corresponding to the sites of the tip Voronoi cell. It has been shown (Ferretti, 2003) that the tension points correctly describe the Mohr's circle in the Mohr-Coulomb plane, for every rotation of the hexagonal element around the tip.

The propagation direction is then derived as the direction of the line joining the tangent point to the Mohr's pole (Figure 1). The crack propagation technique is the nodal relaxation intra-element, with subsequent re-meshing (Figure 3).

3 Constitutive Assumption

The concrete constitutive law here adopted is monotonically non-decreasing, in accordance with the identification procedure for concrete in mono-axial load provided by Ferretti (2001) (Figure 4). It was shown (Ferretti, 2001) how this constitutive law turns out to be size effect insensitive and failure mechanism insensitive for mono-axial compressive load. This result has made it possible to formulate a new concrete law in mono-axial loading, the effective law, which can be considered more representative of the material physical properties than the softening laws are. The tensile branch was here identified

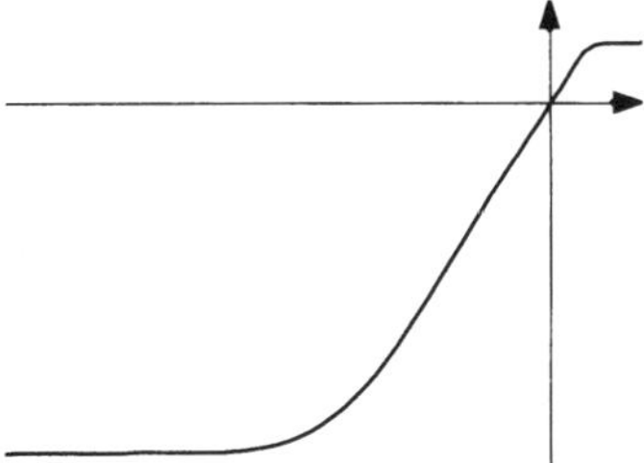

Figure 4. Adopted constitutive law for concrete in mono-axial load.

starting from the compressive one, in the assumption that an homothetic relationship exists between the two branches (Figure 4). A ratio between tensile and compressive strength of 1/12, 1/8, 1/6, 1/4, 1/3 and 1/2 has been considered. The best accordance between analytical and experimental results has been obtained for the ratio equal to 1/8.

4 Numerical Results for Crack Propagation Problems

4.1 Mode I

The code developed by Ferretti (2003) has been extended to provide results in the case of a concrete plate (Figure 5) tensioned at infinity by a load of intensity $p_x = kp_0$ parallel to the x–axis, and $p_y = p_0$ parallel to the y–axis. The plate has an initial straight crack of length $2l_0$ oriented at an angle α_0 to the x–axis (β_0 to the y–axis). Numerical results are here provided for various values of k and α_0.

The p_y analysis for $k = 0$ and $\alpha_0 = \pi/4$, plotted on the deformed configuration of a finite area around the crack, is shown in Figure 6 for the initial straight crack. In this Figure, the darker red color corresponds to the maximal tensile stress, while the darker

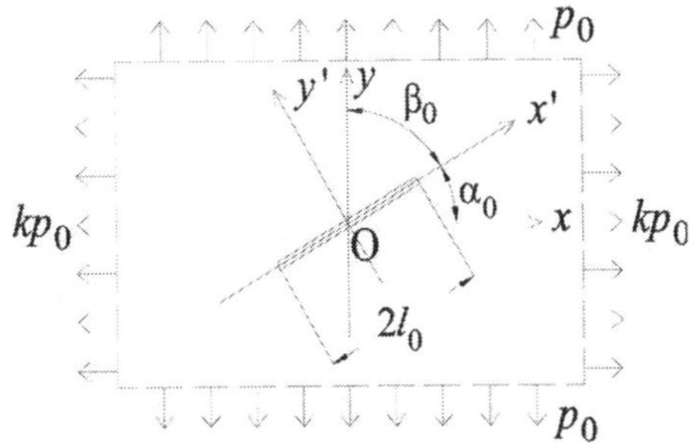

Figure 5. Load and geometrical set-up of the cracked plate.

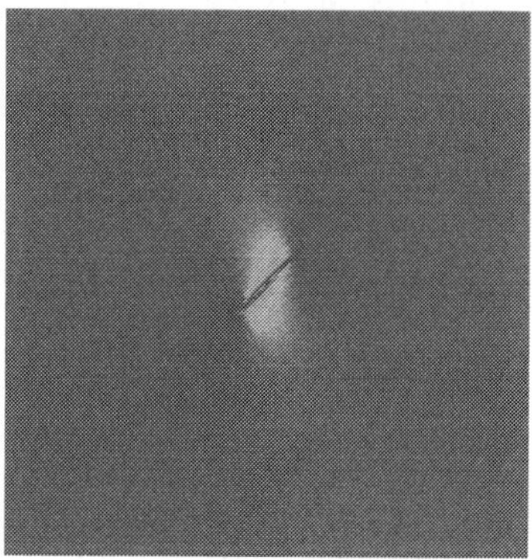

Figure 6. 2D p_y analysis for the initial straight crack with $k = 0$, $\alpha_0 = \pi/4$.

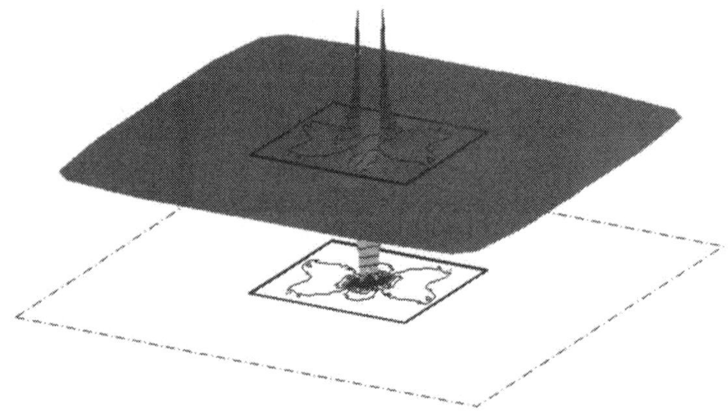

Figure 7. 3D p_y analysis and lines of equal p_y for the initial straight crack with $k = 0$, $\alpha_0 = \pi/4$.

green color corresponds to $p_y = 0$.

In Figure 7, the p_y analysis is performed in 3D, with the level lines plotted in the plane $p_y = 0$. The significance of colors is the same as in Figure 6. The 3D plot correctly shows how p_y reaches a value numerically close to 0 in correspondence with the crack surface. The boxed area in the plane $p_y = 0$ of Figure 7 is the area of Figure 6. As all the level lines are internal to the boxed area, it can be assumed with good approximation that this area represents the stress extinction zone for the initial straight crack.

In Figure 8 and Figure 9, the 2D and 3D p_y analysis for $k = 0$ and $\alpha_0 = \pi/4$ is shown for a generic crack propagation step. The lighter colors of Figures 8 and 9 with respect to those of Figures 6 and 7 are representative of the progressive plate downloading

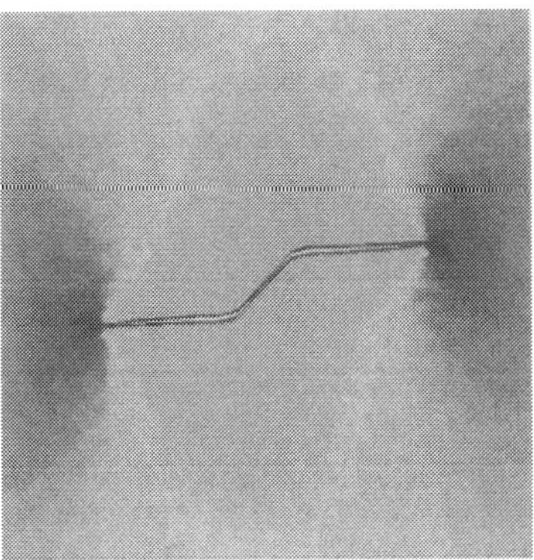

Figure 8. 2D p_y analysis for a generic propagation step with $k = 0$, $\alpha_0 = \pi/4$.

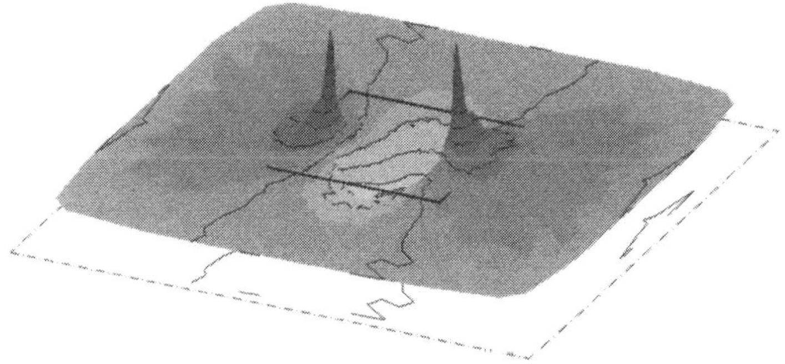

Figure 9. 3D p_y analysis and lines of equal p_y for a generic propagation step with $k = 0$, $\alpha_0 = \pi/4$.

corresponding to a crack propagation. For the same reason, the values in the third axis of Figure 9 are smaller than those of Figure 7. Moreover, in Figure 8 it can be observed how a mono-axial compressive state of stress of small entity (cyan color) arises in the plate, due to the geometrical effect of the non-straight crack deformation in Mode I.

The crack trajectory is plotted in Figure 10 for the value of the angle α_0 equal to $\pi/6$, $\pi/4$ and $\pi/3$, and for the factor k equal to 0, $1/4$, $1/2$, $3/4$ and 1. Also the two meshes of Delaunay-Voronoi were plotted for $k = 0$, in order to show how the adaptive mesh generator recreates the meshes for a generic crack propagation step (Delaunay mesh in red and Voronoi mesh in blue, as indicated in Figure 2). This Figure shows that the crack trajectory tends to approach an asymptote perpendicular to the external tensile load for all the three cases with $k = 0$. The asymptotic behavior is also evident for $k > 0$, but the asymptote is now oriented at an angle γ to the $x-$axis which depends on k:

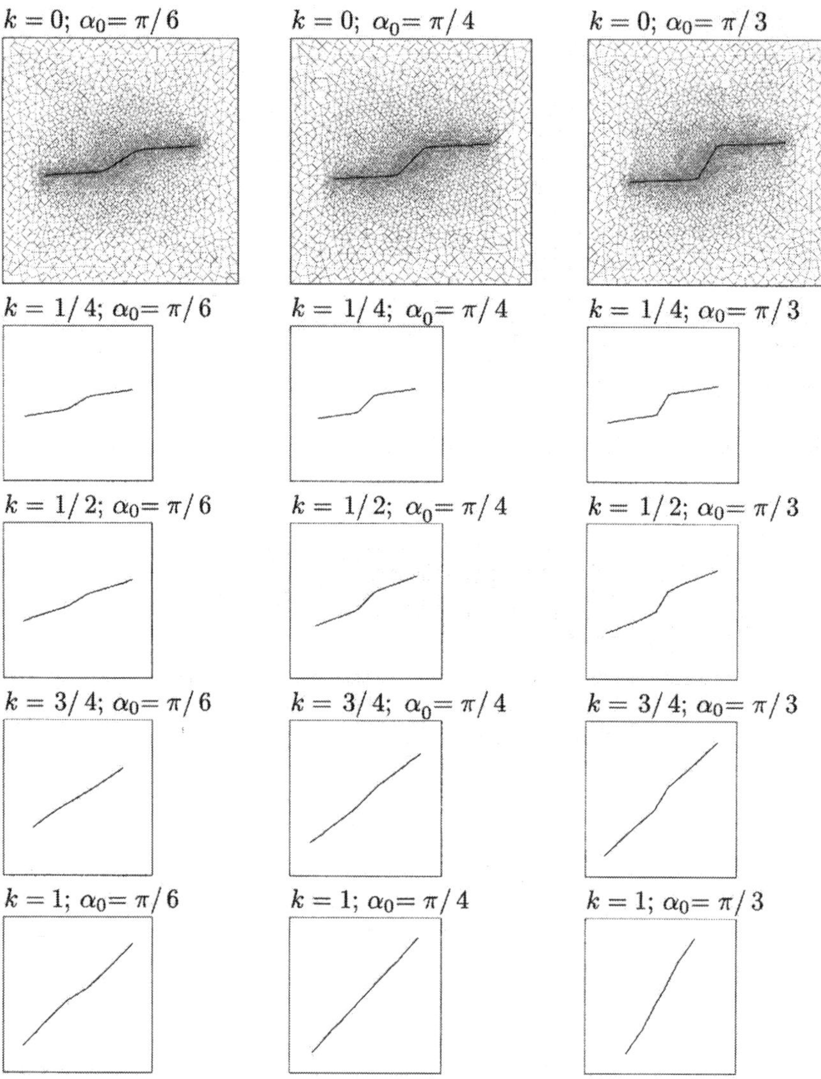

Figure 10. Crack trajectory for $\alpha_0 = \pi/6,\ \pi/4,\ \pi/3$, and $k = 0,\ 1/4,\ 1/2,\ 3/4,\ 1$.

$$\gamma = f(k). \tag{4.1}$$

The angle γ does not depend on the initial inclination α_0, since the crack tends to propagate perpendicularly to the tensile principal direction of the uncracked plate.

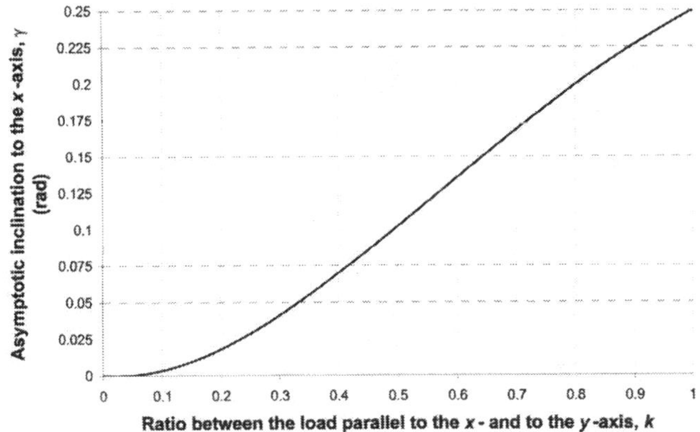

Figure 11. Relation between the angle γ and the ratio k.

For $k = 1$, the angle γ assumes the value $\pi/4$, in good accordance with the homogeneous state of stress represented by this load condition:

$$\gamma|_{k=1} = \frac{\pi}{4}. \tag{4.2}$$

The plot of the function $\gamma = f(k)$ is provided in Figure 11 for the range of values $0 \leq k \leq 1$.

If we know the behavior of the function $\gamma = f(k)$ for $0 \leq k \leq 1$, the behavior of $\gamma = f(k)$ for $1 < k \leq +\infty$ is also known. Actually, the asymptotes of the crack trajectory for a given k and its reciprocal value, $1/k$ (Figure 12), are symmetric with respect to the bisector of the first quadrant in Figure 5, $y = x$. That is to say, the value assumed by γ for a given k is equal to the complementary angle of γ for $1/k$:

$$\gamma\left(\frac{1}{k}\right) = \frac{\pi}{2} - \gamma(k). \tag{4.3}$$

The plot of γ for the complete range of values $0 \leq k \leq +\infty$ can be provided by means of the following change of variable (Figure 13):

$$\psi = 1 - e^{-k}. \tag{4.4}$$

The normalised limiting load in the direction of the y–axis, $p_y \Big/ \max\limits_{\alpha_0,k} p_y$, is plotted in Figure 14.a for each value of the angle α_0 ($\beta_0 = 90 - \alpha_0$) and for the factor k equal, respectively, to 0, $1/4$, $1/2$, $3/4$ and 1. This figure exhibits a p_y limiting load increasing with α_0 in the field $0 \leq k < 1$, stating that $\alpha_0 = 0$ is the critical crack orientation angle for all the biaxial load conditions with the x–component lower than the y–component.

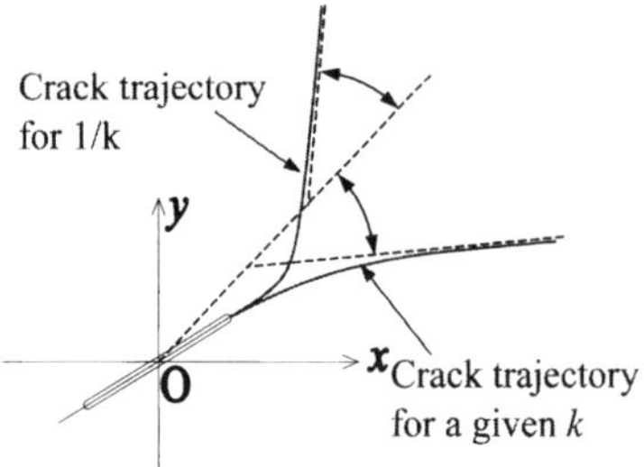

Figure 12. Crack trajectories for a given k and $1/k$.

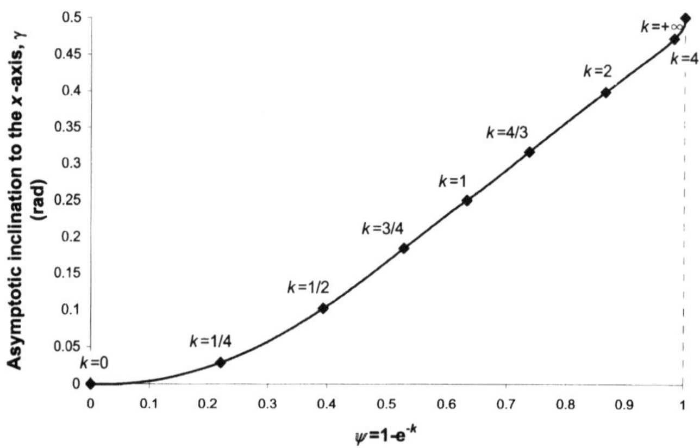

Figure 13. Relation between the angle γ and the variable ψ.

The constant behaviour of the line obtained for $k = 1$ is in good agreement with the homogeneous state of stress represented by this load condition. In this case, all the crack orientations to the $x-$axis return the same limiting load.

The load conditions corresponding to a given k and to its reciprocal, $1/k$, are symmetric with respect to a crack inclination $\alpha_0 = 45°$. In the intent to plot limiting load curves for k and $1/k$ satisfying the symmetry with respect to $\alpha_0 = 45°$, the ideal limiting load p_{id} was defined as:

$$p_{id} = \sqrt{p_x^2 + p_y^2} = p_0\sqrt{k^2 + 1}. \tag{4.5}$$

The limiting load curves in the $\alpha_0 - p_{id} \Big/ \max_{\alpha_0,k} p_{id}$ plane for $0 \leq k \leq +\infty$ are plotted in Figure 14.b. Each line in Figure 14.b represents the function $f_k = p_{id}(\alpha_0)$, returning the value of p_{id} at a given value of k. All the f_k lines together represent the function $f = p_{id}(\alpha_0, k)$, defining the value of the ideal limiting load in function of the crack

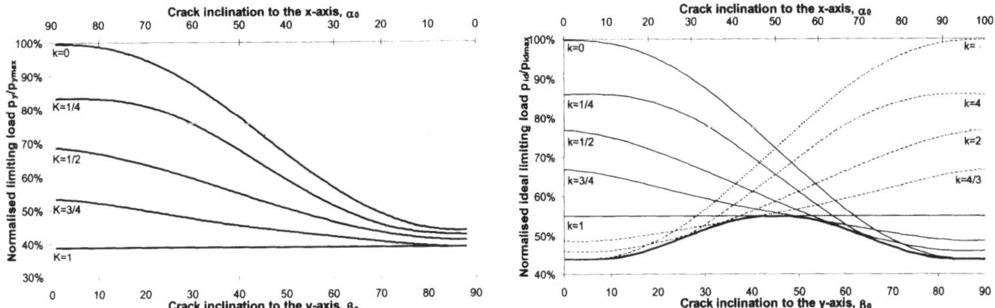

Figure 14. a) Normalised limiting load p_y in function of the crack inclination α_0/β_0 for $0 \leq k \leq 1$; b) Normalised ideal limiting load p_{id} in function of the crack inclination α_0/β_0 for $0 \leq k \leq +\infty$.

orientation and the ratio between the loads in direction of the $x-$ and $y-$axis. The lower envelope of the f_k curves was plotted in thick line in figure Figure 14.b. This line represents the function F defined as:

$$F = F(\alpha_0, k) = \min_{\alpha_0 = const} p_{id}(\alpha_0, k). \tag{4.6}$$

The projection on the $\alpha_0 - k$ plane of the F function, said $\overline{F}$, returns the value of k^{cr} (which is the same, the couple of normalised values p_x^{cr} and p_y^{cr}), corresponding to the minimum value of ideal limiting load p_{id} for a given α_0. In other words, the function $\overline{F}$ can be expressed as the solution of the following differential problem:

$$\overline{F} = k^{cr}(\alpha_0) : \frac{\partial p_{id}(\alpha_0, k)}{\partial k} = 0. \tag{4.7}$$

With the position in Eq. 4.4, it is possible to give a finite 3D representation of p_{id}, in function of α_0 (β_0) and k. Some examples of the 3D surfaces obtained for different values of axis orientation are given in Figure 15. The Figure 15.a is the 3D equivalent representation of the Figure 14.b, while the Figure 15.f is obtained by a 90° rotation of the normalised p_{id} surface around the normalised $p_{id}-$axis. This last plot represents the function $g = p_{id}(k, \alpha_0)$ in the k first independent variable and α_0 second independent variable. The meridians of g give the functions $g_{\alpha_0} = p_{id}(k)$ for a given value of α_0. The lower envelope of the g_{α_0} functions gives the function:

$$G = G(k, \alpha_0) = \min_{k = const} p_{id}(k, \alpha_0). \tag{4.8}$$

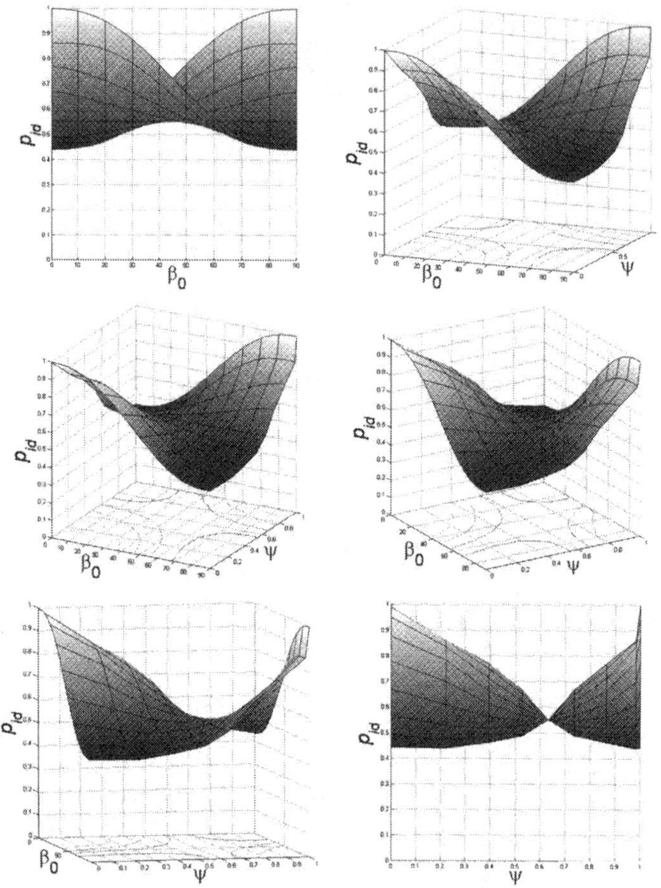

Figure 15. Some examples of finite 3D representation of p_{id}, in function of β_0 and k.

Said $\overline{G}$ the projection on the $k - \alpha_0$ plane of the G function, $\overline{G}$ returns the angle α_0^{cr} minimising p_{id} at a given k. $\overline{G}$ is the solution of the following differential problem:

$$\overline{G} = \alpha_0^{cr}(k) : \frac{\partial p_{id}(\alpha_0, k)}{\partial \alpha_0} = 0. \tag{4.9}$$

As all the g_{α_0} functions intersect for $k = 1$ and they do not have any other point in common, the $\alpha_0 = \alpha_0^{cr}$ minimising p_{id} at a given k results equal to:

$$\alpha_0^{cr} = \begin{cases} 0 & 0 \leq k < 1 \\ [0,\ 90] & k = 1 \\ 90 & 1 < k \leq +\infty \end{cases} \tag{4.10}$$

4.2 Mode II

Results regarding the computation on a geometry of finite dimensions in Mode II are here presented. The computation on finite geometries allows one to avoid the quantitative discrepancy between the experimental results and those calculated via- stress intensity factors (Parton and Morozov, 1978), due to the effect of the specimen boundaries on the stress field around the growing crack. In the case of a beam with two opposite pre-cracks submitted to skew-symmetric load in four-points, the numerical result was so accurate as to allow the complete description of the crack trajectory (Figure 16).

In Figure 16, the darker blue color corresponds to the maximal compressive stress and the darker red color corresponds to the maximal tensile stress.

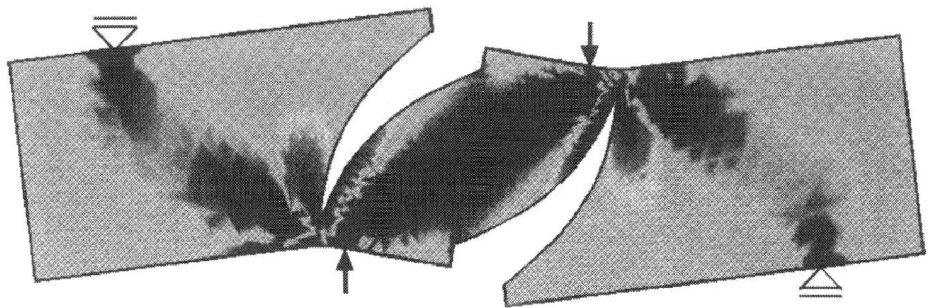

Figure 16. Displacement-stress fields of a beam in skew-symmetric four point bending.

4.3 Mixed Mode

Mixed-Mode crack propagation occurs when the load is applied obliquely to the crack direction and the crack opening direction. If this is the case, the crack can be divided into two parts [Ferretti (2001)]:

part a) Mode I prevails and the two edges of the crack separate;
part b) Mode II prevails and the two edges of the crack slide over one another.

Depending upon the geometry of the domain and the boundary conditions, there may be more than one part a) and more than one part b). Generally speaking, part b) occurs whenever the combination of loading and boundary conditions forces the edges of the crack to close at some point. Numerical simulation is only possible if the position of the point S separating the two parts is known. The dominance of Mode I rather than Mode II crack propagation involves different boundary conditions on the crack surfaces, and it is necessary to specify all the boundary conditions before the simulation starts. In general, S is a function of the load step and crack length, and is, thus, an unknown of the Mixed-Mode problem. To determine S, it is necessary to proceed step-wise.

Numerical results on a concrete compressed cylinder are here presented.

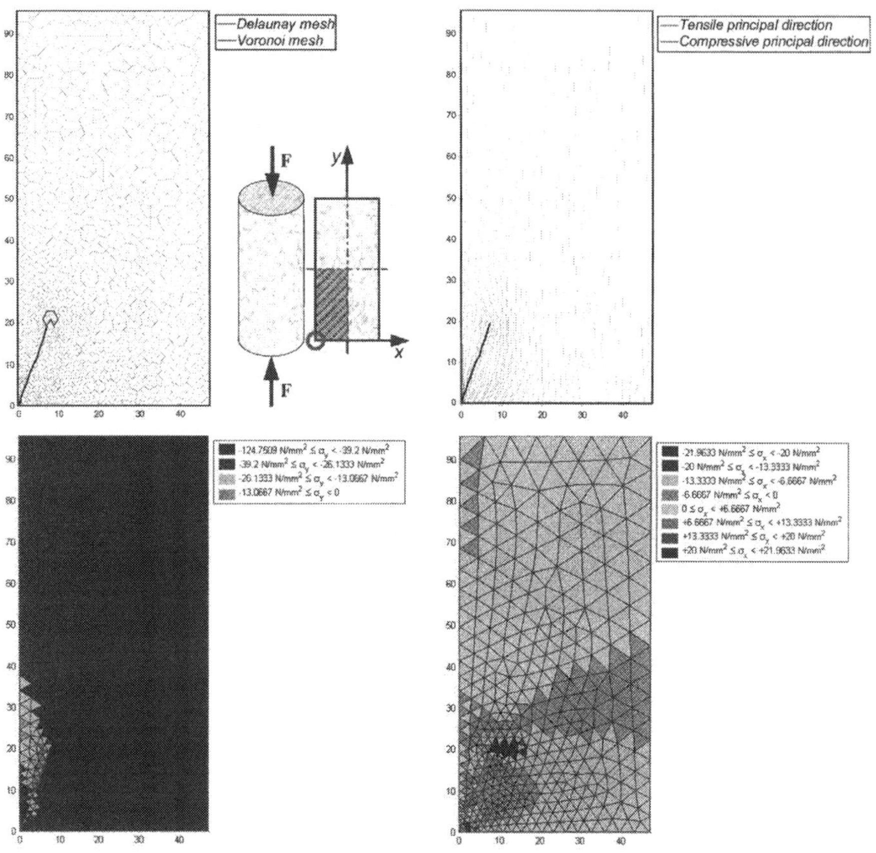

Figure 17. Delaunay and Voronoi discretization of the space domain, computed principal stress directions and stress field in the $y-$ and $x-$axis direction.

5 Conclusions

A crack propagation analysis has been presented for use with the Cell Method, which combines nodal relaxation, intra-element propagation and remeshing. This method permits the mesh dimensions to be refined at specific locations, to improve the solution accuracy. The crack geometry is updated by introducing an extra node at the new crack tip. This is accompanied by a relaxation of the node at the old crack tip.

The crack propagation is studied without using the stress intensity factors and without having to define a model to treat the zone ahead of the crack tip. This allows one to employ the same numerical code for bodies of different dimensions, geometries and boundary conditions, and for materials of different constitutive laws. As an example of the code versatility in front of the geometrical set-up, the interaction between several cracks oriented at any inclination and propagating in plate of finite/infinite dimensions can be easily investigated. As regards the versatility in front of the constitutive law, the numerical analysis can be performed in the linear or non-linear field, with no adjunctive computational burdens. The stress intensity factors can be estimated a-posteriori, since the code allows us to evaluate the compliance decrement following from a crack propagation.

Both Mode I, Mode II, and Mixed-Mode crack propagation have been considered in this paper. The numerical model incorporates an original approach, which automatically estimates which part of the boundary is subjected to Mode I loading, and which part is subjected to Mode II loading. The numerical model subsequently estimates the size of the relative displacements between nodes on the opposite sides of the crack surface subjected to Mode II loading, allowing sliding contact to be described.

The numerical results show that the CM numerical code can give good predictions for fracture mechanics problems, and validate the CM theory for fracture analysis.

The accuracy of the results indirectly validates the adopted innovative constitutive law for concrete, which turned out to be size effect and failure mechanism insensitive.

Finally, the analysis for finite solids is performed directly, without having to apply corrective factors to the solution on an infinite geometry in the same load conditions.

Bibliography

D.S. Dugdale. Yielding of Steel Sheets Containing Slits. *Journal of the Mechanics and Physics of Solids*, 8(2):100–104, 1960.

E. Ferretti. Crack propagation modeling by remeshing using the Cell Method (CM). *To appear in Computer Modeling in Engineering and Science*, 3:, 2003.

E. Ferretti. Identification of Constitutive Law in Mono-Axially Compressed Damaging Materials. *Advanced in Fracture and Damage Mechanics II. Proc. Int. Conf. on Fracture and Damage Mechanics*, pages 379–384, 2001.

A.A. Griffith. The Phenomena of Rupture and Flow in Solids. *Philosophical Transactions of the Royal Society of London*, Ser. A, 221:163–198, 1920.

G.R. Irwin. Analysis of Stresses and Strains Near the Ends of a Crack Traversing a Plate. *Journal of Applied Mechanics*, 24(3):361–364, 1957.

M.Y. Lenov, V.V. Panasyuk. Development of Minute Cracks in Solids. *Prikladnaya Mekhanika*, 5(4):391–401, 1959.

F. McClintock. Ductile Fracture Instability in Shear. *Journal of Applied Mechanics*, 25: 582–588, 1958.

V.V. Novozhilov. On the Foundations of the Theory of Equilibrium Cracks in Elastic Bodies. *Prikladnaja Matematika i Mekhanika*, 33(5):797–812, 1969.

V.Z. Parton, E.M. Morozov. Elastic-Plastic Fracture Mechanics, MIR Publishers, 1978.

E. Tonti. A Direct Discrete Formulation of Field Laws: the Cell Method. *Computer Modeling in Engineering and Science*, 2(2):237–258, 2001.

A.A. Wells. Unstable Crack Propagation in Metals: Cleavage and Fast Fracture. *Proc. Crack Propagation Symposium*, 1:210–230, 1961.

On the Three-Dimensional Vibration Analysis of Rectangular Plates

Cristina Gentilini* and Erasmo Viola*

* DISTART Department, University of Bologna, Bologna, Italy

Abstract This paper is concerned with the free vibration analysis of simply supported rectangular plates with any thickness. The analysis is based on the linear, small-strain, three-dimensional elasticity theory. It aims to raise the quality of investigation beyond that provided by the approximate two- and three-dimensional plate theories, by resorting to an exact three-dimensional analysis. The proposed technique yields the full vibration spectrum of natural frequencies and mode shapes. The vibration response due to the variations of thickness is investigated. Frequency parameters and three-dimensional deformed mode shapes are presented in vivid graphical forms. Finally, some new numerical results are given, which may serve as benchmark solutions for future research on the aforementioned problem and for validating approximate three-dimensional theories and new computational techniques in the future.

1 Introduction

Both thin and thick rectangular plates are widely used in mechanical, civil and aerospace structures. Thus, the knowledge of their free vibration behaviour is very important for the structural designers.

In the classical plate theory (CPT), Leissa (1973), Leissa and Zhang (1983), it is assumed that the straight lines originally normal to the plate median surface will remain straight and normal after deformation (the well-known Kirchhoff normality assumption). This assumption implies that all transverse shear strains are zero in the plate.

However, it is well known that shear deformation becomes increasingly important as the thickness-span ratio of the plate increases. Also, this effect should be considered when a greater accuracy of modes or responses is required, even for thin plates.

Relaxing the normality assumption, Mindlin and co-workers, Mindlin (1951), Mindlin et al. (1956), proposed the so-called first-order shear deformation theory (FSDT) for moderately thick plates. The original normal line has now become a curved line that is closely approximated in an average sense by a straight non-normal line, and as a result, not only the deflection, but also the rotations are required to describe the deformation of the plate. A shear correction factor, κ, is then introduced to compensate the errors resulting from the approximation made on the non-uniform shear strain distribution. Several values of the shear correction factor κ have been suggested. For example, Reissner

(1945), Mindlin et al. (1956) and Uflyand (1948) used values of $5/6$, $\pi^2/12$ and $2/3$, respectively. Nanni (1971), showed that the shear correction factor can be expressed in terms of the Poisson's ratio, ν, and it is given by $\kappa = 20\,(1+\nu)\,/\,\left(24+25\nu+\nu^2\right)$.

It should be realized that when dealing with highly orthotropic or composite plates and in stress analysis, higher-order shear deformation theories (HSDT), Reddy (1984), Whitney and Sun (1973), may need to be used instead of the first-order theories because the latter could be inadequate. The higher-order theories account not only for transverse shear strains, but also for a parabolic variation of the transverse shear strains through thickness, and consequently, there is no need to use shear correction coefficients in computing the shear stresses. This improvement extends the model to account for relatively thick plates where the thickness can be as high as half of the shortest plate planform dimension, Lim et al. (1998a), Lim et al. (1998b).

The two-dimensional plate theories offer a relatively simple mathematical manipulation in analytical or computational implementations. They reduce the dimensions of the plate problem (and thus the determinant size of the eigenvalue equation) from three to two by addressing the quantities of interest, such as membrane forces, bending moments and shear forces, in terms of certain averages over the displacement across the smaller dimension, i.e. the thickness. These simplifications are inherently erroneous, and therefore may lead to unreliable results for relatively thick plates.

In contrast to the two-dimensional theories, no assumptions are made in the three-dimensional elasticity theory that provides, not only reliable solutions, but also brings out the physical characteristics of plates. Three-dimensional elasticity solutions thus provide a real basis for assessing the solutions of the two-dimensional theories. Solutions from the exact three-dimensional elasticity theory have become eminent in recent years. This is due to the rapid advancement in contemporary industries towards high precision engineering applications, which acquires more accurate predictions of the behaviours of mechanical systems. Models established based on the three-dimensional elasticity theory for problems in structural mechanics have several comparative advantages over the simplified and refined plate theories. One of the obvious advantages is that the effects of transverse shear strains, rotatory inertia and in-plane behaviour (such as thickness shearing and thickness twisting motions) are inherently accounted for in the formulation. As a result, in analysing the vibration response of elastic thick plates, the method extracts the full vibration spectrum without any missing modes.

In this context the closed-form solution of three-dimensional eigenvalue problem of simply supported rectangular plates, Srinivas et al. (1970), Gentilini and Viola (2002), Gentilini et al. (2002), assumes a great importance in the dynamical vibrations.

This paper is a complement to the above works by providing three-dimensional elasticity solutions for plates with different thickness ratios, analysing thin, moderately thick and very thick plates. Detailed three-dimensional mode shapes are presented for selected cases to enhance our understanding of the physic of plate vibrations. The present solutions supplement the existing data-base, and further serve as the benchmark for researchers to examine the merits of new development in plate theories, such as finite element methods, finite difference methods, Ritz method, see Viola and Gentilini (2002), and so on.

2 Theoretical Formulation

2.1 Equations of Motion

The geometric configuration of a homogeneous, isotropic and rectangular plate is shown in Figure 1. The plate has length a, width b and uniform thickness h. The plate geometry and dimensions are defined with respect to a Cartesian system $(0;\ x, y, z)$, the origin of which is in a corner of the plate and the axes are parallel to the edges. The corresponding displacement components at a generic point are collected in the following vector:

$$\mathbf{u}(x,y,z,t)=\begin{bmatrix} u(x,y,z,t) & v(x,y,z,t) & w(x,y,z,t) \end{bmatrix}^{\mathrm{T}}.$$

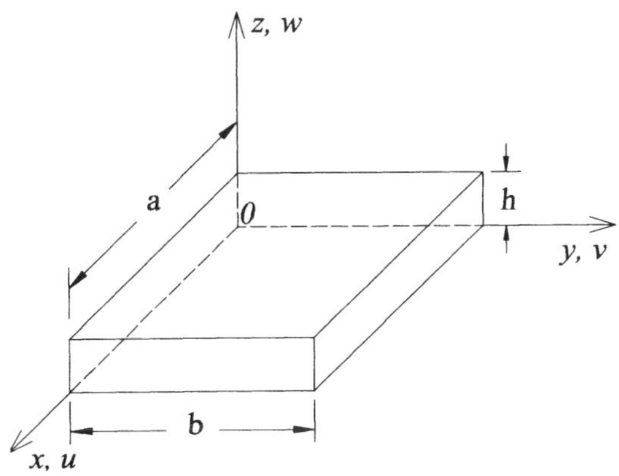

Figure 1. Dimension and geometry of a rectangular plate.

In accordance with the three-dimensional theory, the equations of motion, without considering mass forces, are, Gentilini and Viola (2002):

$$G\nabla^2\mathbf{u}+(\lambda+G)\nabla div\mathbf{u} = \rho\frac{\partial^2\mathbf{u}}{\partial t^2}, \tag{2.1}$$

where

$$\lambda = \frac{2G\nu}{(1-2\nu)},\ G = \frac{E}{2(1+\nu)},$$

E is the Young's modulus, ν is the Poisson's ratio and ρ is the mass density, respectively.

For a plate undergoing free vibration, its periodic displacement components can be expressed in terms of the displacement amplitude functions as follows:

$$u(x,y,z,t) = U(x,y,z)e^{i\omega t}, \tag{2.2}$$

$$v(x,y,z,t) = V(x,y,z)e^{i\omega t}, \tag{2.3}$$

$$w(x,y,z,t) = W(x,y,z)e^{i\omega t}, \tag{2.4}$$

where ω denotes the natural frequency of the plate. The displacement amplitude functions are collected in the following vector

$$\mathbf{U}(x,y,z)=\left[\begin{array}{ccc} U(x,y,z) & V(x,y,z) & W(x,y,z) \end{array}\right]^{\mathrm{T}}$$

and eq. (2.1) can be written as:

$$\nabla^2\mathbf{U}+\frac{1}{1-2\nu}\nabla div\mathbf{U}+\frac{\rho\omega^2\mathbf{U}}{G}=0. \tag{2.5}$$

The edge conditions for a simply supported plate can be specified as

$$\sigma_x = 0,\ v=0,\ w=0 \quad \text{for } x=0,\ a, \tag{2.6}$$
$$\sigma_y = 0,\ u=0,\ w=0 \quad \text{for } y=0,\ b. \tag{2.7}$$

For simplicity and convenience in mathematical formulation, the following non-dimensional coordinate system is introduced:

$$X=\frac{x}{a},\ Y=\frac{y}{b},\ Z=\frac{z}{h}.$$

The edge conditions (2.6) and (2.7) are identically satisfied by setting, Srinivas et al. (1970):

$$\begin{aligned} U(X,Y,Z) &= h\textstyle\sum_{m=1}^{\infty}\sum_{n=1}^{\infty}\Phi(Z)\cos m\pi X\sin n\pi Y,\\ V(X,Y,Z) &= h\textstyle\sum_{m=1}^{\infty}\sum_{n=1}^{\infty}\Psi(Z)\sin m\pi X\cos n\pi Y,\\ W(X,Y,Z) &= h\textstyle\sum_{m=1}^{\infty}\sum_{n=1}^{\infty}\chi(Z)\sin m\pi X\sin n\pi Y, \end{aligned} \tag{2.8}$$

where m and n are natural numbers. By setting

$$M = \frac{m\pi h}{a},\ N=\frac{n\pi h}{b}, \tag{2.9}$$
$$g = \sqrt{M^2+N^2},\ \lambda_{sr}=\omega\sqrt{\frac{\rho h^2}{G}}, \tag{2.10}$$
$$D = \frac{\partial}{\partial Z},\ D^2=\frac{\partial^2}{\partial Z^2}, \tag{2.11}$$

and substituting eqs. (2.8) in eq. (2.5), we obtain the following matricial equation:

$$\begin{bmatrix} D^2+\lambda_{sr}^2-g^2-\frac{M^2}{1-2\nu} & -\frac{MN}{1-2\nu} & \frac{MD}{1-2\nu}\\ -\frac{MN}{1-2\nu} & D^2+\lambda_{sr}^2-g^2-\frac{N^2}{1-2\nu} & \frac{ND}{1-2\nu}\\ -\frac{MD}{1-2\nu} & -\frac{ND}{1-2\nu} & D^2+\lambda_{sr}^2-g^2+\frac{D^2}{1-2\nu} \end{bmatrix}\begin{bmatrix}\Phi(Z)\\ \Psi(Z)\\ \chi(Z)\end{bmatrix}=\begin{bmatrix}0\\0\\0\end{bmatrix}. \tag{2.12}$$

The non-trivial solution of eq. (2.12) yields

$$\begin{bmatrix}\Phi(Z)\\ \Psi(Z)\\ \chi(Z)\end{bmatrix}=\begin{bmatrix} rM & -rM & N & N & M & M\\ rN & -rN & -M & -M & N & N\\ g^2 & g^2 & 0 & 0 & s & -s \end{bmatrix}\begin{bmatrix}Ae^{rZ}\\ Ke^{-rZ}\\ Be^{rZ}\\ R\,e^{-rZ}\\ Ce^{sZ}\\ Se^{-sZ}\end{bmatrix}, \tag{2.13}$$

where A, K, B, R, C, S are six arbitrary constants and r, s can be expressed in the form:

$$r = \sqrt{g^2 - \lambda_{sr}^2}, \quad s = \sqrt{g^2 - \frac{\lambda_{sr}^2 (1 - 2\nu)}{2 - 2\nu}}. \tag{2.14}$$

For all the passages that lead to the expressions (2.13) and (2.14), see Gentilini et al. (2002). Introducing the strain vector $\boldsymbol{\varepsilon}$, the compatibility equation can be expressed in the form:

$$\boldsymbol{\varepsilon} = \mathbf{DU}, \tag{2.15}$$

where $\mathbf{D}$ is the compatibility operator.

In the same way, we introduce the stress vector $\boldsymbol{\sigma}$ and we express the constitutive equations in the following form:

$$\boldsymbol{\sigma} = \mathbf{C}\boldsymbol{\varepsilon}, \tag{2.16}$$

where

$$\mathbf{C} = \begin{bmatrix} 2G+\lambda & \lambda & \lambda & 0 & 0 & 0 \\ \lambda & 2G+\lambda & \lambda & 0 & 0 & 0 \\ \lambda & \lambda & 2G+\lambda & 0 & 0 & 0 \\ 0 & 0 & 0 & G & 0 & 0 \\ 0 & 0 & 0 & 0 & G & 0 \\ 0 & 0 & 0 & 0 & 0 & G \end{bmatrix}. \tag{2.17}$$

is the matrix of elastic stiffness moduli.

Substituting the compatibility equation (2.15) in (2.16), expressing the displacement components through eq. (2.8) and using eq. (2.13), we obtain for the stress components the following expressions:

$$\begin{bmatrix} \sigma_x \\ \sigma_y \\ \sigma_z \\ \tau_{xy} \\ \tau_{xz} \\ \tau_{yz} \end{bmatrix} = G \sum_{m=1}^{\infty} \sum_{n=1}^{\infty} \begin{bmatrix} \sigma_x^* \sin m\pi X \sin n\pi Y \\ \sigma_y^* \sin m\pi X \sin n\pi Y \\ \sigma_z^* \sin m\pi X \sin n\pi Y \\ \tau_{xy}^* \cos m\pi X \cos n\pi Y \\ \tau_{xz}^* \cos m\pi X \sin n\pi Y \\ \tau_{yz}^* \sin m\pi X \cos n\pi Y \end{bmatrix}$$

where:

$$\begin{Bmatrix} \sigma_x^* \\ \sigma_y^* \\ \sigma_z^* \\ \tau_{xy}^* \\ \tau_{xz}^* \\ \tau_{yz}^* \end{Bmatrix} = \begin{bmatrix} -2rM^2 & 2rM^2 & -2MN & -2MN & \frac{2\nu(s^2-g^2)}{1-2\nu}-2M^2 & \frac{2\nu(s^2-g^2)}{1-2\nu}-2M^2 \\ -2rN^2 & 2rN^2 & 2MN & 2MN & \frac{2\nu(s^2-g^2)}{1-2\nu}-2N^2 & \frac{2\nu(s^2-g^2)}{1-2\nu}-2N^2 \\ 2g^2r & -2g^2r & 0 & 0 & r^2+g^2 & r^2+g^2 \\ 2MNr & -2MNr & N^2-M^2 & N^2-M^2 & 2MN & 2MN \\ (r^2+g^2)M & (r^2+g^2)M & Nr & -Nr & 2Ms & -2Ms \\ (r^2+g^2)N & (r^2+g^2)N & -Mr & Mr & 2Ns & -2Ns \end{bmatrix} \begin{Bmatrix} Ae^{rZ} \\ Ke^{-rZ} \\ Be^{rZ} \\ Re^{-rZ} \\ Ce^{sZ} \\ Se^{-sZ} \end{Bmatrix} \tag{2.18}$$

The stress-free conditions for a simply supported plate on the upper and lower surfaces are:

$$\sigma_z = \tau_{xz} = \tau_{yz} = 0 \quad \text{in } Z = 0, 1. \tag{2.19}$$

The satisfaction of (2.19) leads to an homogeneous system of six equations for each (m, n) combination.

A non-trivial solution of the problem is achieved by setting the determinant, Δ, of the coefficient matrix of equation obtained from eqs. (2.19) and (2.18), equal to zero. This yields the characteristic equation:

$$\Delta = \left[8g^2 rs\left(r^2+g^2\right)^2(1-\cosh r\cosh s)+\left(16g^4r^2s^2+\left(r^2+g^2\right)^4\right)\sinh r\sinh s\right]\sinh r=0. \tag{2.20}$$

Solution of this transcendental equation for each combination of m and n yields an infinite sequence of eigenvalues. The eigenvalues depend only on ν and g, so when Poisson's coefficient is fixed, transcendental equation (2.20) represents a surface of type:

$$\Delta = \Delta(g, \lambda_{sr}).$$

This surface can be represented varying g and λ_{sr}, as shown in Figure 2:

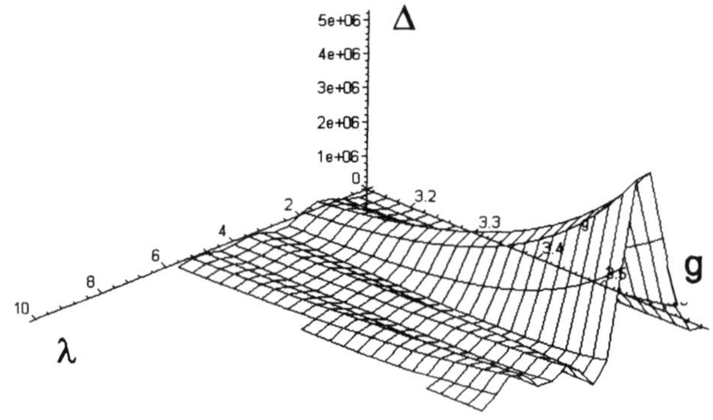

Figure 2. Representation of the transcendental equation.

The sections of the surface with planes characterized by constant values of g, are swinging curves that meet the plane $\Delta = 0$ in infinite values of λ_{sr}. These values are the eigenvalues of the problem. So, for each g value, infinite solutions exist. Some of the eigenvalues may lead to either pure distortional modes in which the cubic dilatation vanishes, or to pure dilatational modes in which all the rigid rotations vanish. Mode shapes (eigenfunctions) are determined by back-substitution of the eigenvalues, one-by-one, in the usual manner.

2.2 Distortional Modes

Distortional modes are characterized by the vanishing of the cubic dilatation

$$I_{1\varepsilon} = \frac{\partial u}{\partial x} + \frac{\partial v}{\partial y} + \frac{\partial w}{\partial z} = 0, \tag{2.21}$$

which involves that the volume of the plate does not change. A consequence of this fact is that the interpolant functions (2.13) assume the following simplified expression

$$\begin{bmatrix} \Phi(Z) \\ \Psi(Z) \\ \chi(Z) \end{bmatrix} = \begin{bmatrix} rM & -rM & N & N \\ rN & -rN & -M & -M \\ g^2 & g^2 & 0 & 0 \end{bmatrix} \begin{bmatrix} Ae^{rZ} \\ Ke^{-rZ} \\ Be^{rZ} \\ R\,e^{-rZ} \end{bmatrix} \tag{2.22}$$

where C, S become zero and the root r takes the value:

$$r = i\alpha\pi. \tag{2.23}$$

with

$$\alpha = 0, 1, 2, \ldots.$$

As a consequence, the expressions of the stress components (2.18) become:

$$\sigma_x^* = -2B'MN\cos(\alpha\pi Z), \tag{2.24}$$

$$\sigma_y^* = 2B'MN\cos(\alpha\pi Z), \tag{2.25}$$

$$\sigma_z^* = 0, \tag{2.26}$$

$$\tau_{xy}^* = B'\left(N^2 - M^2\right)\cos(\alpha\pi Z), \tag{2.27}$$

$$\tau_{xz}^* = -\alpha\pi B'N\sin(\alpha\pi Z), \tag{2.28}$$

$$\tau_{yz}^* = \alpha\pi B'M\sin(\alpha\pi Z), \tag{2.29}$$

where B' is a constant that derived from the calculation, see Gentilini and Viola (2002).

2.3 Dilatational Modes

The dilatational modes are characterized by the vanishing of the rigid rotations:

$$\begin{aligned} \omega_x &= \frac{1}{2}\left(\frac{\partial w}{\partial y} - \frac{\partial v}{\partial z}\right) = 0, \\ \omega_y &= \frac{1}{2}\left(\frac{\partial u}{\partial z} - \frac{\partial w}{\partial x}\right) = 0, \\ \omega_z &= \frac{1}{2}\left(\frac{\partial v}{\partial x} - \frac{\partial u}{\partial y}\right) = 0. \end{aligned}$$

A consequence is that coefficients A, K, B and R have to vanish at the same time, so the interpolant functions (2.13) assume the following simplified expression:

$$\begin{bmatrix} \Phi(Z) \\ \Psi(Z) \\ \chi(Z) \end{bmatrix} = \begin{bmatrix} M & M \\ N & N \\ s & -s \end{bmatrix} \begin{bmatrix} Ce^{sZ} \\ Se^{-sZ} \end{bmatrix}. \tag{2.30}$$

3 Numerical Applications

3.1 Comparison Study

The present formulation is applied to the analysis of simply supported square plates. For simplicity both a and b are taken equal to 1. Poisson's ratio ν is assumed to be 0.3 in all the computation. The solutions obtained from the present study are compared in Figure 3 and in Figure 4 with the solutions obtained from the classical plate theory (CPT).

For plates with thickness ratio of $h/b = 0.05$, the frequency parameters computed by using the CPT are found to deviate slightly from these of the present results. At higher thickness ratio of $h/b = 0.2$, the discrepancies between CPT and 3-D elasticity solutions are significant at higher modes. This is attributed to the simplifying assumption made in the CPT that the straight line originally normal to the plate median surface remains straight and normal after deformation. This leads to neglect the through the thickness shear deformation effect, that instead is implicitly accounted in the present theory.

It is worth mentioning that, when a general 3-D elasticity formulation is considered, a full spectrum of vibration frequencies can be obtained. On the other hand, in most simplified plate theories reported in the literature, which omit the in-plane components, only those vibration frequencies corresponding to the flexural, thickness twist and thickness shear modes are obtainable. It is confirmed that thin plate theory overestimates the frequencies, which it identifies. This is due to the fact that it models a more rigid structure, because it does not take into account all the real degrees of freedom of the plate.

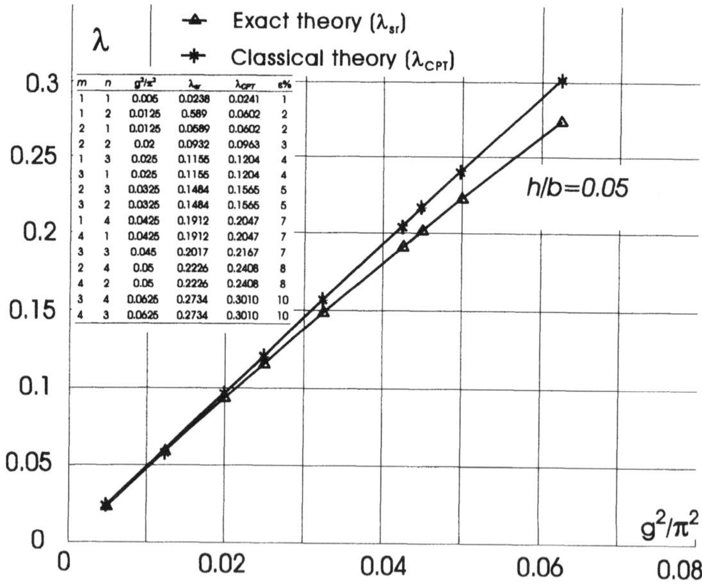

m	n	g^2/π^2	λ_{sr}	λ_{CPT}	ε%
1	1	0.006	0.0238	0.0241	1
1	2	0.0125	0.589	0.0602	2
2	1	0.0125	0.0589	0.0602	2
2	2	0.02	0.0932	0.0963	3
1	3	0.025	0.1155	0.1204	4
3	1	0.025	0.1155	0.1204	4
2	3	0.0325	0.1484	0.1565	5
3	2	0.0325	0.1484	0.1565	5
1	4	0.0425	0.1912	0.2047	7
4	1	0.0425	0.1912	0.2047	7
3	3	0.045	0.2017	0.2167	7
2	4	0.05	0.2226	0.2408	8
4	2	0.05	0.2226	0.2408	8
3	4	0.0625	0.2734	0.3010	10
4	3	0.0625	0.2734	0.3010	10

Figure 3. Comparison between CPT and the present analysis for a square plate with $h/b = 0.05$.

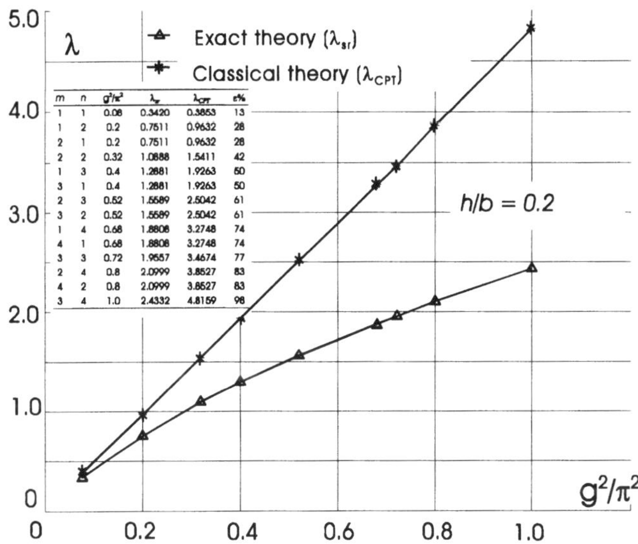

m	n	g^2/π^2	λ_{sr}	λ_{CPT}	$\varepsilon\%$
1	1	0.08	0.3420	0.3853	13
1	2	0.2	0.7511	0.9632	28
2	1	0.2	0.7511	0.9632	28
2	2	0.32	1.0888	1.5411	42
1	3	0.4	1.2881	1.9263	50
3	1	0.4	1.2881	1.9263	50
2	3	0.52	1.5589	2.5042	61
3	2	0.52	1.5589	2.5042	61
1	4	0.68	1.8808	3.2748	74
4	1	0.68	1.8808	3.2748	74
3	3	0.72	1.9557	3.4674	77
2	4	0.8	2.0999	3.8527	83
4	2	0.8	2.0999	3.8527	83
3	4	1.0	2.4332	4.8159	98

Figure 4. Comparison between CPT and the present analysis for a square plate with $h/b = 0.2$.

3.2 Results and Discussions

In order to analyse the vibratory behaviour of thin to very thick plates, frequency parameters, λ_{sr}, and 3-D deformed mode shapes were computed for plates with thickness ratios of 0.05, 0.2, 0.6 and 1.0, respectively.

The first six modes have been depicted in Figure 5. These figures expressed vividly the vibratory motion of elastic plates. It is observed that the frequency parameters, λ_{sr}, increase and for the thinnest plate the first vibration mode is an out-of-plane flexural motion. At higher thickness ratios, it is interesting to observe that the number of flexural modes decrease with increasing plate thickness. Most of the flexural motions in these figures have been replaced by surface-parallel modes as the plate thickness increases. Some typical surface-parallel modes can be seen in the second, third, sixth and eighth modes for the plate with $h/b = 0.2$. For the plates characterized by $h/b = 0.6$ and $h/b = 1.0$, the first mode has become distortional and only the third is an out-of-plane mode. The flexural mode at the first mode of vibration has switched to become the third mode as the thickness h/b changes from 0.05 to 0.6. As the thickness increases, the flexural modes interchange with the surface parallel modes, in fact vibrations outside the plane of the plate become more difficult.

In Tables 1 and 2 the first fifteen eigenvalues are collected of the plates under study. For each eigenvalue the corresponding values of m, n and the kind of mode, if flexural (MF), distortional (MD) or degenerate (MDg) are indicated. If the mode is distortional, also the value of $\alpha-$parameter is indicated between brackets. It is worth underlining the meaning of this parameter.

From Figure 5 it is possible to notice that, for example, if we consider the plate with $h/b = 0.6$, the first distortional mode is characterized by $\alpha = 0$, while the eighth mode is characterized by $\alpha = 1$. It could be seen that in the case where $\alpha = 0$, through the thickness there are no waves, while for $\alpha = 1$ we have an half-wave through the thickness. So the α–parameter represents the number of half-waves through the thickness for the distortional modes.

Figure 6 shows the distribution of the non-dimensional stress components, σ_x^*, σ_z^*, τ_{xy}^* and τ_{xz}^*, through the thickness, for plates with $h/b = 0.2$, 0.6 and 1.0. It could be seen that σ_x^* stress component distribution is almost linear for plates with $h/b = 0.2$, but for higher thickness ratios this assumption of the classical plate theory is not respected. For σ_z^* stress component, the assumption of Mindlin to retain this component on average zero through the thickness is satisfied. Moreover, for what it concerns with the distribution of τ_{xz}^*, it is clearly seen that the assumption made by Mindlin to consider its value constant across the thickness is not respected, in fact this stress component has a parabolic distribution. In particular, in Figure 7 the stress components distributions for the eighth and the ninth distortional mode of a plate with $h/b = 1.0$ are depicted. In conclusion, stress and displacement distributions across thickness in both thin plate and Mindlin theories are much different from the true distributions.

Table 1. Frequency parameters for plates with $h/b = 0.05$ and $h/b = 0.2$.

N°	m	n	λ_{sr}	MODE
1	1	1	0.0238	MF
2	1	2	0.0589	MF
3	2	1	0.0589	MF
4	2	2	0.0932	MF
5	1	3	0.1155	MF
6	3	1	0.1155	MF
7	2	3	0.1484	MF
8	3	2	0.1484	MF
9	0	1	0.1570	MD ($\alpha = 0$)
10	1	0	0.1570	MD ($\alpha = 0$)
11	1	4	0.1912	MF
12	4	1	0.1912	MF
13	3	3	0.2017	MF
14	1	1	0.2221	MD ($\alpha = 0$)
15	2	4	0.2226	MF

N°	m	n	λ_{sr}	MODE
1	1	1	0.3420	MF
2	0	1	0.6283	MD ($\alpha = 0$)
3	1	0	0.6283	MD ($\alpha = 0$)
4	1	2	0.7511	MF
5	2	1	0.7511	MF
6	1	1	0.8885	MD ($\alpha = 0$)
7	2	2	1.0888	MF
8	0	2	1.2566	MD ($\alpha = 0$)
9	2	0	1.2566	MD ($\alpha = 0$)
10	1	3	1.2881	MF
11	3	1	1.2881	MF
12	1	2	1.4050	MD ($\alpha = 0$)
13	2	1	1.4050	MD ($\alpha = 0$)
14	1	1	1.4922	MDg
15	2	3	1.5589	MF

Table 2. Frequency parameters for plates with $h/b = 0.6$ and $h/b = 1.0$.

N°	m	n	λ_{sr}	MODE
1	0	1	1.8849	MD ($\alpha = 0$)
2	1	0	1.8849	MD ($\alpha = 0$)
3	1	1	1.9557	MF
4	1	1	2.6657	MD ($\alpha = 0$)
5	1	2	3.5133	MF
6	2	1	3.5133	MF
7	0	1	3.6637	MD ($\alpha = 1$)
8	1	0	3.6637	MD ($\alpha = 1$)
9	0	2	3.7698	MD ($\alpha = 0$)
10	2	0	3.7698	MD ($\alpha = 0$)
11	1	1	4.0423	MDg
12	1	1	4.1201	MD ($\alpha = 1$)
13	1	2	4.2147	MD ($\alpha = 0$)
14	2	1	4.2147	MD ($\alpha = 0$)
15	2	2	4.6279	MF

N°	m	n	λ_{sr}	MODE
1	0	1	3.1416	MD ($\alpha = 0$)
2	1	0	3.1416	MD ($\alpha = 0$)
3	1	1	3.7419	MF
4	1	1	4.4428	MD ($\alpha = 0$)
5	0	1	4.4428	MD ($\alpha = 1$)
6	1	0	4.4428	MD ($\alpha = 1$)
7	1	1	5.2013	MDg
8	1	1	5.4414	MD ($\alpha = 1$)
9	0	2	6.2830	MD ($\alpha = 0$)
10	2	0	6.2830	MD ($\alpha = 0$)
11	1	2	6.2943	MF
12	2	1	6.2943	MF
13	1	2	6.9519	MD ($\alpha = 0$)
14	2	1	6.9519	MD ($\alpha = 0$)
15	0	1	7.0248	MD ($\alpha = 2$)

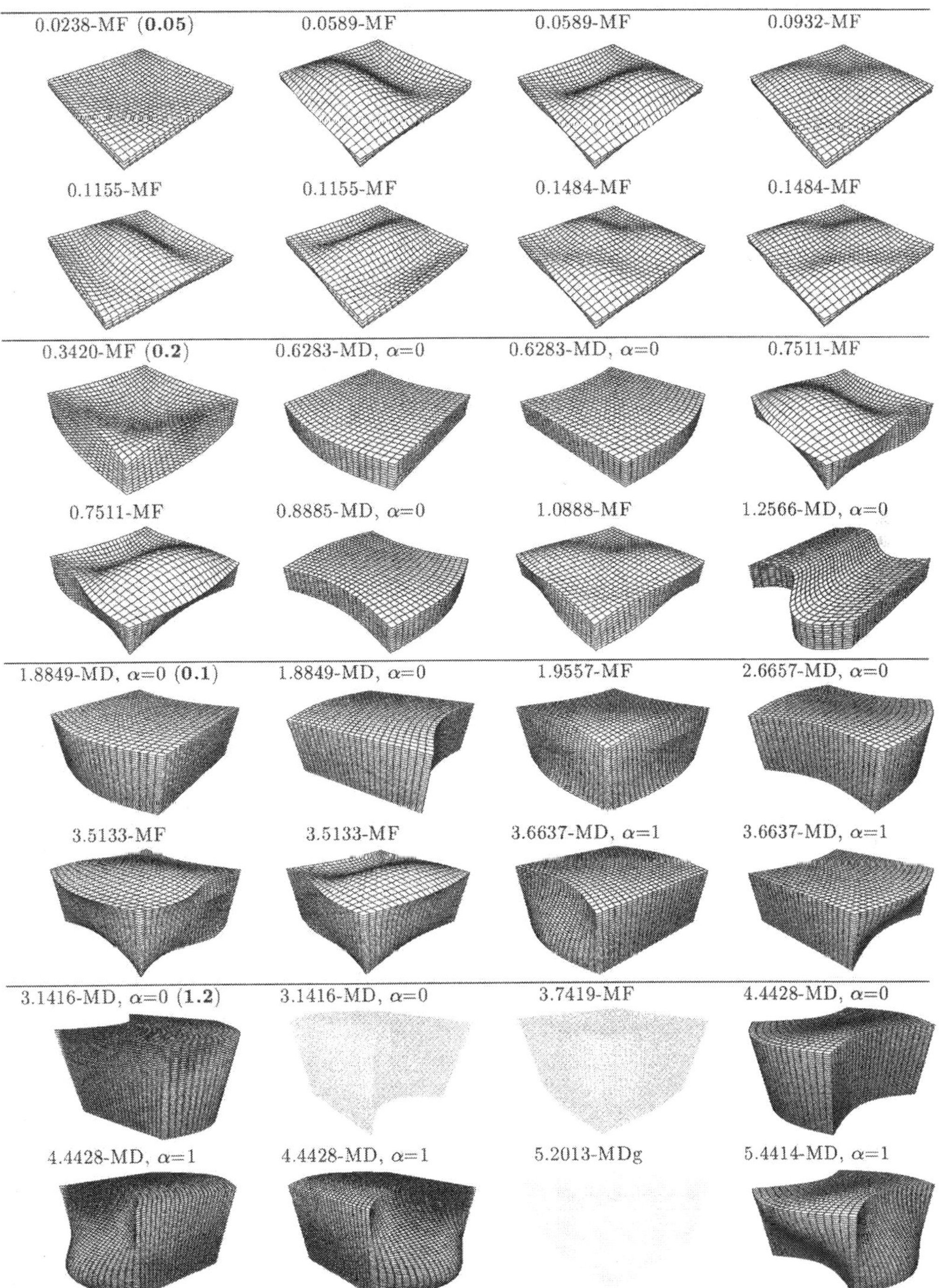

Figure 5. Mode shapes and eigenvalues λ_{sr} for plates with $h/b = 0.05$, 0.2, 0.6 and 1.0.

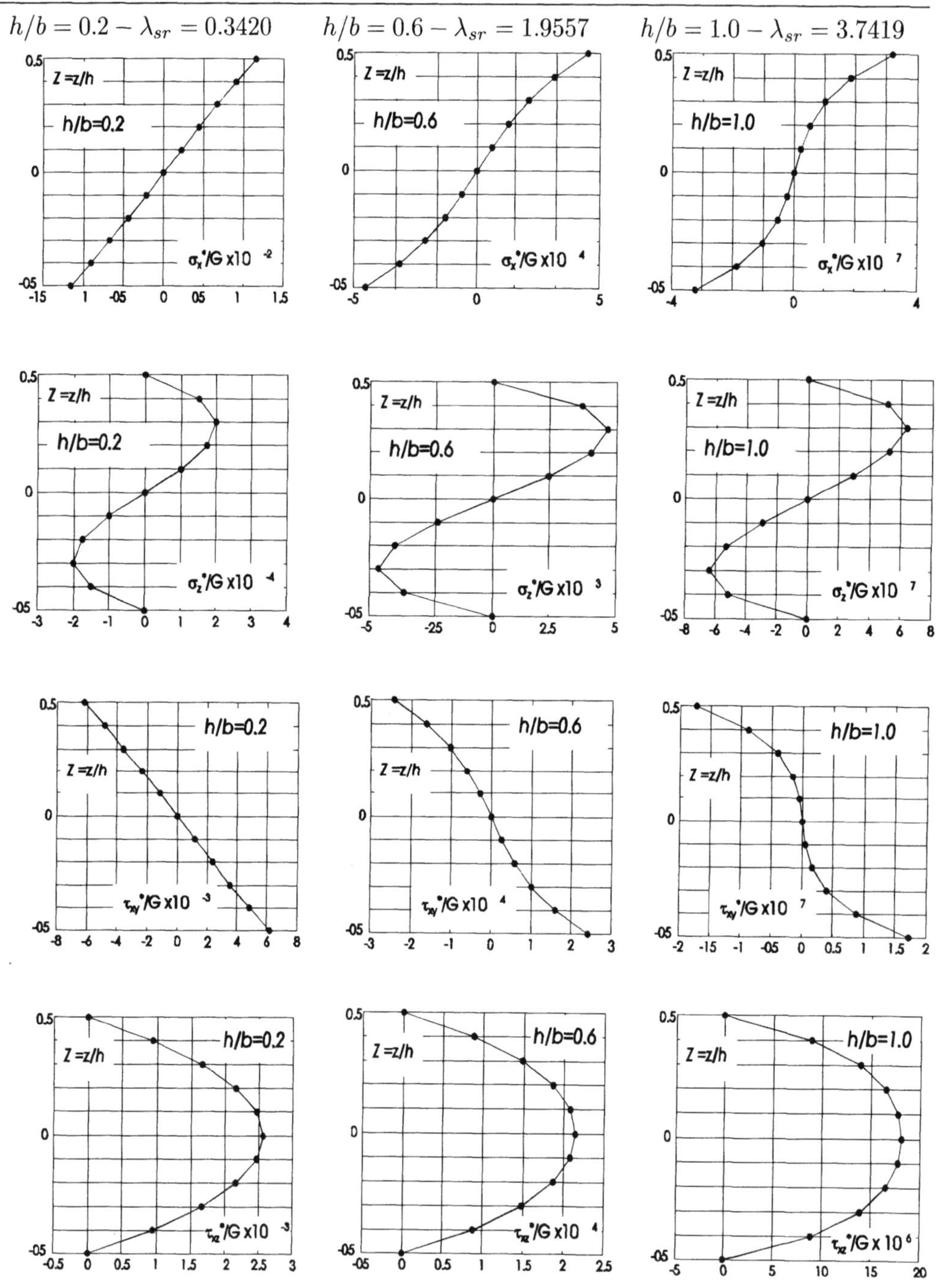

Figure 6. Distibution of stress components through the thickness, for plates with $h/b = 0.2$, 0.6 and 1.0.

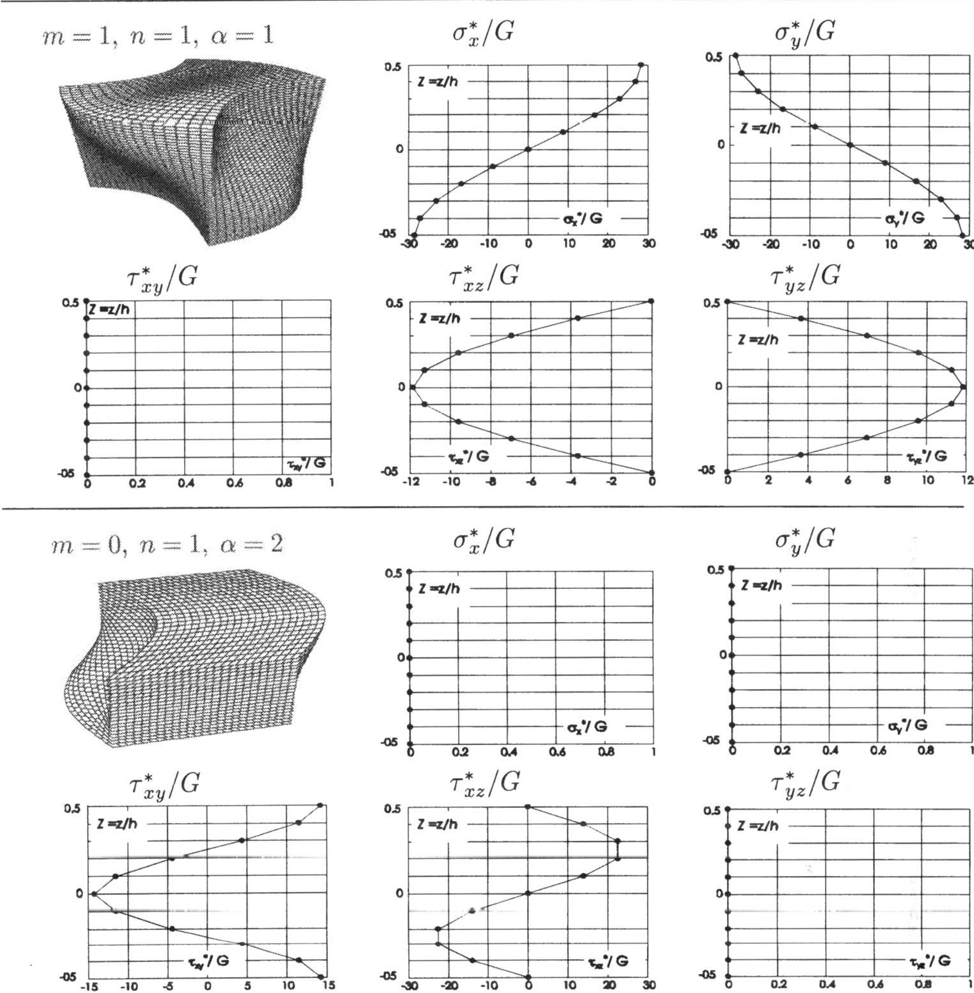

Figure 7. Distribution of stress components through the thickness, for the eighth and the ninth distortional mode of a plate with $h/b = 1.0$.

4 Conclusions

The free vibration characteristics of simply supported rectangular plates have been investigated. Since no simplifying assumption has been made about the variation of stresses and displacements through the thickness, the solution is exact. The linear small-strain three dimensional elasticity theory adopted in this derivation allows computation of the full vibration spectrum for the plates. A transcendental eigenvalue equation system was obtained. This was used to determine the vibration frequencies and mode shapes of the plates. Vibration behaviours of rectangular plates with various thickness ratios were investigated. Vivid graphical representations of the vibration modes were manifested in three-dimensional deformed mesh geometry. The three-dimensional mode shapes encompass flexural, thickness twist and thickness shear motions, and in particular the twist and shear modes which cannot be predicted by two dimensional plate theories.

The thickness variations of stresses have been discussed and shown through some graphics.

This study has yielded new results, which may serve as benchmarks for future research on the aforementioned problem that can be used to validate results obtained by other approximate approaches such as the FEM, FDM and the BEM.

Bibliography

C. Gentilini, A. Mazza, and E. Viola. Analisi dinamica esatta delle piastre spesse. Technical Report 84, DISTART, University of Bologna, 2002.

C. Gentilini and E. Viola. Risposta in termini di tensione nell'analisi dinamica di piastre. Technical Report 85, DISTART, University of Bologna, 2002.

A.W. Leissa. The free vibration of rectangular plates. *Journal of Sound and Vibration*, 31(3):257–293, 1973.

A.W. Leissa and Z. Zhang. On the three-dimensional vibrations of a cantilevered rectangular parallelepiped. *Journal of Acoustic of America*, 73:2013–2021, 1983.

C.W. Lim, S. Kitipornchai, and K.M. Liew. Numerical aspects for free vibration of thick plates part I: Formulation and verification. *Computer Methods in Applied Mechanics and Engineering*, 156:15–29, 1998a.

C.W. Lim, S. Kitipornchai, and K.M. Liew. Numerical aspects for free vibration of thick plates part II: Numerical efficiency and vibration frequencies. *Computer Methods in Applied Mechanics and Engineering*, 156:31–44, 1998b.

R.D. Mindlin. Influence of rotatory inertia and shear on flexural motion of isotropic elastic plates. *ASME Journal of Applied Mechanics*, 73:31–38, 1951.

R.D. Mindlin, A. Schacknow, and H. Deresiewickz. Flexural vibrations of rectangular plates. *ASME Journal of Applied Mechanics*, 18:430–436, 1956.

J. Nanni. Das eulersche knickproblem unter berucksichtigung der querkrafte. *Zeitschrift fur Angewandte Mathematik und Phisysik*, 22, 1971.

J.N. Reddy. A simple high-order theory for laminated composite plates. *Journal of Applied Mechanics*, 51:745–752, 1984.

E. Reissner. The effect of transverse shear deformation on the bending of elastic plates. *ASME Journal of Applied Mechanics*, 12(2):69–77, 1945.

S. Srinivas, C.V. Joga Rao, and K. Rao. An exact analysis for vibration of simple-supported homogeneous and laminated thick rectangular plates. *Journal of Sound and Vibration*, 12:187–199, 1970.

Y.S. Uflyand. The propagation of waves in the transverse vibrations of bars and plates. *Akademia Nauk SSSR, Prikladnaya Matematika i Mekhanika*, 12, 1948.

E. Viola and C. Gentilini. Influenza dello spessore sul comportamento dinamico delle piastre. Technical Report 86, DISTART, University of Bologna, 2002.

J.M. Whitney and C.T. Sun. A higher-order theory for extensional motion of laminated composites. *Journal of Sound and Vibration*, 30(1), 1973.

The Crack Detection Problem in Vibrating Beams

Antonino Morassi
Department of Civil Engineering, University of Udine, Udine, Italy

Abstract. This paper deals with a class of inverse problems concerning the identification of cracks in elastic beams from vibration measurements. A review of some recent results is given.

1 Introduction

Dynamic tests are commonly used as a diagnostic tool to detect structural modifications that may have occurred in mechanical systems during service. The final aim is to interpret the changes induced by a possible damage on the dynamic behavior of a system, using them to predict location and degree of severeness of the degradation, see Hearn and Testa (1991).

Within the class of diagnostic problems in Structural Mechanics, the crack detection problem in vibrating beams has received great attention in the scientific community in last two decades. There are good reasons for this interest: firstly, the mechanical system consisting of a single beam describes the behavior of many structures which are important both for the civil and mechanical engineering field; secondly, the problem of identifying a crack in a beam gives rise to the basic diagnostic problem and, therefore, it represents an important benchmark to test the accuracy of identification techniques.

Most of the diagnostic techniques for vibrating beams are formulated as an optimality criterion, where the stiffness distribution of a chosen reference configuration of the beam is updated so that the first few natural frequencies and mode shapes closely match the measured ones at a certain level of deterioration, see, for example, Shen and Taylor (1991), Liang et al. (1992), Vestroni and Capecchi (1996), Sinha et al. (2002). On the one hand, these optimization techniques have the advantage of allowing for investigation of quite general classes of systems. On the other hand, the lack of satisfactory framework of general properties gives rise to several indeterminacy, which, in some cases, may obstruct applications to practical problems, see Davini et al. (1995), Morassi and Rovere (1997).

In this paper the crack detection problem in beams is investigated from a different point of view, namely attention is focused on finding conditions which allow for a rigorous, e.g. mathematically proved, identification of the damage. Despite very extensive literature on damage detection in beams, few results of this kind are available. Concerning identification from frequency measurements, for instance, in a relatively recent paper by Narkis (1994) it was shown that if the damaged beam is a perturbation of a virgin beam, namely whenever the crack is very small, then the only information required for crack localization is the variation of the first two natural frequencies caused by the crack. The results were proved for uniform free-free vibrating rods and for uniform simply supported beams in bending. See also Cerri and Vestroni (2000) for detection of damage in beams subjected to diffused, and not necessarily small, cracking. In Section 2 of the present work it is shown how Narkis results can be extended to derive simple closed-form expressions for damage location and severity for a large class of vibrating beams. Related results concerning identification of multiple small cracks and damage detection in dissipative systems are collected in Section 4. In Section 3 the changes in the nodes of the modes shapes are used to identify localized damage in a beam. All the proposed diagnostic techniques were tested on a basis of an extensive series of dynamic tests carried out on steel cracked beams, see references for more details.

2 Crack Identification from Changes in Natural Frequencies

2.1 Axially Vibrating Beams

It is assumed that the spatial variation of the infinitesimal free vibrations of an undamaged straight rod of length L is governed by the differential equation

$$(au')' + \lambda\rho u = 0 \quad \text{in } (0, L), \tag{2.1}$$

where $u = u(x)$ describes the mode and $\sqrt{\lambda}$ is the associated natural frequency. In this Section, the rod is assumed to have no material damping. The quantities $a = a(x) \equiv EA(x)$ and $\rho = \rho(x)$ denote the axial stiffness and the linear-mass density of the rod. E is Young's modulus of the material and $A(x)$ the cross-section area of the rod. This analysis is concerned with rods for which the axial stiffness is strictly positive and continuously differentiable function; ρ will be assumed to be continuous and strictly positive function. The end conditions are taken as

$$a(0)u'(0) - hu(0) = 0, \tag{2.2}$$

$$a(L)u'(L) + Hu(L) = 0, \tag{2.3}$$

where $h \geq 0$ and $H \geq 0$ are the stiffnesses of the elastic end supports. Three important cases can be distinguished:

$$\text{Supported (S)}: \quad h = \infty = H, \quad u(0) = 0 = u(L), \tag{2.4}$$

$$\text{Free (F)}: \quad h = 0 = H, \quad u'(0) = 0 = u'(L), \tag{2.5}$$

$$\text{Cantilever (C)}: \quad h = \infty, H = 0, \quad u(0) = 0 = u'(L). \tag{2.6}$$

Modes and frequencies are the eigensolutions of the boundary value problem (2.1)–(2.3), and the mth eigenpair of the undamaged rod, $m \geq 0$, is denoted by $\{u_m, \lambda_m\}$. It is well known that for such a and ρ, and end conditions (2.2), (2.3), there is an infinite sequence $\{\lambda_m\}_{m=0}^{\infty}$ such that $0 \leq \lambda_0 < \lambda_1 < ...$ with $\lim_{m\to\infty}\lambda_m = \infty$.

Suppose that a crack appears at the cross-section of abscissa $s \in (0, L)$. Assuming that the crack remains always open during the longitudinal vibration, by modelling it as a massless translational spring, at $x = s$, see Freund and Herrmann (1976) and Cabib et al. (2001), the eigenvalue problem for the damaged rod is the following

$$(au_d')' + \lambda_d \rho u_d = 0 \quad \text{in } (0, s) \cup (s, L), \tag{2.7}$$

where, in addition to the boundary conditions (2.2), (2.3) it is necessary to consider the jump conditions

$$[u_d'(s)] = 0, \qquad K_A[u_d(s)] = a(s)u_d'(s), \tag{2.8}$$

that are to hold at the cross-section where the crack occurs. In equations (2.8), $[\phi(s)] \equiv (\phi(s^+) - \phi(s^-))$ denotes the jump of the function $\phi(x)$ at $x = s$. The expression K_A is the stiffness of the spring simulating the damage, and it can be related to the crack geometry as suggested, for example, by Freund and Herrmann (1976) or by Dimarogonas and Paipetis (1983). The undamaged system corresponds to $K_A \to \infty$ or $\epsilon \equiv 1/K_A \to 0$.

The variational formulation of (2.1)–(2.3) shows that eigenvalues of the system are increasing functions of h, H and K_A, and thus decreasing functions of ϵ, so that

$$\lambda_{dm} \leq \lambda_m, \qquad \text{m=0,1,2,...} \tag{2.9}$$

Moreover, as for any constrained system, the following interlacing result holds

$$\lambda_{m-1} \leq \lambda_{dm} \leq \lambda_m, \qquad \text{m=1,2,...} \tag{2.10}$$

The first order variation of the natural frequencies with respect to ϵ may be found by taking

$$u_d = u + \epsilon\widetilde{u}, \qquad \lambda_d = \lambda + \epsilon\widetilde{\lambda} \tag{2.11}$$

in equations (2.7), (2.8) and (2.2), (2.3). It follows that

$$(a\widetilde{u}')' + \lambda\rho\widetilde{u} + \widetilde{\lambda}\rho u = 0 \quad \text{in } (0, s) \cup (s, L), \tag{2.12}$$

$$[\widetilde{u}'(s)] = 0, \qquad [\widetilde{u}(s)] = a(s)u'(s). \tag{2.13}$$

Multiplying (2.1) by $\widetilde{u}$, (2.12) by u and subtracting, and then integrating by parts in the interval $(0, L)$, it follows that

$$(a(s)u'(s))^2 + \widetilde{\lambda} \int_0^L \rho u^2 = 0, \tag{2.14}$$

which, with the normalizing condition $\int_0^L \rho u^2 = 1$, gives

$$\delta\lambda_m \equiv \epsilon\widetilde{\lambda_m} = -\frac{(a(s)u_m'(s))^2}{K_A}, \qquad \text{m=0,1,2,....} \tag{2.15}$$

Inequalities (2.9), (2.11) and (2.15) show that the eigenvalues change with the damage. In particular, the change in a natural frequency produced by a single crack may be expressed as the product of two terms, the first of which is proportional to the severity and the second depends only on the location of the damage. This second term is the square of the axial force

$$N_m(s) \equiv a(s)u_m'(s) \tag{2.16}$$

in the mth mode shape of the undamaged rod, evaluated at the cracked cross-section.

Equation (2.15) has an important consequence: the ratios of the change in two different natural frequencies depend only on the damage location, not on its severity. This fact was probably first observed by Adams et al. (1978). That is

$$\frac{\delta\lambda_n}{\delta\lambda_m} = \left(\frac{N_n(s)}{N_m(s)}\right)^2 \equiv f(s), \tag{2.17}$$

where $s \in (0, L)$ and $\delta\lambda_m < 0$ (if $\delta\lambda_m = 0$ the possible crack locations coincide in the node points of the axial force N_m resulting from the mth vibrating mode of the undamaged rod). A relation similar to (2.17) has been achieved in Narkis (1994) by considering a linearization of the characteristic polynomial explicitly referred to the cracked rod, in case the damaged system should be a perturbation of the undamaged one. Adapting such a method to the analysis of non-uniform beams is likely to be difficult, while the use of the expression (2.17) makes it possible to overcome such hindrance.

The problem related to the crack location from frequency data lies in determining the solutions of the equation (2.17) for a measured value of the ratio $\delta\lambda_n/\delta\lambda_m$. It follows from equation (2.17) that all, and the only possible, locations of the crack are the abscissas of the points of the $f(x) = (N_n(x)/N_m(x))^2$ diagram intersecting with the horizontal straight line drawn parallel to the abscissa axis at a distance equal to the ratio $\delta\lambda_n/\delta\lambda_m$. On a practical level, once the free vibration problem related to the integral rod is solved, the behavior of $f(x)$ is known and therefore it is possible, via numerical method for example, to determine the solutions of equation (2.17).

Some qualitative properties of the crack location problem for initially non-uniform rods were investigated in detail in Morassi (2001). In brief, it was shown that the problem of locating a crack in a vibrating rod from knowledge of the damage-induced shifts in a pair of natural frequencies is generally ill-posed: if the system is symmetrical, then a crack at any one of a set of symmetrically placed points will produce identical changes in natural frequencies. Even if the system is not symmetrical, cracks in different locations can still produce identical changes in a pair of natural frequencies. In spite of this ill-posedeness of the diagnostic problem, it will be shown in the remainder of this Section that there are certain situations in which the effects of the non-uniqueness of the solution may be considerably reduced by means of a careful choice of the data.

The simple but very common case of uniform rods, e.g. rods for which $a = a(x)$ and $\rho = \rho(x)$ are constants on the interval $(0, L)$ will be considered here. The most exhaustive results concern free vibrating rods (F). C_m^F is denoted by

$$C_m^F \equiv -\frac{\delta\lambda_m^F}{Bm^2}, \tag{2.18}$$

where m is a positive integer and B is the constant

$$B = \left(a\sqrt{\frac{2}{\rho L}}\frac{\pi}{L}\right)^2. \tag{2.19}$$

It can be proved that, *if $C_m^F > 0$, the measurement of the pair $\{C_m^F, C_{2m}^F\}$, $m \geq 1$, uniquely determines the severity of the damage, e.g. the spring stiffness K_A, and the variable $S = \cos(2m\pi s/L)$ of the damage location s.*

The eigenpairs of a free uniform rod (F) are given by

$$\lambda_m^F = \frac{a}{\rho}\left(\frac{m\pi}{L}\right)^2, \qquad u_m^F(x) = \sqrt{\frac{2}{\rho L}}\cos(\frac{m\pi x}{L}), \tag{2.20}$$

$m = 0, 1, 2, \ldots$. The rigid mode $u_0^F(x)$ obviously is always insensitive to damage. Putting the expressions of λ_m^F and $u_m^F(x)$ for $m \geq 1$ into equation (2.15) gives

$$C_m^F = \frac{1}{K_A}\sin^2(\frac{m\pi s}{L}) \tag{2.21}$$

and, by using standard trigonometric identities, gives

$$K_A(4C_m^F - C_{2m}^F) = 4K_A^2(C_m^F)^2. \tag{2.22}$$

Since $C_1^F > 0$, from the identity above it follows that

$$4C_1^F - C_2^F > 0, \tag{2.23}$$

and

$$\text{if} \quad C_m^F > 0 \quad \text{then} \quad 4C_m^F - C_{2m}^F > 0, \quad \text{for m} \geq 2. \tag{2.24}$$

Let it be assumed that $C_m^F > 0$. Equation (2.22) can be solved for the damage severity

$$K_A = \frac{1 - \frac{C_{2m}^F}{4C_m^F}}{C_m^F}. \tag{2.25}$$

Note that conditions (2.23), (2.24) guarantee that K_A takes positive values. By inserting the expression (2.25) of K_A into equation (2.21) the damage can be localized

$$S = \frac{C_{2m}^F}{2C_m^F} - 1, \tag{2.26}$$

where $S \in [-1, 1)$ because of inequalities (2.23), (2.24). Note that the ratio of changes in the first two natural frequency is sufficient to localize the damage (except for symmetrical positions). Finally, if $C_m^F = 0$ for a certain $m \geq 2$, then from equation (2.21) it follows that $S = 1$, that is the crack is located in one of the points of zero-sensitivity of the mth mode, and K_A remains undetermined. This establishes the assertion.

The preceding result improves known results about crack localization in different directions. One of the results by Narkis (1994) is a particular case concerning the unique localization of the damage based on the knowledge of the first two frequencies of a free-free vibrating rod.

The expressions (2.25) and (2.26) for the damage parameters indicate that the pair of natural frequencies mth and $2m$th plays a crucial role when localizing the damage. In fact, provided that the mth frequency proves to be sensitive to damage, that is $C_m^F > 0$, the pair $\{C_m^F, C_{2m}^F\}$ uniquely determines the damage severity, namely the stiffness K_A. Quite surprisingly the expression for K_A turns out to be the same for all pairs of values $\{C_m^F, C_{2m}^F\}$. Finally, it is shown that the number of the possible crack locations, corresponding to the same ratio C_{2m}^F/C_m^F, increases as the order m of the modes assessed increases, which accounts for the recourse to "low" frequencies for the problem of damage localization.

The preceding result does not consider the problem that occurs when a pair of values such as $\{C_k^F, C_l^F\}$, with $l \neq 2k$, is chosen as data. It can be shown that in these cases the solution of the inverse problem is generally non-unique (even by leaving symmetrical positions aside), see Morassi (2001) for more details.

The analysis has hitherto been related to uniform beams under axial vibration with free ends and it has been borne out that the first two frequencies allow the crack (except for symmetrical positions) to be uniquely identified. Such a result does not prove true if different boundary conditions, for example (C) or (S), are being considered. For instance, in case (C), when the crack is located in the half of the rod adjacent to the fixed end, there are two distinct locations corresponding to the same ratio between the variations in the first two frequencies, as already asserted by Narkis (1994). In these situations, an alternative method of proceeding lies in resorting at the same time to frequency measurements on the cracked rod which derive from different boundary conditions.

The attention is concentrated again on initially uniform rods and the data resulting from boundary conditions of the types (S) and (F) are first considered. It can be proved that *the measurement of the $(m+1)$th frequency in the cracked rod under boundary conditions of the type (F) and of the mth frequency under boundary conditions of type (S), for $m \geq 0$, determines uniquely the severity of the crack and the location variable $S = cos(2(m+1)\pi s/L)$, where s stands for the abscissa of the cracked cross-section.* Defining $C_m^S \equiv -\delta\lambda_m^S/B(m+1)^2$, $m = 0, 1, 2, \ldots$, it follows that

$$K_A = \frac{1}{C_{m+1}^F + C_m^S}, \qquad S = -1 + \frac{2}{1 + \frac{C_{m+1}^F}{C_m^S}}. \tag{2.27}$$

(If $C_m^S = 0$ then $S' = -1$.) In particular, it turns out that the damage is uniquely determined (except for symmetrical positions) by the measurement of the pair $\{C_0^S, C_1^F\}$.

A similar result holds true also when frequency measurements are derived from (F) and (C) boundary conditions. In this case it can be shown that *from the knowledge of the mth frequency in the cracked rod under boundary conditions (C) and of the $(1+2m)$th frequency under boundary conditions (F) it is possible to uniquely determine the stiffness K_A and the position variable $S = cos((1+2m)\pi s/L)$* , $m = 0, 1, 2, \ldots$. By proceeding as exemplified above, if $C_m^S > 0$ then it follows that

$$K_A = \frac{1 - \frac{C_{1+2m}^F}{4C_m^C}}{C_m^C}, \qquad S = 1 - \frac{C_{1+2m}^F}{2C_m^C}, \tag{2.28}$$

for $m = 0, 1, 2, \ldots$.

2.2 Bending Vibrating Beams

In the previous section the inverse problem of identifying a crack in an axially vibrating beam from frequency measurements has been discussed. Here a cracked beam in bending will be considered. To start, a simply supported uniform Euler-Bernoulli beam with an open crack at the cross-section of abscissa s is considered. According to Freund and Herrmann (1976) the crack is represented by the insertion of a massless rotational elastic spring at the damaged cross-section. The stiffness K_B of the spring may be related in a precise way to the geometry of the damage. Denoting as usual the Young's modulus of the material by E and the linear mass-density by ρ, the mth eigenpair $\{w_{dm}(x), \lambda^S_{dm}\}$ of the bending vibrations of the cracked beam satisfies the following boundary value problem

$$(EIw_d'')'' = \lambda_d^S \rho w_d \quad \text{in } (0,s) \cup (s,L), \tag{2.29}$$

$$w_d = 0, \quad w_d'' = 0 \quad \text{at } x = 0 \text{ and } x = L, \tag{2.30}$$

where the jump conditions

$$[w_d(s)] = [w_d''(s)] = [w_d'''(s)] = 0, \qquad EIw_d''(s) = K_B[w_d'(s)], \tag{2.31}$$

hold at the cross-section where the crack occurs. In the equations above I and A represent the moment of inertia and the area of the cross-section, respectively.

If the crack is small, namely K_B is large enough, on proceeding as in Section 2.1 and with the above notation, the first order variation of the mth eigenvalue with respect to $1/K_B$ is given by

$$\delta\lambda_m^S = -\frac{(M_m^S(s))^2}{K_B}, \qquad \text{m=0,1,2,....,} \tag{2.32}$$

where $M_m^S(s) \equiv -EIw_m''(s)$ is the bending moment at the cross-section of abscissa s in the mth (normalized) bending mode of the undamaged beam.

At this stage the problem of identifying the position and severity of the crack from the knowledge of the changes in a pair of natural frequencies can be posed. In the above mentioned paper by Narkis (1994), it was shown that knowledge of the first two frequency changes induced by the damage suffices to identify uniquely the crack location (except for symmetrical positions). Here an improvement of such result is presented. Denote by C_m^S the quantity

$$C_m^S \equiv -\frac{\delta\lambda_m^S}{B(1+m)^2}, \tag{2.33}$$

where $m \geq 0$ and B is the constant

$$B = \left(EI\sqrt{\frac{2}{\rho L}}\left(\frac{\pi}{L}\right)^2\right)^2. \tag{2.34}$$

It can be proved that, *if* $C_m^S > 0$*, the measurement of the pair* $\{C_m^S, C_{2m}^S\}$, $m \geq 0$, *determines uniquely the severity of the damage, e.g. the spring stiffness* K_B*, and the variable* $S = cos(2(m+1)\pi s/L)$ *of the damage location* s.

A result similar to what attained at the end of Section 2.1 is valid for cracked beams in bending as well, namely *the measurement of the* m*th frequency of the cracked beam under simply supported boundary conditions* (S) *and of the* $(m+1)$*th frequency for sliding-sliding boundary conditions* (Sl) *(e.g.,* $w' = w''' = 0$ *at* $x = 0$ *and* $x = L$*) determines uniquely the stiffness* K_B *of the rotating spring and the position variable* $S = cos(2(m+1)\pi s/L)$.

Defining $C_m^{Sl} \equiv -\delta\lambda_m^{Sl}/Bm^4$ and keeping the conventional meaning of the symbols, $m = 0, 1, 2$, it is possible to attain the following expressions for the damage parameters

$$K_B = \frac{1}{C_{m+1}^{Sl} + C_m^S}, \qquad S = -1 + \frac{2}{1 + \frac{C_m^S}{C_{m+1}^{Sl}}}. \tag{2.35}$$

The latter expression is valid if $C_{m+1}^{Sl} > 0$. If $C_{m+1}^{Sl} = 0$ then $S = 1$.

3 Crack Identification from Changes in Node Positions

3.1 Axially Vibrating Beams

In this Section an investigation of the effect of a crack on the nodes of free axial vibration modes of a beam is presented. The spatial variation of the infinitesimal undamped free vibration of the undamaged and cracked rod are governed by the eigenvalue problems (2.1)–(2.3) and (2.7), (2.2), (2.3), (2.8), respectively.

The eigenfunctions of the undamaged system exist only for certain values of λ, the eigenvalues. Denote a generic eigenvalue by λ_0. However, for any (real) values of λ it may be found the solution, $\vartheta(x) = \vartheta(x, \lambda)$, of (2.1) subject only to (2.2). The solution will be determined apart from an arbitrary multiplicative constant which, if necessary, may be chosen to satisfy some norming condition. Similarly, the solution $\varphi(x) = \varphi(x, \lambda)$ of (2.1), (2.3) may be found. The eigenvalues λ_0 of the undamaged system are the values of λ for which $\vartheta(x, \lambda)$, $\varphi(x, \lambda)$ are multiples of each other, i.e. $\vartheta(x, \lambda_0) = c\varphi(x, \lambda_0)$, where $c \neq 0$ is a real constant.

Again, the eigenfunctions of the damaged rod will exist only for certain values of λ, the eigenvalues of the damaged system. To the left of the damaged section, the eigenfunction $u_d(x)$ will be proportional to $\vartheta(x, \lambda)$, and to the right it will be proportional to $\varphi(x, \lambda)$. Thus to see how the nodes of the eigenfunctions to the left of s change, the zeros of $\vartheta_0(x, \lambda_0) = \vartheta_0(x)$ and $\vartheta(x, \lambda) = \vartheta(x)$ can be compared; and to see how the nodes to the right of s change, the zeros of $\varphi(x, \lambda_0)$ and $\varphi(x, \lambda)$ can be compared. *It will be proved that nodes move toward the damaged location.* To show this, adaptations of two Sturm's theorems are needed.

Lemma 3.1. *Let $\vartheta_0(x)$, $\vartheta(x)$ be solutions of* (2.1) *with parameter λ_0 and λ, respectively, where $\lambda_0 > \lambda$. Between two consecutive zeros of $\vartheta(x)$, there is at least one zero of $\vartheta_0(x)$.*

Lemma 3.2. *Let $\vartheta_0(x)$, $\vartheta(x)$ be the solutions of* (2.1), (2.2) *with parameter λ_0 and λ, respectively, where $\lambda_0 > \lambda$. If $\vartheta(x)$ has m zeros in an interval $(0, b]$, $\vartheta_0(x)$ has at least n zeros in the same interval, and the mth zero of $\vartheta_0(x)$ is less than the nth zero of $\vartheta(x)$.*

The analysis of Section 2.1 shows that the damage cannot increase the natural frequencies of the system. *If the damage actually does decrease a natural frequency, and this will certainly happen if $u'(s) \neq 0$ for that mode, then, from Lemma 3.1 and Lemma 3.2, the nodes of the modes to the left of the damage, at s, like the zeros of $\vartheta(x)$, will move to the right. A similar argument shows that the nodes to the right of s will move to the left.* This means that for every principal mode, which has at least two nodes, there is exactly one neighboring pair which move toward each other, and the crack is located between them.

If the damage is small, i.e. $\epsilon = 1/K_A$ is small enough, the first order change in the position of the nodes with respect to ϵ may be determined. To do this, the first order change in the zeros of $\vartheta(x)$ to the left of s will be estimated. If x_0 is a zero of ϑ_0, then, to first order in ϵ, the corresponding zero of ϑ will be $x = x_0 + \epsilon\widetilde{x}$. By using a perturbation argument it can be shown that the change in position of the nodes is given by

$$\epsilon\widetilde{x} = \frac{\epsilon(a(s)u_0'(s))^2 \int_0^{x_0} \rho u_0^2}{a(x_0)(u_0'(x_0))^2}, \tag{3.1}$$

where u_0 is the eigenfunction of the undamaged rod. The corresponding result for the change in position of nodes to the right of s is

$$\epsilon\widetilde{x} = \frac{-\epsilon(a(s)u_0'(s))^2 \int_{x_0}^{L} \rho u_0^2}{a(x_0)(u_0'(x_0))^2}. \tag{3.2}$$

Experimentally, nodal positions are easier to measure than are mode shapes; in fact, their require, roughly speaking, just a detection of modal component sign, rather than a set of measurements of magnitudes, see Gladwell and Morassi (1999) for some experimental applications.

3.2 Bending Vibrating Beams

The result obtained in previous Section left an important question unsolved, and namely the capability to extend the identification procedure to cracked beams in bending vibration. Although desirable for the application of diagnostic methods, the existence of a similar property is all but obvious. In fact, the monotonicity property shown in Section 3.1 is a consequence of the oscillation and separation theorems for solutions of the Sturm-Liouville operator governing the axial vibration of a thin rod. General properties of this kind become more involved for the fourth-order differential operator of the bending vibrations of a beam, see, for instance, the classical paper by Leighton and Nehari (1958). In particular, on adapting the Comparison Theorem 5.1 of Leighton and Nehari (1958), one can show that the bending analogue of the monotonicity property found for the second-order case involves conjugate points (associated to one end of the beam) and crack location, that is *conjugate points move toward the crack*. Conjugate points are points of the beam axis given by the intersection of two special solutions of the differential operator governing the bending vibrations. Unfortunately, unlike nodal points, conjugate points lack a clear physical interpretation; therefore, since their variations cannot be measured, damage location cannot be deduced. Considering the effect of a crack on the nodal points of flexural modes, the insertion of a crack in some areas of the beam axis was found to cause some nodes to shift away from the damaged area. This effect does not apply to axial modes. Despite no monotonicity property is found to exist between nodes and crack location, a parametrical analysis of the bending problem proved useful to characterize the areas of the beam axis where a crack causes nodes always shift along the same direction, for varying damage severity, in originally uniform beams with general boundary conditions. In the examined cases, the signs of nodal shifts in the first three/four vibration modes proved essential to assess crack location with good accuracy, see Dilena and Morassi (2002a) for more details.

4 Extensions

4.1 Crack Identification from Antiresonance Measurements

Cracks at any one of a set of symmetrically placed points of a symmetrical structure produce identical changes to natural frequencies; so, as it was shown in Section 2.1, the measurement of a (suitable) pair of natural frequencies cannot eliminate symmetrical solutions in the damage location problem. Here it will be shown how an appropriate use of natural frequencies and antiresonance frequencies may overcome the non-uniqueness of the damage location problem due to structural symmetry.

To show this, attention will be focused on axially vibrating beams. The flexural case and some experimental applications are fully discussed in Dilena and Morassi (2003b). To start, a free uniform rod with a small open crack at the cross-section of abscissa s is considered. First order variations of natural frequencies with respect to the damage were derived in Section 2.1, equation (2.18). Antiresonances correspond to zeros of frequency response functions (frf) $H(\sqrt{\lambda}, x_i, x_o)$, where x_i, x_o are the abscissas of the excitation point and measurement point, respectively. When $x_i = x_o$, the zeros of the frf $H(\sqrt{\lambda}, x_i, x_i)$ are the frequencies of a beam in which the longitudinal displacement at the cross-section of abscissa x_i is hindered. In particular, if $x_i = x_o = 0$, then the antiresonances of the frf $H(\sqrt{\lambda}, 0, 0)$ are the square root of the eigenvalues λ_m^C of the cantilever rod (2.6). Therefore, under the assumption of small crack, namely K_A large enough, on proceeding as in Section 2.1 and with the same notation, the first order variation of the (square of the) mth antiresonance of the frf $H(\sqrt{\lambda}, 0, 0)$ with respect to $1/K_A$ may be evaluated by equation (2.15).

Denoting $C_m^C \equiv -\delta\lambda_m^C/B(1+2m)^2$, with $B = (a\pi\sqrt{2/\rho L}/2L)^2$, $m = 0, 1, 2, ...$, it turns out that $C_m^C = cos^2((1+2m)\pi s/2L)/K_A$ and therefore equation (2.28) gives a closed form expression of the stiffness K_A and of the position variable $S = cos((1+2m)\pi s/L)$ of the damage location s. Considering $m = 0$, *it turns out that the damage is uniquely determined by the measurement of the first resonance of the rod and the first antiresonance of the frf* $H(\sqrt{\lambda}, 0, 0)$. It is worth noticing that the required data can be extracted from the frf measurement of $H(\sqrt{\lambda}, 0, 0)$ only.

A similar result holds true also for a uniform cantilever rod with a small open crack at the cross-section of abscissa s. In this case, the antiresonances of the frf $H(\sqrt{\lambda}, L, L)$ are the square roots of the eigenvalues λ_m^S of the supported rod (2.4). Denoting $C_m^S \equiv -\delta\lambda_m^S/Bm^2$, with $B = (a\pi\sqrt{2/\rho L}/L)^2$, $m = 1, 2, ...$, it turns out that knowledge of the mth natural frequency of the cracked cantilever and of the $(1+2m)$th antiresonance of the frf $H(\sqrt{\lambda}, L, L)$ suffice to uniquely determine the severity of the damage and the position variable $S = cos((1+2m)\pi s/L)$, $m = 0, 1, 2, ...$,

$$K_A = \frac{1}{2C_m^C - C_{1+2m}^S}, \qquad S = \frac{1}{1 - \frac{C_{1+2m}^S}{2C_m^C}} - 1. \tag{4.1}$$

In particular, *for* $m = 0$, *the damage is uniquely determined from the first resonance of the cantilever and the first antiresonance of the frf* $H(\sqrt{\lambda}, L, L)$.

4.2 Multiple cracks

The physical model, which will be mainly investigated in this Section, is a simply supported uniform Euler-Bernoulli beam with two cracks of equal severity located at cross sections of abscissa s_1 and s_2, with $0 < s_1 < s_2 < L$, where L is the beam length.

Assuming that cracks remain always open during the flexural vibration, the infinitesimal bending vibrations of the cracked beam are governed by the eigenvalue value problem (2.29), (2.30), where the jump conditions (2.31) hold at the cross sections of abscissa $x = s_1$ and $x = s_2$ where cracks occur. As in Section 2.2, every crack is represented by inserting a massless rotational spring of stiffness K_B.

If cracks are small, namely $\epsilon \equiv 1/K_B$ is small enough, then the first order variation of the natural frequencies with respect to ϵ may be evaluated. On proceeding as in Section 2.2 and with the above notation one has

$$\delta\lambda_m^S = -\frac{(M_m(s_1))^2}{K_B} - \frac{(M_m(s_2))^2}{K_B}, \qquad \text{m=1,2,...,} \tag{4.2}$$

where $M_m(x) \equiv -EIw_m''(x)$ is the bending moment at the cross-section of abscissa x in the mth (normalized) bending mode of the undamaged beam.

Since only three parameters need to be determined, namely s_1, s_2 and K_B, it will be investigated to what extent the measurement of the first three natural frequencies may be useful for identifying the damage.

The system is symmetrical with respect to $x = L/2$ and therefore a crack located at any one of a set of symmetrically placed points will produce identical changes in natural frequencies. Without affecting the character of generality of the analysis, it can be assumed that $0 < s_1 < s_2 \le L/2$. By inserting the expressions of the eigenmodes $w_m(x)$ of the undamaged beam, for m=1,2,3, into equation (4.2) and by using standard trigonometric identities, the following system of three non-linear equations is obtained

$$\alpha + \beta = 2 - 2KC_1^S, \quad \alpha^2 + \beta^2 = 2 - KC_2^S, \quad \alpha^3 + \beta^3 = 2 - \frac{K}{2}(3C_1^S + C_3^S), \tag{4.3}$$

where

$$\alpha \equiv \cos\left(\frac{2\pi s_1}{L}\right) \in [-1, 1), \quad \beta \equiv \cos\left(\frac{2\pi s_2}{L}\right) \in [-1, 1) \tag{4.4}$$

and

$$C_m^S \equiv -\frac{\delta\lambda_m^S}{\frac{2}{\rho L}\lambda_m^S}, \qquad \text{m=1,2,3.} \tag{4.5}$$

Since $0 < s_1 < s_2 \le L/2$, it turns out that $C_m > 0$ (to simplify notation, the superscript S will be omitted) and functions $\alpha = \alpha(s_1)$, $\beta = \beta(s_2)$ are one-to-one correspondences, with $\alpha \neq \beta$. By using algebraic identities in equations (4.3), the following polynomial equation in the unknown K_B follows

$$K_B\left(4C_1^3K_B^2 + 3C_1(C_2 - 4C_1)K_B + \left(\frac{15}{2}C_1 - 3C_2 + \frac{C_3}{2}\right)\right) = 0. \tag{4.6}$$

Apart from the trivial solution $K_B = 0$, which clearly does not correspond to the assumption of small damage, it can be proved that the polynomial of second degree enclosed in square brackets in equation (4.6) always has two real positive (possibly equal) roots for every set of data $\{C_1, C_2, C_3\}$, see Morassi and Rollo (2001), that is there exist two values of the stiffness K_B of the spring simulating the damage

$$K_{B1,2} = \frac{1}{8C_1^2}\left[-3(C_2-4C_1) \pm \left(9(C_2-4C_1)^2 - 16C_1\left(\frac{15}{2}C_1 - 3C_2 + \frac{C_3}{2}\right)\right)^{\frac{1}{2}}\right], \tag{4.7}$$

where indexes 1 and 2 correspond to + sign and - sign respectively.

By inserting the expression (4.7) of K_B into first two equations of (4.3), the location of the damage may be determined. Note that the role of variables α and β in system (4.3) is completely interchangeable, namely, if the ordinate triple $\{K, \alpha, \beta\}$ is a solution of the diagnostic problem, also $\{K, \beta, \alpha\}$ is a solution. Then, it is enough to determine the position variable α. By using equations (4.3) the following polynomial equation of second degree in the variable α is obtained

$$g(\alpha) \equiv 2\alpha^2 - 2\alpha(2 - 2K_BC_1) + (2 - 2K_BC_1)^2 - (2 - K_BC_2) = 0. \tag{4.8}$$

It was shown in Morassi and Rollo (2001) that the polynomial $g(\alpha)$ in (4.8) always has two distinct real roots α_1, α_2 belonging to the interval $[-1, 1)$

$$\alpha_{1,2} = \frac{1}{2}\left[(2 - 2K_BC_1) \pm \left(-4K_B^2C_1^2 + 8K_BC_1 - 2K_BC_2\right)^{\frac{1}{2}}\right] \tag{4.9}$$

where indexes 1 and 2 represent + sign and - sign before the square root respectively.

Finally, the complete set of solutions of the system (4.3) with reference to cracks located on the left half of the beam is given by $\{K_{B1}, s_1(K_{B1}), s_2(K_{B1})\}$, $\{K_{B2}, s_1(K_{B2}), s_2(K_{B2})\}$. That is, two cracks of same severity K_{B1} (evaluated via expression (4.7) with + sign) located at the cross-sections of abscissa $s_1(K_{B1})$, $s_2(K_{B1})$ (evaluated via expressions (4.9) and (4.4)) produce changes in the first three natural frequencies identical to those induced by two cracks of same severity K_{B2} (expression (4.7) with - sign) located at the cross-sections of abscissa $s_1(K_{B2})$, $s_2(K_{B2})$ (expressions (4.9) and (4.4)).

The present method can be adapted to identify two small cracks of the same severity in an axially vibrating beam with free ends, see Morassi and Rollo (2001).

4.3 Crack Identification in Dissipative Systems

Consider a thin straight elastic rod of length L with a crack at the cross section of abscissa $s \in (0, L)$. The beam has constant cross-section of area A, Young's modulus E and uniform linear mass density ρ. For definiteness, the left end of the beam is assumed to be free, whereas its right end is constrained by a lumped linear damper of viscous damping coefficient $c > 0$. Assuming that the crack remains always open during motion and modelling it as a massless translational spring at $x = s$ of stiffness K_A, the infinitesimal free longitudinal vibrations of the damaged rod (F-D) are governed by the following dimensionless eigenvalue problem

$$u_d'' - \lambda_d u_d = 0 \qquad \text{in } (0,\sigma) \cup (\sigma, 1), \tag{4.10}$$

$$u'_d(0) = 0, \qquad u'_d(1) + \gamma\sqrt{\lambda_d}u_d(1) = 0, \tag{4.11}$$

$$[u_d(\sigma)] = \epsilon u'_d(\sigma), \qquad [u'_d(\sigma)] = 0, \tag{4.12}$$

where $u_d = u_d(z)$ and where the dimensionless quantities z, σ, γ and ϵ are defined as follows

$$z = \frac{x}{L}, \quad \sigma\frac{s}{L}, \quad \gamma = c(EA\rho)^{-\frac{1}{2}}, \quad \epsilon = \frac{EA}{K_A L}. \tag{4.13}$$

The undamaged system corresponds to $K_A \to \infty$ or $\epsilon \to 0$. Under the assumption that $\gamma < 1$, the mth eigenpair $\{u_m, \lambda_m^2\}$ of the undamaged rod is given by $\sqrt{\lambda_m} = \ln\left((1-\gamma)/(1+\gamma)\right)/2 + im\pi \equiv \xi + im\pi$, $u_m(x) = A_m\cosh(\sqrt{\lambda_m}z)$, m $= 0, \pm1, \pm2, ...$, where i is the imaginary unit and A_m is an arbitrary complex number. For $\gamma > 1$ the mth eigenpair is given by $\sqrt{\lambda_m} = \ln((\gamma-1)/(1+\gamma))/2 + \mathrm{i}(1+2m)\pi/2$, $u_m(z) = A_m\cosh(\sqrt{\lambda_m}z)$, $m = 0, \pm1, \pm2, ...$ Note that if $\gamma = 1$ there is no spectrum at all. To fix ideas, and considering that this is a common situation in practice, the condition $\gamma < 1$ will be assumed hereafter. It can be shown that also for the eigenvalue problem (4.10), (4.11), (4.12) there is an infinite sequence of simple eigenvalues.

Similarly to the ideal undamped case, see Section 2.1, the first order change in an eigenvalue produced by a single small crack may be expressed as the product of two terms, the first of which is proportional to the severity of damage and the second only depends on the location of damage

$$\delta\lambda_m = \epsilon(u'_m(\sigma))^2, \quad \mathrm{m} = 0, \pm1, \pm2, ..., \tag{4.14}$$

where the normalizing condition $\int_0^1 u_m^2 + \gamma u_m^2(1)/2\sqrt{\lambda_m} = 1$ was taken into account. However, concerning the problem of damage identification, there is a substantial difference between the undamped case and the dissipative case considered here. In fact, the mode shape of the undamaged rod is now a complex value function and this fact has a strong influence on the damage detection procedure, as it will be shown in the following.

Again, under the assumption that the undamaged system is completely defined and with damage simulated as above, only the stiffness K_A and the abscissa s of the cracked cross-section need to be determined. Therefore, changes in a suitable pair of eigenvalues will be considered in identifying damage. By introducing the expression of the mth eigenpair of the undamaged rod into equation (4.14) gives

$$C_m = \epsilon\left[\sinh^2(\xi\sigma)\cos^2(m\pi\sigma) - \cosh^2(\xi\sigma)\sin^2(m\pi\sigma) + \mathrm{i}\cdot\sinh(\xi\sigma)\cosh(\xi\sigma)\sin(2m\pi\sigma)\right] \tag{4.15}$$

where $C_m \equiv \delta\lambda_m/\lambda_m$ and $\xi \equiv \mathrm{Re}\sqrt{\lambda_m}$.

In order to identify damage, it is convenient to deal with the imaginary part of C_m's, $m \neq 0$. Let it be assumed that $\mathrm{Im}C_m \neq 0$ for a certain $m \neq 0$. Then, by using standard trigonometric identities, from expression (4.15) it follows that

$$S \equiv \cos(2m\pi\sigma) = \frac{\mathrm{Im}C_{2m}}{2\mathrm{Im}C_m}, \tag{4.16}$$

namely, *if* $\mathit{Im}C_m \neq 0$, *the measurement of the pair* $\{\mathit{Im}C_m, \mathit{Im}C_{2m}\}$, $m \neq 0$, *uniquely determines the variable* $S \equiv \mathit{cos}(2m\pi\sigma)$ *of the normalized damage location* σ. Note that the sign of m does not affect the localization result. Moreover, for those values of σ such that $|S| = 1$, it turns out that $\text{Im}C_m = 0$, and then $S \in (-1, 1)$. Therefore there are exactly $2|m|$ possible crack locations σ_k, $k = 1, ..., 2|m|$, corresponding to the same ratio $\text{Im}C_{2m}/\text{Im}C_m$. These damage locations are symmetrically placed with respect to the mid-point of the rod, namely to every generic crack location σ_k corresponds its symmetric $\sigma_k^{sym} = \sigma_{2|m|-k+1}$, $k = 1, ..., 2|m|$. Note that, for $|m| = 1$, crack location can be uniquely determined from equation (4.16), except for a symmetrical position.

By inserting the expression of a possible damage location, let say σ_k, $k = 1, ..., 2|m|$, into the expression of $\text{Im}C_m$, the corresponding damage severity ϵ_k can be determined

$$\text{Im}C_m = \epsilon_k \sinh(\xi\sigma_k)\cosh(\xi\sigma_k)\sin(2m\pi\sigma_k). \tag{4.17}$$

Since $\sin(2m\pi\sigma_{2|m|-k+1}) = -\sin(2m\pi\sigma_k)$, $k = 1, ..., 2|m| - 1$, and $\xi < 0$, equation (4.17) gives $2|m|$ values of σ_k, half of them are negative, the remaining half being positive. More precisely, $\epsilon_k\epsilon_{k+1} < 0$, $k = 1, ..., 2|m| - 1$. Then, since the only physically plausible values of ϵ_k are the positive ones, half of the possible $2|m|$ damage locations identified from equation (4.16) can be discarded. This fact has an important consequence in the simplest situation where $|m| = 1$, because it implies the unique determination of crack location from the knowledge of ratio $\text{Im}C_2/\text{Im}C_1$. *It is worth noticing that there is no direct analogue of this property in the undamped case.*

Suppose now that $\text{Im}C_m = 0$ for a certain $m \neq 0$. From expression (4.15) it easily follows that $S = 1$, i.e. the possible damage locations are $\sigma_k = k/2|m|$, $k = 1, ..., 2|m| - 1$. Damage severity can be identified by substituting $\sigma = \sigma_k$, $k = 1, ..., 2|m| - 1$, into the expression of the real part of C_m. A direct calculation shows that

$$\text{Re}(C_m) = \frac{\epsilon_k}{2}((-1)^k \cosh(2\xi\sigma_k - 1)). \tag{4.18}$$

The expression within brackets is positive for even k and negative for odd k. Then, $\epsilon_k\epsilon_{k+1} < 0$, $k = 1, ..., 2|m| - 2$, and reasoning as before $(|m| - 1)$ or $|m|$, depending on the case, of the $(2|m| - 1)$ possible damage locations $\sigma_k = k/2|m|$ can be discarded.

The case when both ends of the rod are constrained by a lumped linear damper of equal viscous coefficient c can be investigated by a similar procedure. For clamped-damped boundary conditions, a quite common situation in the applications, the simultaneous use of eigenvalues related to different boundary conditions enables for the identification of the crack. Finally, it can be shown that similar results hold true for the case $\gamma > 1$, see Dilena and Morassi (2003a).

Bibliography

R.D. Adams, P. Cawley, C.J. Pye, and B.J. Stone. A Vibration Technique for Non-Destructively Assessing the Integrity of Structures. *Journal of Mechanical Engineering Science*, 20:93–100, 1978.

E. Cabib, L. Freddi, A. Morassi, and D. Percivale. Thin Notched Beams. *Journal of Elasticity*, 64:157–178, 2001.

M.N. Cerri and F. Vestroni. Detection of Damage in Beams Subjected to Diffused Cracking. *Journal of Sound and Vibration*, 234:259–276, 2000.

C. Davini, A. Morassi, and N. Rovere. Modal Analysis of Notched Bars: Tests and Comments on the Sensitivity of an Identification Technique. *Journal of Sound and Vibration*, 179:513–527, 1995.

M. Dilena and A. Morassi. Identification of Crack Location in Vibrating Beams from Changes in Node Positions. *Journal of Sound and Vibration*, 255:915–930, 2002.

M. Dilena and A. Morassi. Detecting Cracks in a Longitudinally Vibrating Beam with Dissipative Boundary Conditions. *Journal of Sound and Vibration*, 2003 (to appear).

M. Dilena and A. Morassi. The Use of Antiresonances for Crack Detection in Beams. *Journal of Sound and Vibration*, 2003 (to appear).

A.D. Dimarogonas and S.A. Paipetis. *Analytical Methods in Rotor Dynamics*, London: Applied Science, 1983.

L.B. Freund and G. Herrmann. Dynamic Fracture of a Beam or Plate in Plane Bending. *Journal of Applied Mechanics*, 76:112–116, 1976.

G.M.L. Gladwell and A. Morassi. Estimating Damage in a Rod from Changes in Node Positions. *Inverse Problems in Engineering*, 7:215–233, 1999.

G. Hearn and R.B. Testa. Modal Analysis for Damage Detection in Structures. *ASCE Journal of Structural Engineering*, 117:3042–3063, 1991.

W. Leighton and Z. Nehari. On the Oscillation of Solutions of Self-Adjoint Linear Differential Equations of the Fourth Order. *Transactions of the American Mathematical Society*, 89:325–377, 1958.

R.Y. Liang, J. Hu, and F. Choy. Theoretical Study of Crack-Induced Eigenfrequency Changes on Beam Structures. *ASCE Journal of Engineering Mechanics*, 118:384–396, 1992.

A. Morassi. Crack-Induced Changes in Eigeparameters of Beam Structures. *ASCE Journal of Engineering Mechanics*, 119:1798–1803, 1993.

A. Morassi. Identification of a Crack in a Rod Based on Changes in a Pair of Natural Frequencies. *Journal of Sound and Vibration*, 242:577–596, 2001.

A. Morassi and M. Rollo. Identification of Two Cracks in a Simply Supported Beam from Minimal Frequency Measurements. *Journal of Vibration and Control*, 7:729–739, 2001.

A. Morassi and N. Rovere. Localizing a Notch in a Steel Frame from Frequency Measurements. *ASCE Journal of Engineering Mechanics*, 123:422–432, 1997.

Y. Narkis. Identification of Crack Location in Vibrating Simply Supported Beams. *Journal of Sound and Vibration*, 172:549–558, 1994.

M.-H.H. Shen and J.E. Taylor. An Identification Problem for Vibrating Cracked Beams. *Journal of Sound and Vibration*, 150:457–484, 1991.

J.K. Sinha, M.I. Friswell, and S. Edwards. Simplified Models for the Location of Cracks in Beam Structures Using Measured Vibration Data. *Journal of Sound and Vibration*, 251:13–38, 2002.

F. Vestroni and D. Capecchi. Damage Evaluation in Cracked Vibrating Beams Using Experimental Frequencies and Finite Element Models. *Journal of Vibration and Control*, 2:69–86, 1996.

Formulation and identification problems for cracked orthotropic materials

Patrizia Ricci[‡] , Erasmo Viola[‡], Aldino Piva[†], Lucio Nobile[‡]
[‡] DISTART-Department, Faculty of Engineering, Bologna
[†] Department of Physics, University of Bologna, Bologna

Abstract The purpose of the present paper is to identify the elastic parameters of an orthotropic material from measured displacements of crack edges. This paper consists of two parts. In the first part, the expressions for stresses and displacements for plane strain of crack edges in an orthotropic material are derived through the complex variable approach. Mode I crack problems on three mutually orthogonal planes are considered and solved in a unified form. In the second part, an iterative statistical identification method, based on Taylor expansion of experimental data in the neighbourhood of initial parameters, is used to identify the elastic constants of an orthotropic medium from the Mode I crack displacements measured from the mid-point of the crack. This method is based on a Bayesian approach. The convergence of the identified parameters and the approach of the analytical displacements to their experimental values is illustrated. Some graphics are provided to illustrate the sensitivity of the solution with the initial values parameters.

1 Introduction

Inverse problems have been extensively treated in several branches of science. The approach is to determine the optimal estimates of the model parameters by minimizing a selected measure-of-fit between the responses of the system and the model.

Numerical problems arising in inverse elasticity problems were studied (Zabaras 1989 and Maniatty 1994). Zabaras et al. (1989) analysed the solution of a class of inverse linear elastic problems using a spatial regularization method. They used an algorithm which employs a sensitivity analysis and a least-squares minimization of the difference between calculated and measured displacements. Maniatty and Zabaras (1994) presented a technique to estimate the errors in the solution of inverse elasticity problems based on statistical analysis. Worthwhile noting is the theoretical work of Tanuma (1995), in which he calculated a formula for the relative displacements of crack edges in an infinite anisotropic elastic material containing a plane crack. Besides, he considered the inverse problem of determining the elastic constants from measured relative displacements on the crack. More recently, the statistical identification method was applied (Hasan 1998) to identify the elastic constants of an isotropic medium with Mode I displacements due to applied load. Use was made of the Mode I and Mode III displacement data, measured from the mid-point of the crack.

Federici (1999) proposed a statistical approach to identify the elastic constants of an orthotropic material from the measured Mode I crack edge displacements, extending the work in (Hasan 1998) to an orthotropic elastic material. Analytical expressions for Mode I crack edge displacements on three orthogonal planes $x-y$, $y-z$ and $x-z$ of a Cartesian orthogonal system were derived.

The objective of this paper is to obtain analytical expressions of stresses and displacements for plane strain using a complex variable approach solving in a unified form Mode I crack problems on three mutually orthogonal planes. An iterative statistical identification method, sensitivity-based, is used to identify the elastic constants of an orthotropic medium from the Mode I crack displacements measured from the mid-point of the crack. The iterative feature of the method is based on a Taylor series expansion up to the first order of the experimental data in the neighbourhood of the initial parameters. The experimental data are those of Mode I crack edge displacements obtained experimentally. Some graphics illustrate the convergence of the estimated elastic parameters to the respective exact values. The influence of a confidence coefficient on the convergence velocity is also given. It is shown, by the study of sensitivity, that the solution is a function of the initial parameter values.

2 Basic equation for an orthotropic material

Let an orthotropic elastic medium be referred to a Cartesian coordinate system whose axes x, y and z coincide with the three mutually orthogonal principal directions of the material. The elastic constants referring to the system are given in matrix form:

$$C = \begin{bmatrix} c_{11} & c_{12} & c_{13} & 0 & 0 & 0 \\ c_{21} & c_{22} & c_{23} & 0 & 0 & 0 \\ c_{31} & c_{32} & c_{33} & 0 & 0 & 0 \\ 0 & 0 & 0 & c_{44} & 0 & 0 \\ 0 & 0 & 0 & 0 & c_{55} & 0 \\ 0 & 0 & 0 & 0 & 0 & c_{66} \end{bmatrix} \tag{2.1}$$

Let u, v and w be displacement components in Cartesian coordinates. The stress components are represented by the following equations (see Table 1)

$$\sigma_1 = a\frac{\partial u_1}{\partial x_1} + b\frac{\partial u_2}{\partial x_2} \tag{2.2}$$

$$\sigma_2 = b\frac{\partial u_1}{\partial x_1} + c\frac{\partial u_2}{\partial x_2} \tag{2.3}$$

$$\tau_{12} = d\left(\frac{\partial u_1}{\partial x_2} + \frac{\partial u_2}{\partial x_2}\right) \tag{2.4}$$

where u_1, u_2 are the auxiliary components. For plane strain, the equilibrium equations are:

$$a\frac{\partial^2 u_1}{\partial x_1^2} + d\frac{\partial^2 u_1}{\partial x_2^2} + (b+d)\frac{\partial^2 u_2}{\partial x_1 \partial x_2} = 0 \tag{2.5}$$

$$c\frac{\partial^2 u_2}{\partial x_2^2}+d\frac{\partial^2 u_1}{\partial x_1^2}+(b+d)\frac{\partial^2 u_1}{\partial x_1 \partial x_2}=0 \tag{2.6}$$

Plane	x_1	x_2	u_1	u_2	σ_1	σ_2	τ_{12}	a	b	c	d
$(x-y)$	x	y	u	v	σ_x	σ_y	τ_{xy}	c_{11}	c_{12}	c_{22}	c_{66}
$(z-x)$	z	x	w	u	σ_z	σ_x	τ_{zx}	c_{33}	c_{13}	c_{11}	c_{55}
$(y-z)$	y	z	v	w	σ_y	σ_z	τ_{yz}	c_{22}	c_{23}	c_{33}	c_{44}

Table 1. Symbols for the unified formulation of the problem

3 Crack problems

In order to identify the elements of the matrix in Eq.(2.1) of the elastic constants of an orthotropic medium under plane strain, three Mode I crack problems for a line crack of length $2a$ will be considered. They are illustrated in Fig. 1(a)-(c). The problems in Fig.(1) correspond to a crack edge subjected to a uniform tensile stress $T(x_1)$ symmetric compared to the x-axis, such that the stress vanishes at infinity. The boundary value problems corresponding to Fig.(1) may be formulated as:

$$\sigma_2(x_1,0)=T(x_1), \qquad |x_1|\leq l \tag{3.1}$$

$$\tau_{12}(x_1,0)=0, \qquad |x_1|\leq \infty \tag{3.2}$$

$$u_2(x_1,0)=0, \qquad |x_1|>l \tag{3.3}$$

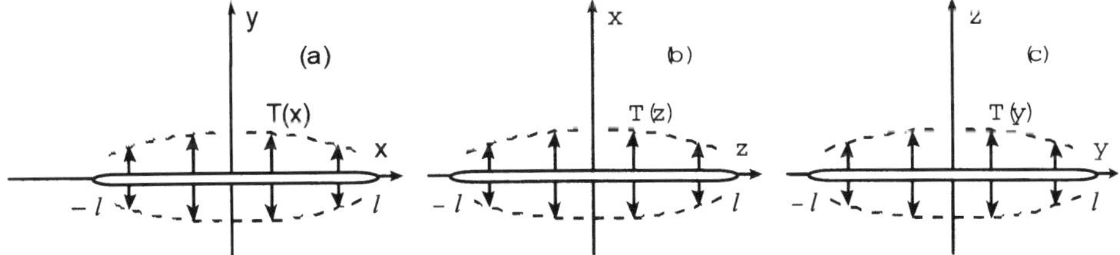

Figure 1. Problems for a linear crack

4 The boundary value problem solutions for complex variable

In terms of complex variables, following the work in Viola (1986) and Piva (1987), it may be shown that the stress and the displacement components are given by (see Table 1):

$$\sigma_1=d\,\mathrm{Im}\left[\lambda_1\Omega_1(z_1)+\lambda_2\Omega_2(z_2)\right] \tag{4.1}$$

$$\sigma_2 = d\,\mathrm{Im}\left[q_1^2\lambda_3\Omega_1(z_1) + q_2^2\lambda_4\Omega_2(z_2)\right] \tag{4.2}$$

$$\tau_{12} = -d\,\mathrm{Re}\left[q_1\lambda_3\Omega_1(z_1) + q_2\lambda_4\Omega_2(z_2)\right] \tag{4.3}$$

and

$$u_1(x_1,x_2) = \frac{\frac{b+d}{a}q_1^2}{\frac{d}{a}-q_1^2}\,\mathrm{Im}\,\omega_1(z_1) + \frac{\frac{b+d}{a}q_2^2}{\frac{d}{a}-q_2^2}\,\mathrm{Im}\,\omega_2(z_2) \tag{4.4}$$

$$u_2(x_1,x_2) = -\,\mathrm{Re}\left[q_1\omega_1(z_1) + q_2\omega_2(z_2)\right] \tag{4.5}$$

where $\Omega_j(z_j)$, $\omega_j(z_j)$ and $\omega_j'(z_j) = \Omega_j(z_j)$ are holomorphic functions of the complex variables $z_j = x + i\frac{y}{q_j}$, $j = 1,2$. The above expressions, in terms of the following expressions $q_j, \lambda_1, \lambda_2, \lambda_3, \lambda_4$, are obtained solving in a unified form Mode I crack problems on three mutually orthogonal planes:

$$q_1 = \left[\frac{1}{2}\left(\frac{d}{a}+\frac{c}{d}-\frac{(b+d)^2}{ad}\right) - \left(\left(\frac{1}{2}\left(\frac{d}{a}+\frac{c}{d}-\frac{(b+d)^2}{ad}\right)\right)^2 - \frac{c}{a}\right)^{\frac{1}{2}}\right]^{\frac{1}{2}}$$

$$q_2 = \left[\frac{1}{2}\left(\frac{d}{a}+\frac{c}{d}-\frac{(b+d)^2}{ad}\right) + \left(\left(\frac{1}{2}\left(\frac{d}{a}+\frac{c}{d}-\frac{(b+d)^2}{ad}\right)\right)^2 - \frac{c}{a}\right)^{\frac{1}{2}}\right]^{\frac{1}{2}}$$

$$\lambda_1 = \frac{\left(\frac{b+d}{a}\right)\left[\frac{1}{2}\left(\frac{d}{a}+\frac{c}{d}-\frac{(b+d)^2}{ad}\right) + \left(\left(\frac{1}{2}\left(\frac{d}{a}+\frac{c}{d}-\frac{(b+d)^2}{ad}\right)\right)^2 - \frac{c}{a}\right)^{\frac{1}{2}}\right]}{\frac{d}{a}\left[\frac{d}{a} - \left(\frac{1}{2}\left(\frac{d}{a}+\frac{c}{d}-\frac{(b+d)^2}{ad}\right) - \left(\left(\frac{1}{2}\left(\frac{d}{a}+\frac{c}{d}-\frac{(b+d)^2}{ad}\right)\right)^2 - \frac{c}{a}\right)^{\frac{1}{2}}\right)\right] + \frac{b}{d}}$$

$$\lambda_2 = \frac{\left(\frac{b+d}{a}\right)\left[\frac{1}{2}\left(\frac{d}{a}+\frac{c}{d}-\frac{(b+d)^2}{ad}\right) + \left(\left(\frac{1}{2}\left(\frac{d}{a}+\frac{c}{d}-\frac{(b+d)^2}{ad}\right)\right)^2 - \frac{c}{a}\right)^{\frac{1}{2}}\right]}{\frac{d}{a}\left[\frac{d}{a} - \left(\frac{1}{2}\left(\frac{d}{a}+\frac{c}{d}-\frac{(b+d)^2}{ad}\right) + \left(\left(\frac{1}{2}\left(\frac{d}{a}+\frac{c}{d}-\frac{(b+d)^2}{ad}\right)\right)^2 - \frac{c}{a}\right)^{\frac{1}{2}}\right)\right] + \frac{b}{d}}$$

$$\lambda_3 = 1 - \frac{\frac{b+d}{a}}{\left(\frac{d}{a} - \left(\frac{1}{2}\left(\frac{d}{a}+\frac{c}{d}-\frac{(b+d)^2}{ad}\right) - \left(\left(\frac{1}{2}\left(\frac{d}{a}+\frac{c}{d}-\frac{(b+d)^2}{ad}\right)\right)^2 - \frac{c}{a}\right)^{\frac{1}{2}}\right)\right)}$$

$$\lambda_4 = 1 - \frac{\frac{b+d}{a}}{\left(\frac{d}{a} - \left(\frac{1}{2}\left(\frac{d}{a} + \frac{c}{d} - \frac{(b+d)^2}{ad}\right) + \left(\left(\frac{1}{2}\left(\frac{d}{a} + \frac{c}{d} - \frac{(b+d)^2}{ad}\right)\right)^2 - \frac{c}{a}\right)^{\frac{1}{2}}\right)\right)}$$

Taking into account the symmetry of the elastic potentials $\Omega_j(z_j)$ with respect to the x-axis, the boundary conditions in Eqs.(3.1)-(3.3) reduce to the following Hilbert problems:

$$\Omega_1^+(x_1) + \Omega_1^-(x_1) = -\frac{2iT}{dq_1\lambda_3(q_1 - q_2)} \qquad |x_1| < l \tag{4.6}$$

$$\Omega_2^+(x_1) + \Omega_2^-(x_1) = -\frac{2iT}{dq_2\lambda_4(q_1 - q_2)} \qquad |x_1| < l \tag{4.7}$$

Under the assumption that displacements are single-valued functions, the solution to the Eqs.(4.6)-(4.7) is given by:

$$\Omega_1(z_1) = \frac{iT}{dq_1\lambda_3(q_1 - q_2)}\left[\frac{z_1}{X(z_1)} - 1\right], \quad \Omega_2(z_2) = \frac{iT}{dq_2\lambda_4(q_1 - q_2)}\left[1 - \frac{z_2}{X(z_2)}\right] \tag{4.8}$$

where $X(z_j) = \sqrt{z_j^2 - l^2}$ are the Plemelj function of the problems. Integrating Eq.(4.8) these result

$$\omega_1(z_1) = -\frac{iT}{dq_1\lambda_3(q_1 - q_2)}G(z_1), \omega_2(z_2) = -\frac{iT}{dq_2\lambda_4(q_1 - q_2)}G(z_2) \tag{4.9}$$

in which $G(z_1) = z_j - \sqrt{z_j^2 - l^2}$, $j = 1, 2$.

5 Displacement discontinuities

Substituting Eq.(4.9) into Eqs.(4.4)-(4.5), respectively, the following expressions of the displacement components are obtained:

$$u_1(x_1, x_2) = \frac{T}{d(q_1 - q_2)}\mathrm{Re}\left[\frac{\frac{b+d}{a}q_2}{\lambda_4\left(\frac{d}{a} - q_2^2\right)}G(z_2) - \frac{\frac{b+d}{a}q_1}{\lambda_3\left(\frac{d}{a} - q_1^2\right)}G(z_1)\right] \tag{5.1}$$

$$u_2(x_1, x_2) = \frac{T}{d(q_1 - q_2)}\mathrm{Im}\left[\frac{G(z_2)}{\lambda_4} - \frac{G(z_1)}{\lambda_3}\right] \tag{5.2}$$

By defining $\Delta u_2(x_1)$ as the displacement discontinuities across the crack for the problems in Fig. 1, it is found that

$$\Delta u_2(x_1) = \frac{2T}{d(q_1 - q_2)}\left(\frac{1}{\lambda_4} - \frac{1}{\lambda_3}\right)\sqrt{l^2 - x_1^2} \tag{5.3}$$

The above-mentioned discontinuities $\Delta v_{(x-y)}$,$\Delta u_{(z-x)}$,$\Delta w_{(y-z)}$ may be calculated at the half-length of the crack and represented as functions of the elastic constants reported in appendix A.

6 The identification method

Let

$$\mathbf{r} = (r_1 r_2 .....r_m)^T \tag{6.1}$$

be a vector of unknown parameters r_i $(i = 1, 2...m)$ that need to be identified and let

$$\mathbf{r}_a = (r_{1a}\ r_{2a}.....r_{ma})^T \tag{6.2}$$

be a vector of the corresponding estimated parameters. The unknown parameters in Eq.(6.2) are also referred to analytical or initial parameters.

Let $y = y\,(r)$ be a vector of regular functions

$$\mathbf{y} = \mathbf{y}\,(r) = (y_1\,(r)\ \ y_2\,(r)\,.....y_m\,(r))^T \tag{6.3}$$

where each element $y_i\,(r)$ is, in general, a function of all parameters in Eq.(6.1).

Furthermore, introduce the vector

$$\mathbf{y}_s = (y_{1s}\ y_{2s}.....y_{ms})^T \tag{6.4}$$

where $y_{is}\,(r)$ represents the experimental determination of the analytical quantities $y_i\,(r)$, for $(i = 1, 2...m)$. By considering the Taylor series expansion of the regular functions $y_i\,(r)$ in the neighbourhood of the estimated parameters and retaining terms up to the first order, the result is

$$\mathbf{y}\,(\mathbf{r}) = \mathbf{y}\,(\mathbf{r}_a) + \left.\frac{\partial \mathbf{y}}{\partial \mathbf{r}}\right|_{r=r_a} (\mathbf{r} - \mathbf{r}_a) \tag{6.5}$$

By appropriate substitutions, Eq.(6.5) leads to

$$\Delta\mathbf{y} = \mathbf{S}\Delta\mathbf{r} \tag{6.6}$$

where $\mathbf{S}$ is the sensitivity matrix. From Eq.(6.6) the vector

$$\widehat{\mathbf{r}} = (\widehat{r}_1 \widehat{r}_2 ...\widehat{r}_m)^T \tag{6.7}$$

of correct parameters is obtained. They are of the form

$$\mathbf{r} = \mathbf{r}_a + \mathbf{H}\Delta\mathbf{y} \tag{6.8}$$

where $\mathbf{r}_a$ is the vector of estimated parameters at the ith iteration; $\widehat{\mathbf{r}}$ the same vector at the (i+1)th iteration; $\mathbf{H}$ is an estimator matrix whose algorithm depends on the identification method adopted.

For the improved statistical identification method, $\mathbf{H}$ assumes the following expression:

$$\mathbf{H} = \beta^{-1}\mathbf{C}_{rr}\mathbf{S}^T\left(\beta^{-1}\mathbf{S}\mathbf{C}_{rr}\mathbf{S}^T + \mathbf{C}_{\varepsilon\varepsilon}\right)^{-1} \tag{6.9}$$

In Eq.(6.9), $\mathbf{C}_{\varepsilon\varepsilon}$ is the diagonal covariance matrix of errors on the measured data, $\mathbf{C}_{rr}$ is the diagonal covariance matrix of the priori parameters and β is a confidence coefficient.

7 Identification of elastic constants of an orthotropic medium

The previously illustrated iterative method of identification is now applied to identify the nine elastic constants for an orthotropic medium from measured Mode I crack displacement discontinuities given in appendix A.

The vector of the elastic parameters is defined as

$$\mathbf{r} = \left(C_{11}C_{22}C_{33}C_{44}C_{55}C_{66}C_{12}C_{13}C_{23}\right)^T \tag{7.1}$$

and the vector of the experimental data is

$$\mathbf{y}_s = \begin{pmatrix} \underset{(x-y)}{\Delta v_s} & \underset{(y-z)}{\Delta w_s} & \underset{(z-x)}{\Delta u_s} \end{pmatrix}^T \tag{7.2}$$

The Taylor series expansions of $\Delta v_{(x-y)}$,$\Delta w_{(y-z)}$,$\Delta u_{(z-x)}$, up to the first order terms, lead to

$$\begin{pmatrix} \underset{(x-y)}{\Delta v_s} \\ \underset{(y-z)}{\Delta w_s} \\ \underset{(z-x)}{\Delta u_s} \end{pmatrix} = \begin{pmatrix} \underset{(x-y)}{\Delta v} \\ \underset{(y-z)}{\Delta w} \\ \underset{(z-x)}{\Delta u} \end{pmatrix} + \mathbf{S}\left(\mathbf{r}^{(i+1)} - \mathbf{r}^{(i)}\right) \qquad i = 0, 1, 2\ldots \tag{7.3}$$

where the matrix $\mathbf{S}$ is given by

$$\mathbf{S} = \left[\frac{\partial\left(\underset{(x-y)}{\Delta v}\right)}{\partial \mathbf{r}} \quad \frac{\partial\left(\underset{(y-z)}{\Delta w}\right)}{\partial \mathbf{r}} \quad \frac{\partial\left(\underset{(z-x)}{\Delta u}\right)}{\partial \mathbf{r}} \right]^T \tag{7.4}$$

From Eq.(7.2) the vector of the estimated parameters at the (i+1)*th* iteration is obtained as follows:

$$\mathbf{r}^{(i+1)} = \mathbf{r}^{(i)} + \mathbf{H}\left(\begin{pmatrix} \underset{(x-y)}{\Delta v_s} \\ \underset{(y-z)}{\Delta w_s} \\ \underset{(z-x)}{\Delta u_s} \end{pmatrix} - \begin{pmatrix} \underset{(x-y)}{\Delta v} \\ \underset{(y-z)}{\Delta w} \\ \underset{(z-x)}{\Delta u} \end{pmatrix}\right) \tag{7.5}$$

where the matrix **H** is given in Eq.(6.9).

8 Results and comments

The identification method is used to identify the elastic constants of Boron-Epoxy which are reported in Table 2. The experimental data correspond to Mode I displacements $\Delta v_{(x-y)}$,$\Delta w_{(y-z)}$,$\Delta u_{(z-x)}$, measured in the mid-point of the crack with length $2a$=80 mm. To simulate the experimental data, a sample of 20 measurements for each of the above displacements due to T= 145 N/mm^2 has been considered and their mean together with the coefficient of variations are reported in Table 3:

Constant	C_{11}	C_{22}	C_{33}	C_{44}	C_{55}	C_{66}	C_{12}	C_{13}	C_{23}
Actual values (N/mm^2)	210345	22298	22298	4137	10342	10342	7012	7012	5750
A priori coefficient of variation	0.0012	0.0015	0.0012	0.004	0.002	0.002	0.018	0.018	0.006

Table 2:Identified parameters for Boron-Epoxy

Experimental data	Mean (mm)	Variation Coefficient
Δv_s $(x-y)$	0.597	0.02
Δw_s $(y-z)$	0.981	0.02
Δu_s $(z-x)$	0.597	0.02

Table 3: Experimental mean values and variation coefficients

Fig.2 shows the influence of the coefficient β on the convergence velocity of the identified parameters through the convergence of the analytical displacements to the experimental displacements. The more the coefficient β decreases, the more the convergence velocity increases. Fig. 2 illustrates the convergence of displacements for three different values of the coefficient β.

The convergence of the analytical displacements to the experimental displacements, when the variation of the only one coefficient Cij is possible, shows that displacements $\Delta v_{(x-y)}$, $\Delta w_{(y-z)}$, $\Delta u_{(z-x)}$ allow identification of some constants only, as shown in Fig. 3 for the elastic coefficients C_{33},C_{44},C_{55} and in Fig. 4 for the elastic coefficient C_{11},C_{12},C_{23} for $\beta = 0.01$. In Fig. 3, the coefficient C_{33} is identified through the Δu and Δw displacements, the coefficient C_{44} is only identified through the Δu displacement as the coefficient C_{55} through the Δw displacement. In Fig. 4, C_{11} is identified through the Δv and Δw displacements and C_{12} through the Δv displacement, but the variation of this displacement is almost constant so the identification is very difficult. The coefficient C_{23} is identified through the Δu displacement. In Fig. 5, the sensitivity diagrams show the elastic coefficients C_{ij} as a function of the displacements Δv, Δu and Δw, for the coefficient $\beta = 1$, $\beta = 0.1$ and $\beta = 0.01$, respectively.

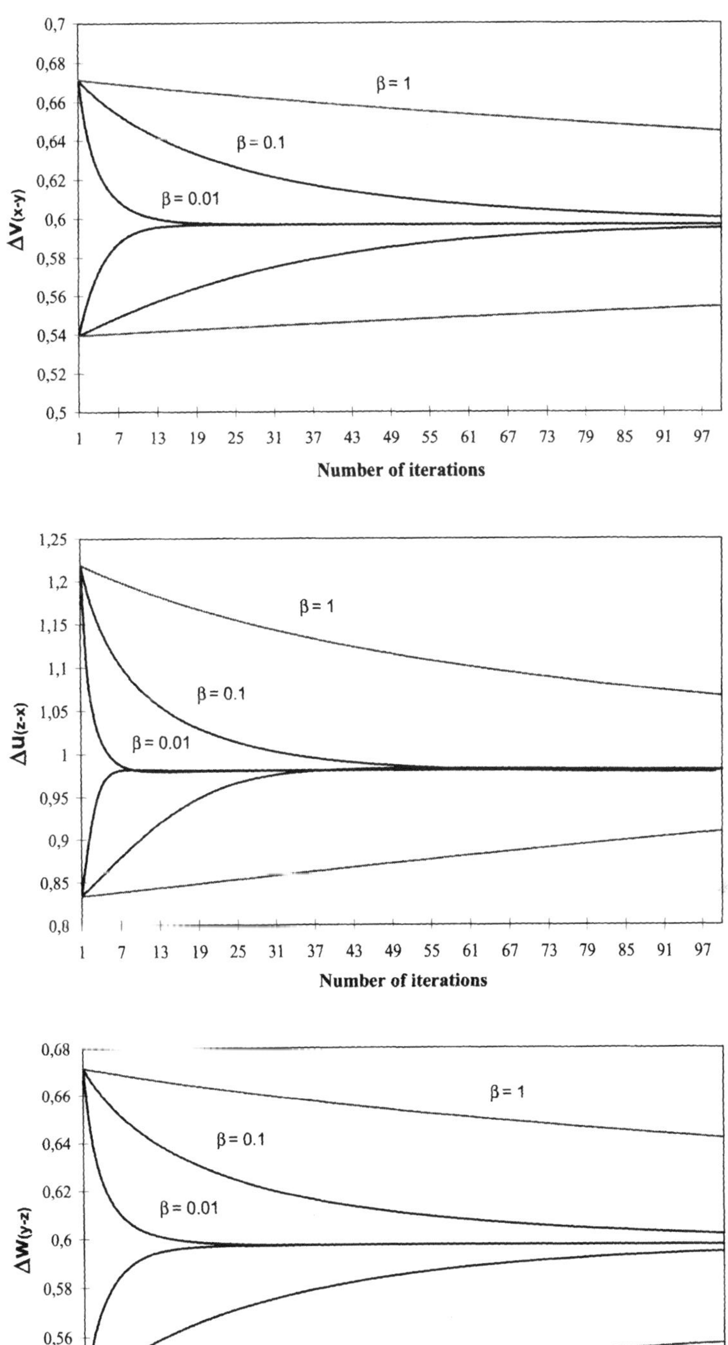

Figure 2: Convergence of analytical displacement Δv, Δu, Δw to their experimental values and influence of coefficient β on convergence.

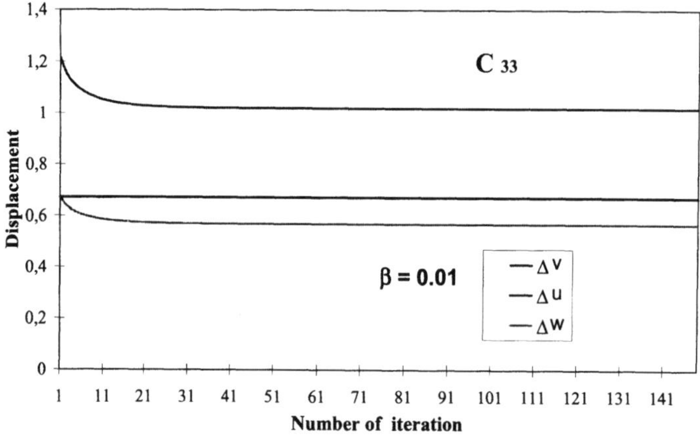

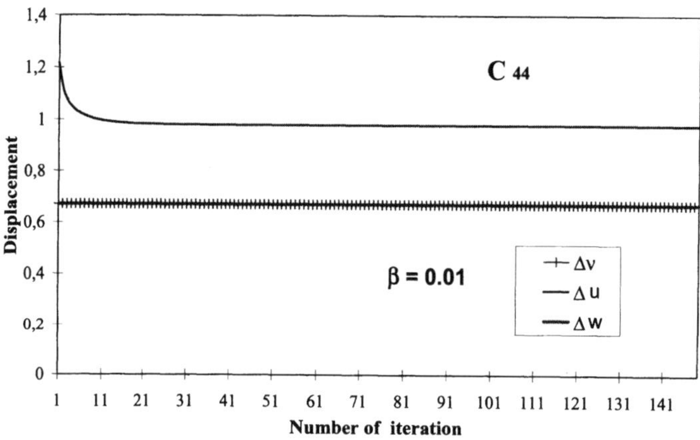

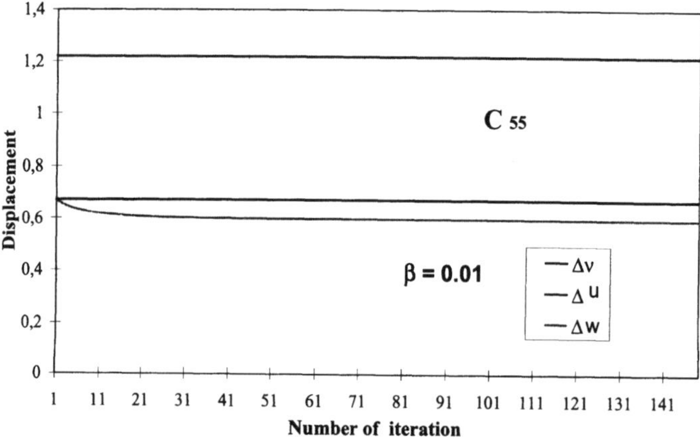

Figure 3: Convergence of analytical displacements as a function of the elastic coefficients

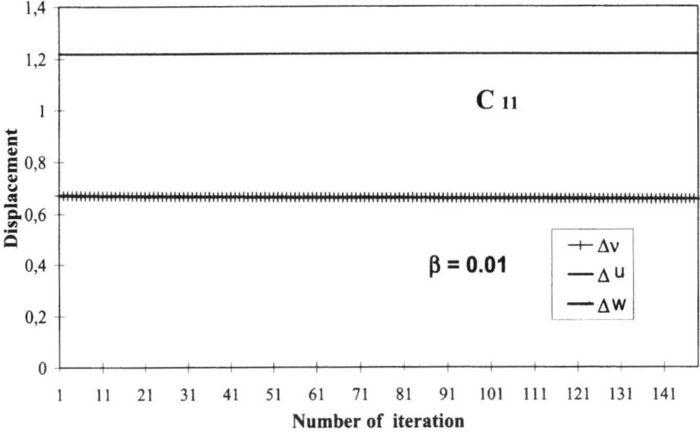

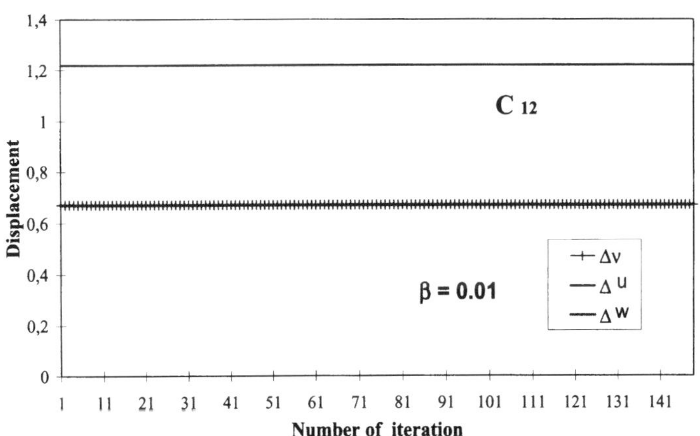

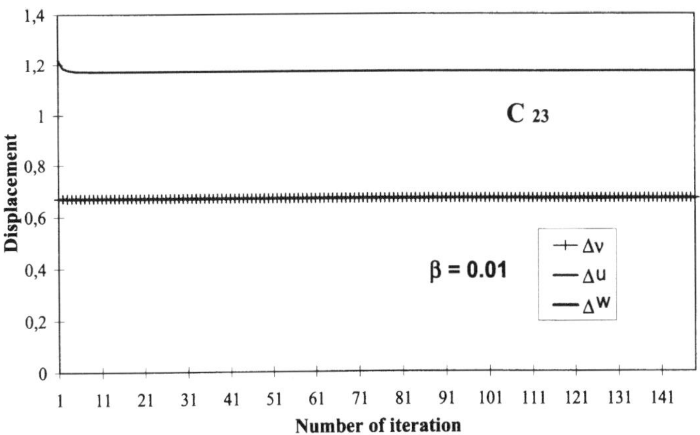

Figure 4: Convergence of analytical displacements as a function of the elastic coefficients

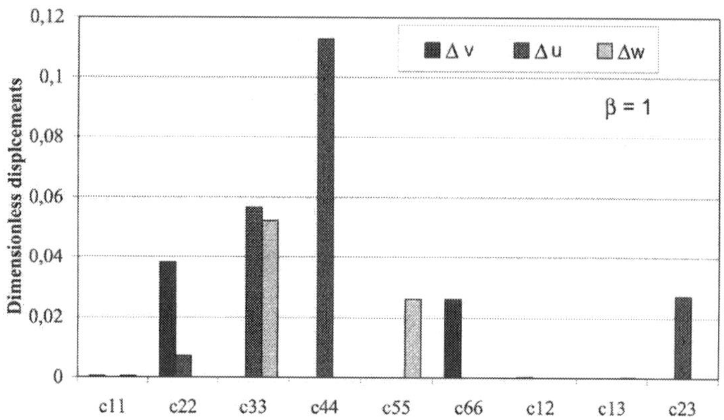

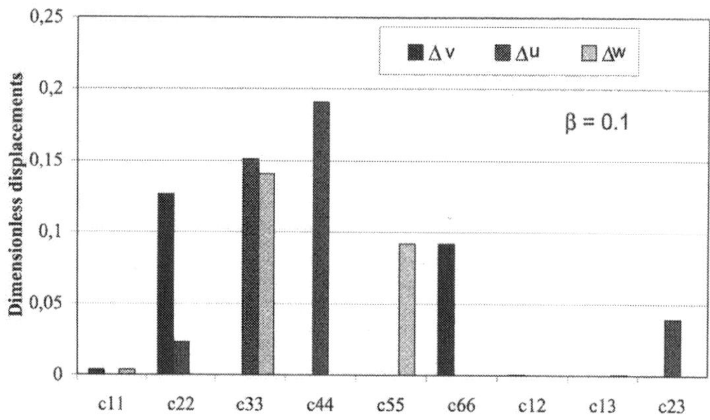

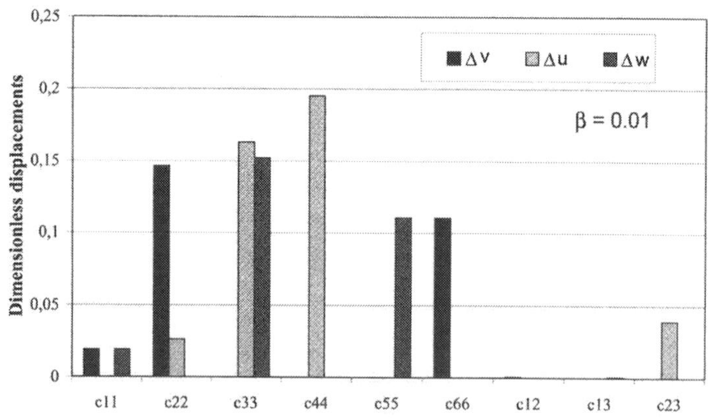

Figure 5: Sensitivity diagrams

Bibliography

J.D.C. Collins, G.C. Hurt, T.K. Hasselman, B. Kennedy. Statistical identification of the structures. *AAIAA Journal*, 12(2):185–190, 1974.

J. Cunha, S. Cogan,C. Berthod. Application of genetic algorithms for the identification of elastic constants of composite materials from dynamic test. *International Journal for Numerical Methods in Engineering*, 45:891–900, 1999.

P. Eychoff. System identification. John Wiley, New York, 1974.

L. Federici, A. Piva,E. Viola. Crack edge displacement and elastic constant determination for an orthotropic material, *Theoretical and Applied Fracture Mechanics*, 31:173–187, 1999.

F. Hartmann. The mathematical foundation of structural mechanics. Spring-Verlag, Berlin, 1985.

W.M. Hasan,A. Piva,E. Viola. Use of singular value decomposition method to detect ill-conditioning of structural identification problems. *Computers and Structures*, 2(63): 267–275, 1997.

W.M. Hasan,A. Piva,E. Viola. Parameter estimation from measured displacements of crack edges in an isotropic materials, *Engineering Fracture Mechanics*, 59(5):697–712, 1998.

G.C. Hurt, J.T.P. Yao. System identification in structural dynamics. *Journal Engineering Mechanical Division ASCE*, 103:1089–1104, 1977.

A. Maniatty,N. Zabaras, K. Stelson. Finite element analysis of some inverse elasticity problems. *Journal of Engineering Mechanics ASCE*, 115:1303–1317, 1989.

A. Maniatty,N. Zabaras. Investigation of regularization parameters and error estimating in inverse elasticity problems. *Int. J. Num. Meth. Engng.*, 37:1039–1052, 1994.

L. Nobile, P. Ricci,E. Viola. Elastic parameter estimation from measured crack edge displacement in orthotropic materials, *Second European Conference on Computational Mechanics*, June 26-29, Cracow-Poland, 2001.

L. Nobile,A. Piva, P. Ricci,E. Viola. Sull'identificazione dei parametri elastici di un solido fessurato di materiale ortotropo. *XV Congresso AIMETA di Meccanica Teorica e Applicata*, 26-29 Settembre 2001, Taormina.

A. Piva. An alternative approach to elastodynamic crack problems in an orthotropic medium, *Quarter.J.Appl.Math.*, 45:97–104, 1987.

D.S. Schnur, N. Zabaras. An inverse method for determining elastic materials properties and a material interface. *International Journal for Numerical Methods in Engineering*, 33:2039–2057, 1992.

K. Tanuma. Relative displacements of the two sides of a plane crack, *Journal of Elasticity*, 36:257–270, 1995.

E. Viola, A. Piva. Effect of orthotropic on elastodynamic crack behaviour, *Proceeding of the VIII AIMETA Conference*, vol.1, Torino, 39(4):155–158, 1986.

N. Zabaras,V. Morellas,D. Schnur. Spatially regularized solution of inverse elasticity problems using the BEM. *Commun. Appl. Num. Meth.*, 5:547–553, 1989.

O.C. Zienkiewicz,R.L. Taylor. The finite element model. Volume 1,2, McGraw-Hill Book Company, London, 1989.

Appendix A:

Displacement discontinuities $\Delta v_{(x-y)}$,$\Delta w_{(y-z)}$,$\Delta u_{(z-x)}$ of cross the crack for the problems in Fig. 1

$$\underset{(x-y)}{\Delta v} = \left\{ \frac{2Ta}{1-\dfrac{(c_{12}+c_{66})}{c_{11}\left(\dfrac{c_{66}}{c_{11}}+\dfrac{c_{12}^2-c_{11}c_{22}+2c_{12}c_{66}}{2c_{12}c_{66}}-\sqrt{-\dfrac{c_{22}}{c_{11}}+\dfrac{\left(-c_{12}^2+c_{11}c_{22}-2c_{12}c_{66}\right)^2}{4c_{11}^2c_{66}^2}}\right)}} - \right.$$

$$\left. + \frac{2Ta}{1-\dfrac{(c_{12}+c_{66})}{c_{11}\left(\dfrac{c_{66}}{c_{11}}+\dfrac{c_{12}^2-c_{11}c_{22}+2c_{12}c_{66}}{2c_{12}c_{66}}+\sqrt{-\dfrac{c_{22}}{c_{11}}+\dfrac{\left(-c_{12}^2+c_{11}c_{22}-2c_{12}c_{66}\right)^2}{4c_{11}^2c_{66}^2}}\right)}} \right\} /$$

$$\left\{ c_{66}\left[-\sqrt{\frac{-c_{12}^2+c_{11}c_{22}-2c_{12}c_{66}}{2c_{11}c_{66}}-\sqrt{-\frac{c_{22}}{c_{11}}+\frac{\left(-c_{12}^2+c_{11}c_{22}-2c_{12}c_{66}\right)^2}{2c_{11}^2c_{66}^2}}} - \right.\right.$$

$$\left.\left. +\sqrt{\frac{-c_{12}^2+c_{11}c_{22}-2c_{12}c_{66}}{2c_{11}c_{66}}+\sqrt{-\frac{c_{22}}{c_{11}}+\frac{\left(-c_{12}^2+c_{11}c_{22}-2c_{12}c_{66}\right)^2}{2c_{11}^2c_{66}^2}}}\right]\right\}$$

$$\underset{(z-x)}{\Delta u} = \left\{ \frac{2Ta}{1-\dfrac{(c_{13}+c_{55})}{c_{11}\left(\dfrac{c_{55}}{c_{11}}+\dfrac{c_{13}^2-c_{11}c_{33}+2c_{13}c_{55}}{2c_{11}c_{55}}-\sqrt{-\dfrac{c_{33}}{c_{11}}+\dfrac{\left(-c_{13}^2+c_{11}c_{33}-2c_{13}c_{55}\right)^2}{4c_{11}^2c_{55}^2}}\right)}} - \right.$$

$$\left. + \frac{2Ta}{1-\dfrac{(c_{13}+c_{55})}{c_{11}\left(\dfrac{c_{55}}{c_{11}}+\dfrac{c_{13}^2-c_{11}c_{33}+2c_{13}c_{55}}{2c_{11}c_{55}}+\sqrt{-\dfrac{c_{33}}{c_{11}}+\dfrac{\left(-c_{13}^2+c_{11}c_{33}-2c_{13}c_{55}\right)^2}{4c_{11}^2c_{55}^2}}\right)}} \right\} /$$

$$\left\{ c_{55}\left[-\sqrt{\frac{-c_{13}^2+c_{11}c_{33}-2c_{13}c_{55}}{2c_{11}c_{55}}-\sqrt{-\frac{c_{33}}{c_{11}}+\frac{\left(-c_{13}^2+c_{11}c_{33}-2c_{13}c_{55}\right)^2}{2c_{11}^2c_{55}^2}}} - \right.\right.$$

$$+\sqrt{\frac{-c_{13}^2+c_{11}c_{33}-2c_{13}c_{55}}{2c_{11}c_{55}}+\sqrt{-\frac{c_{33}}{c_{11}}+\frac{\left(-c_{13}^2+c_{11}c_{33}-2c_{13}c_{55}\right)^2}{2c_{11}^2c_{55}^2}}}\Bigg]\Bigg\}$$

$$\underset{(y-z)}{\Delta w}=\left\{\frac{2Ta}{1-\dfrac{(c_{23}+c_{44})}{c_{22}\left(\dfrac{c_{44}}{c_{22}}+\dfrac{c_{23}^2-c_{22}c_{33}+2c_{23}c_{44}}{2c_{22}c_{44}}-\sqrt{-\dfrac{c_{33}}{c_{22}}+\dfrac{\left(-c_{23}^2+c_{22}c_{33}-2c_{23}c_{44}\right)^2}{4c_{22}^2c_{44}^2}}\right)}}-\right.$$

$$\left.+\frac{2Ta}{1-\dfrac{(c_{23}+c_{44})}{c_{22}\left(\dfrac{c_{44}}{c_{22}}+\dfrac{c_{23}^2-c_{22}c_{33}+2c_{23}c_{44}}{2c_{22}c_{44}}+\sqrt{-\dfrac{c_{33}}{c_{22}}+\dfrac{\left(-c_{23}^2+c_{22}c_{33}-2c_{23}c_{44}\right)^2}{4c_{22}^2c_{44}^2}}\right)}}\right\}/$$

$$\Bigg\{c_{44}\Bigg[-\sqrt{\frac{-c_{23}^2+c_{22}c_{33}-2c_{23}c_{44}}{2c_{22}c_{44}}-\sqrt{-\frac{c_{33}}{c_{22}}+\frac{\left(-c_{23}^2+c_{22}c_{33}-2c_{23}c_{44}\right)^2}{2c_{22}^2c_{44}^2}}}-$$

$$+\sqrt{\frac{-c_{23}^2+c_{22}c_{33}-2c_{23}c_{44}}{2c_{22}c_{44}}+\sqrt{-\frac{c_{33}}{c_{22}}+\frac{\left(-c_{23}^2+c_{22}c_{33}-2c_{23}c_{44}\right)^2}{2c_{22}^2c_{44}^2}}}\Bigg]\Bigg\}$$

Interaction of BEM analysis and experimental testing on Pile-Soil Systems

Francesco Vinciprova[1], Orlando Maeso Fortuny[2], Juan José Aznárez[2] and Giuseppe Oliveto[1]

[1] Dipartimento di Ingegneria Civile e Ambientale, Università di Catania, Catania, Italia
[2] Departamento de Ingeniería Civil, Universidad de Las Palmas de Gran Canaria, Las Palmas, España

Abstract. This work considers some important modelling problems for soil-pile systems via application of the Boundary Element Method (BEM). Among these are the correct definition of the number and shape of elements to be used in the discretisation of the pile surface and of the parameters which influence them, the number and shape of elements needed for the discretisation of the soil free surface and of the soil layer's interfaces. The results are used to model a pile-soil system in relation to a full scale static and dynamic experimental test to be conducted on site. The results of the modelling and subsequent analyses have been used in the planning of the experimental testing procedure, while the results of the experimental tests will be used later to validate the mathematical model and the assumptions on which this is based. Several open problems, which deserve the attention of the scientific community, have been implicitly or explicitly outlined within the treatment of the problem.

1 Introduction

The effects of soil-structure interaction are of paramount importance in several static and/or dynamic problems of structural engineering. With the exception of particularly firm soils or of considerably flexible structures, the interaction between soil and structure always plays an important role with reference to both the structural behaviour and the effects on the soil itself.

Dynamic soil-structure interaction has been studied since the second half of last century in relation to the development in many countries of peaceful applications of nuclear energy and the subsequent construction of nuclear power stations.

Both surface and deep foundations have been studied. Initially foundations have been considered as rigid and mass less; subsequently the foundation flexibility has also been included. Many papers were written on the advancement of knowledge and on the state of the art as soon as these reached several stages of maturation. Another work on the state of the art (Hall, W. S., Oliveto, G., Editors, 2003) has been recently published by Kluwer Academic with contributions on several aspects of soil-structure interaction.

The volume, which is divided in two parts, one specifically devoted to soil-structure interaction problems and the other to strictly related ones, approaches the subject via the boundary element method (BEM). The first part is composed of five chapters devoted to general aspects of soil structure interaction problems, formulation and computational aspects, simplified analysis

procedures, probabilistic methods of analysis and to the procedures embodied in the codes of practice.
The second part contains four chapters, one devoted to theoretical formulations, two others to computational aspects of fracture dynamics and the final one to the definition of the seismic motion on the surface of soft soil with complex geometries via application of the BEM.
Each chapter, written by world renowned experts in the specific field of interest, presents an up to date review of the state of knowledge and art on the considered sector.

The present work originated while the volume previously described took shape and came to a conclusion. Among the many soil-structure interaction problems which are of interest the choice fell on the study of the soil-pile system for two fundamental reasons. The first reason comes from the experience acquired by the research group in Catania on dynamical testing of foundation rafts on piles (Buda, G., Oliveto, G., 2001). The second is related to the fact that, among all foundation systems, the dynamical behaviour of structures founded on piles is the least known, especially with reference to the correspondence of experimental and numerical results.

In view of the difficulty of interpretation of experimental results on complex structures, the present work focusses on the behaviour of a single pile-soil system. The results obtained for the single pile may be useful in the interpretation of the dynamical behaviour of foundation systems on piles. At the levels of dynamical identification and modelling, the problem is considerably simplified. Therefore this study is specifically addressed to the interpretation of experimental results that will be obtained by the experimental tests described in a parallel contribution to the present volume, (Buda, G. Sciacca, P., Oliveto, G., 2003).

2 Preliminary Considerations

An important part of the research programme, as already pointed out in the Introduction, is the modelling of the pile-soil system. In the initial planning stage the problem arose of whether to formulate and develop a computer programme from scratch or to build the model on the basis of results and software available elsewhere. After a careful analysis, with the fundamental contribution of international participants to the present research, it was concluded that it would have been more effective to refer to the already available material. Within the framework of a network of international cooperation developed and consolidated through the realisation of important events and the achievement of significant results (Euromech Colloquium 414, 2000), (Hall, W. S., Oliveto, G., Editors, 2001), (Interaction Problems in Structural Mechanics, AIMETA 2001), the most suitable location for the development of the present work was identified. On suggestion of José Dominguez, whose contribution within the volume edited by W.S. Hall and G. Oliveto (Kluwer Academic, 2003) appeared the most adherent to the research interests pursued in Catania, Francesco Vinciprova, a post-graduate student at the University of Catania, has joined the research group at the University of Las Palmas, Canary Islands, where has worked onto the modelling of the pile-soil system under the supervision of Orlando Maeso Fortuny. The present work reports most of the results obtained in Las Palmas and further developments from Catania.

3 Boundary Element Analysis Programs

The numerical investigation described in the following were performed by means of two boundary element structural analysis programs. The first of such programs, known by the acronym MECIMEN, considers rigid surface and embedded foundations. More specifically it allows for the definition of the dynamical stiffness matrix for such foundation systems and the soil. In such an analysis the foundation is considered rigid and mass less. The mass of the foundation may be accounted for when the dynamic stiffness matrix, also known as the impedance matrix, is coupled to the superstructure. The programme takes advantage of the Kelvin-Stokes fundamental solution of elastodynamics which is referred to the three-dimensional space or full space as opposed to the halfspace. This requires the discretisation not only of the soil foundation interface, but also of the free surface where stress-free conditions need to be enforced. In the literature on the subject it is now a tradition to discretise only a small part of the free surface close to the considered foundation; in fact, many numerical experiments show that results do not change when a sufficient distance from the foundation-soil interface has been reached. However the extension of the free surface to be discretised and the density of the discretisation may change from case to case. In the present study optimal conditions have been identified with reference to mesh extension and density as functions of several parameters such as the width and the depth of the foundation. Moreover the density of the discretisation of the soil-foundation interface also plays a dominant role on the results and needs to be optimised.

The second computer programme that has been used is known under the acronym of PILOT and allows for the analysis of foundation rafts on piles. In this case the problems of discretisation of the pile surface and of the free surface need to be addressed. The mesh extension and density optimisation depends in this case upon pile diameter and length, pile arrangement and distance, and on the geometry of the raft.

Both computer programmes can take advantage of geometrical and structural symmetries to minimise computer storage and time.

4 Theoretical Background

The boundary element method is based on the resolution of a boundary integral equation derived via application of the elastic reciprocity theorem. As such the method is mainly applicable to the solution of linear problems, although several applications to non-linear problems are possible and have been performed by means of appropriate techniques which may vary from case to case.

A typical application of the BEM to non-linear problems is in the study of fracture propagation in elastic solids (Aliabadi, M. H., 2003; Frangi A. et al., 2003).

The integral equation at the basis of the method, for linear problems, may be written as follows

$$\mathbf{C}(\mathbf{y})\cdot\mathbf{u}(\mathbf{y})+\iint_S \hat{\mathbf{T}}(\mathbf{x}-\mathbf{y})\cdot\mathbf{u}(\mathbf{x})\,dS(\mathbf{x})=\iint_S \hat{\mathbf{U}}(\mathbf{x}-\mathbf{y})\cdot\mathbf{t}^n(\mathbf{x})\,dS(\mathbf{x}) \qquad \mathbf{y}\in S \qquad (1)$$

where S represents the boundary of the solid considered, $\mathbf{y}$ defines the point of application of the harmonic load which defines the fundamental solution, obviously known, $\hat{\mathbf{T}}(\mathbf{x}-\mathbf{y})$ and $\hat{\mathbf{U}}(\mathbf{x}-\mathbf{y})$ are second order tensors related to the fundamental solution and define the effects in terms of stress and displacement at point $\mathbf{x}$. Finally $\mathbf{u}(\mathbf{x})$ and $\mathbf{t}^n(\mathbf{x})$ define the unknown displacement and stress fields, while tensor $\mathbf{C}(\mathbf{y})$ depends on the boundary geometry at point $\mathbf{y}$. Further details on the above subject may be found in the specialised literature (Domínguez, J., 1993; Wrobel, L. C., Aliabadi, M. H., 2002).

Equation (1) applies to points **y** belonging to the boundary S of the body considered. In actual problems the solution at all points in a body is required, therefore also at points in the interior of the body. The integral equation that defines the displacement field in the interior of a body is the same as equation (1) with the boundary tensor $\mathbf{C}(\mathbf{y})$ replaced by the identity tensor **I**.

The numerical solution to equation (1) is pursued via the boundary element method through an appropriate discretisation of the boundary surface S. By subdividing the latter in M elements equation (1) may be written as follows:

$$\mathbf{C}(\mathbf{y})\cdot\mathbf{u}(\mathbf{y})+\sum_{m=1}^{M}\iint_{S_m}\hat{\mathbf{T}}(\mathbf{x}-\mathbf{y})\cdot\mathbf{u}(\mathbf{x})\,dS(\mathbf{x})=\sum_{m=1}^{M}\iint_{S_m}\hat{\mathbf{U}}(\mathbf{x}-\mathbf{y})\cdot\mathbf{t}(\mathbf{x})\,dS(\mathbf{x}) \tag{2}$$

In this equation the tensor functions $\hat{\mathbf{T}}(\mathbf{x}-\mathbf{y})$ and $\hat{\mathbf{U}}(\mathbf{x}-\mathbf{y})$ are known functions of $(\mathbf{x}-\mathbf{y})$, nevertheless it is necessary to describe points **x** and **y** on S. For this purpose, for each element S_m, the position vector **x** is defined in terms of nodal coordinates by means of suitable shape functions as follows:

$$\mathbf{x}(\xi_1,\xi_2)=\sum_{c=1}^{N}N_c(\xi_1,\xi_2)\mathbf{x}_c \tag{3}$$

The shape functions $N_c(\xi_1,\xi_2)$ define a mapping between the actual physical element and a square or triangular plane domain. As well as the boundary surface S the unknown functions $\mathbf{u}(\mathbf{x})$ and $\mathbf{t}^n(\mathbf{x})$ need to be discretised. For the latter the same nodal points and shape functions used for the discretisation of the boundary surface may be used (isoparametric description). In this way the following discretised form of equation (1) may be found:

$$\begin{aligned}&\mathbf{C}(\mathbf{y})\cdot\mathbf{u}(\mathbf{y})+\sum_{m=1}^{M}\int_{-1}^{1}\int_{-1}^{1}\hat{\mathbf{T}}(\mathbf{x}(\xi_1,\xi_2)-\mathbf{y})\cdot\sum_{c=1}^{N}N_c(\xi_1,\xi_2)\mathbf{u}_c\,\frac{\partial S(\mathbf{x}(\xi_1,\xi_2))}{\partial\xi_1\partial\xi_2}d\xi_1 d\xi_2=\\&=\sum_{m=1}^{M}\int_{-1}^{1}\int_{-1}^{1}\hat{\mathbf{U}}(\mathbf{x}(\xi_1,\xi_2)-\mathbf{y})\cdot\sum_{c=1}^{N}N_c(\xi_1,\xi_2)\mathbf{t}_c^n\,\frac{\partial S(\mathbf{x}(\xi_1,\xi_2))}{\partial\xi_1\partial\xi_2}d\xi_1 d\xi_2\end{aligned} \tag{4}$$

The only unknowns of the problem appearing in equation (4) are the nodal parameters $\mathbf{u}_c$ and $\mathbf{t}_c^n$ of the unknown solution. Altogether these are equal in number to six times the number of nodes used in the discretisation. It is therefore necessary to devise a procedure starting from (4) which allows the development of as many algebraic equations as is needed to solve the problem. This can be done in several ways, but historically the collocation method was developed first and is the one used in the previously quoted computer programmes. This method consists in the association of the loading point **y** of the fundamental solution with one of the nodal points $\mathbf{y}_c$. In such a way a vector algebraic equation corresponding to three scalar equation is obtained. By repeating the collocation for each nodal point, a system of $3N$ equations in $3N$ unknown is found of the type:

$$\mathbf{H}\cdot\mathbf{u}=\mathbf{G}\cdot\mathbf{t} \tag{5}$$

where N is the number of nodal points. Vectors **u** and **t** are obtained by assembling vectors $\mathbf{u}_c$ and $\mathbf{t}_c^n$ corresponding to the nodal points. Because each of these possesses 3 components, the vector which is obtained by assembling them will have $3N$ components. On the other hand, as was mentioned above, matrices **H** and **G** will have dimensions $3N\times3N$. The components of matrices **H** and **G** derive from the integration of the stress and displacement kernels. In some cases the stress vectors $\mathbf{t}_c^n$ are associated with elements rather than nodal points. In these cases, by setting $J = N_c \cdot M$, the matrix **G** and the vector **t** will have dimensions $3N \times J$ and $J \times 1$ respectively. In general some components of **u** and some components of **t** are known a priori from the boundary conditions of the problem. The remaining ones, which are always equal in number to $3N$, represent the unknowns of the problem. By separating the known from the unknown components of vectors **u** and **t**, the system of equations (5) may be written in the form:

$$\mathbf{A} \cdot \mathbf{x} = \mathbf{f} \tag{6}$$

The coefficient matrix **A** is generally fully populated and unsymmetrical. This is one of the main limits of the collocation method, a limitation which may be overcome at the expense of a greater initial computational effort by using the so-called symmetrical Galerkin BEM (Bonnet, M. et al., 1998).

The solution of equation (6) provides the boundary unknowns. The solution in the interior of the body may be evaluated on the basis of the boundary solution. For this equation (4) is used with tensor $\mathbf{C}(\mathbf{y})$ replaced by the identity tensor **I** and attributing to vector **y** the position of the point were the displacement $\mathbf{u}(\mathbf{y})$ needs to be evaluated. In equation (4) the nodal values of the boundary displacement and stress vectors are already known while the stress and displacement kernels, which are dependant on point **y** where is required the displacement $\mathbf{u}(\mathbf{y})$, must be evaluated and integrated. The stress tensor is evaluated by using the boundary integral equation

$$\mathbf{T}(\mathbf{y}) = \iint_S \mathbf{D}^*(\mathbf{x}-\mathbf{y}) \cdot \mathbf{t}^n(\mathbf{x})\, dS(\mathbf{x}) - \iint_S \mathbf{S}^*(\mathbf{x}-\mathbf{y}) \cdot \mathbf{u}(\mathbf{x})\, dS(\mathbf{x}) \tag{7}$$

where $\mathbf{D}^*$ and $\mathbf{S}^*$ are third order tensors obtained by differentiation of the stress and displacement kernels (Domínguez, J., 1993). It is of interest to notice that all functions in the integrals of equation (7) are already known; nevertheless it is necessary to use a discretised form of equation (7) for the numerical evaluation of $\mathbf{T}(\mathbf{y})$.

5 Discretisation Problems

Some discretisation problems will be considered in the following, with separate treatments for two-dimensional surface and embedded foundations and for foundations on piles. In both cases problems concerning the discretisation of the foundation itself and of the free surface will be addressed. Both aspects are in fact complementary and of considerable importance for the accuracy of the solution.

5.1 Rigid Plane Extended Foundations.

In principle foundations of any shape are never rigid. However in many applications it may not only be convenient but even sufficiently accurate to consider foundations as rigid, at least with reference to stress transmission from the structure to the soil and vice-versa. To validate such a hypothesis concurs not only the thickness of the foundation plate, sometimes really relevant, but often and predominantly the stiffening effect provided by the superstructure. In practical applications only the soil-foundation interface is considered as rigid, while in structural analyses of the foundation structures these are usually considered in every respect as deformable bodies.

Given this premise it is evident that the kinematics of such a foundation is fully described by only six degrees of freedom. However, while these six degrees of freedom allow for the complete characterisation of the displacement field, the same is not true as far as the stress field is concerned. In fact the stress field at the soil-foundation interface depends not only on the local displacement field, but also on the overall displacement field in the soil. While it is true that such interface forces are globally characterised by only six parameters, their effective distribution may only be captured with sufficient accuracy by an appropriately dense discretisation of the interface surface. The problems associated to such a discretisation will be shown in the following through significant numerical applications.

Besides the interface characterisation, two additional discretisation problems arise with reference to extended plane foundations. One concerns the discretisation of the free surface of the halfspace while the other, arising only for embedded foundations, is related to the discretisation of the surfaces arising from the excavation needed to create the foundation plane. In what follows some applications concerning surface and embedded foundations will be briefly considered.

Surface Foundations. In this case the discretisation problems concern on one hand the soil foundation interface and on the other the free surface of the half-space. The first problem has been considered briefly earlier on. For the second problem the discretisation of the free surface of the half-space is required because of the use of the Kelvin-Stokes fundamental solution referred to the whole space. The boundary conditions on the free surface are not satisfied automatically and must therefore be enforced. The availability of a fundamental solution for the half-space, such as the Mindlin static solution, would avoid the discretisation of the free surface. In the present work the importance of the discretisation of the free surface is firstly emphasised. In fact, with reference to a square foundation of width $2B$, three cases are considered where an extension of the free surface respectively equal to $2B$, $5B$ and $10B$ is discretised. It must be observed that, in view of symmetry, only one quarter of the foundation is considered and the extension of the free surface should each time be compared with B rather than with $2B$. In each of the three cases considered, shown in Figure 1, the number of elements is the same for the foundation-soil interface while it gradually increases for the free surface.

The number of elements considered in each case is given in Table 1.

Table 1. Number of elements considered for three different extensions of the free surface.

Extension of the Free Surface	Number of elements	
	Soil-Foundation Interface	Free Surface
$2B$	16	40
$5B$	16	64
$10B$	16	96

Why such discretisations were used and why in the most comprehensive case the extension of the free surface discretisation was limited to $10B$ are obvious questions. The answers are strongly dependant on the objectives of the analyses. In the present case the objective consisted in the evaluation of the coefficients of the dynamic stiffness matrix in the frequency range 0-12 *rad/s*. In the case where B=25 m and $c_s = 150\ m/s$ this requires a dimensionless frequency interval $0 \le a_0 \le 2$ with $a_0 = \omega B / c_s$.

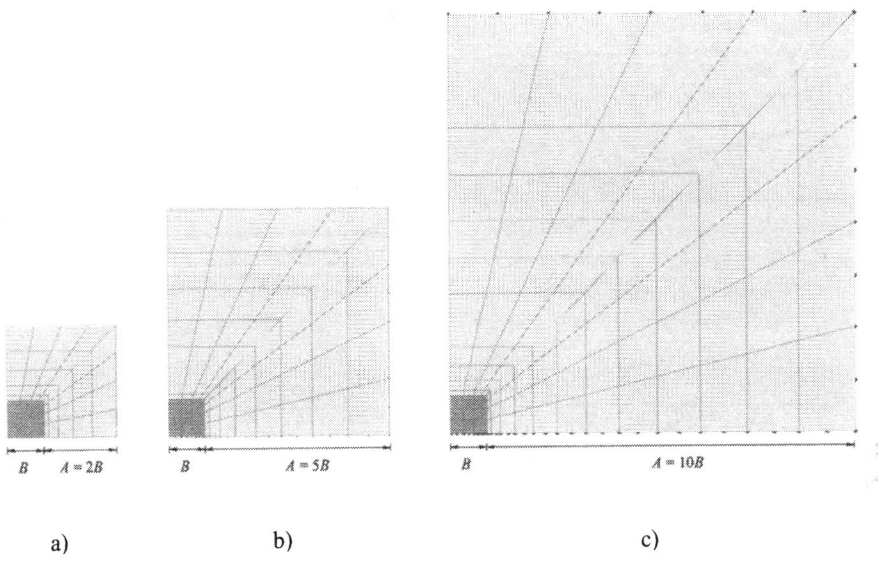

Figure 1. Three different discretisations of the half-space free surface; a) A=2B; b) A=5B; c) A=10B.

In such a frequency interval the coefficients of the dynamic stiffness matrix must be defined with sufficient accuracy. This accuracy depends, however, not only on the extension of the discretised free surface, but also on the maximum element size and consequently on the number of elements considered which also becomes an important parameter. In fact, because a quadratic representation is used, in order for the displacement field to be described with sufficient accuracy the maximum element size should not exceed half a wavelength. Therefore it is the minimum wavelength which controls the maximum element size. The wavelength is related to the wave velocity through the relationship

$$c_s = \frac{\lambda}{T} = \frac{\omega}{2\pi}\lambda$$

where T is the period and ω is the frequency of the wave. In terms of dimensionless frequency it follows that

$$a_0 = \frac{\omega B}{c_s} = \frac{2\pi}{\lambda}B$$

Therefore it must be the case that

$$d_{\max} \le \frac{\lambda}{2} = \frac{\pi}{a_{0\max}}B$$

and for $a_{0\max} = 2$, $d_{\max} \le \pi B/2$. For the numerical values previously considered it follows that $d_{\max} \le 1.57B$.

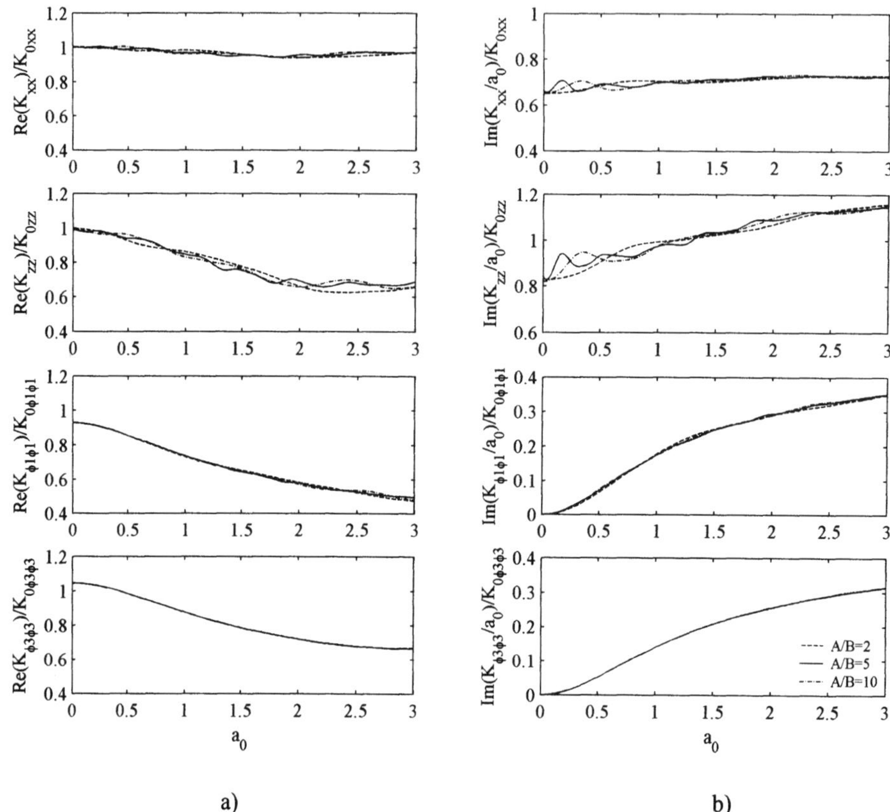

Figure 2. Dynamic stiffness coefficients for square surface foundations; a) real part, b) imaginary part.

This condition on d_{max} is generally satisfied for the elements in the discretisations given in Figure 1; with wide margins for the elements closer to the soil-foundation interface and approaching the limit as the distance from the foundation becomes larger. It may easily be seen the difference in the results is rather small and even the least extended discretisation may provide acceptable results, at least within the considered frequency range. The results for the three models of Figure 1 are shown in Figure 2 as the diagonal terms of the dynamic stiffness matrix.

Embedded Foundations. The investigation presented so far shows how the extension of the free surface to be discretised depends essentially on the wavelength considered and consequently on the corresponding frequency. This may be emphasised even more if the subject is widened by including among the variables the depth E of the foundation plane. In the present study three levels of foundation depth have been considered, from the surface foundation, E/B=0, to an intermediate case characterised by E/B=2/3, to the extreme case E/B=2. In the last case the depth of the foundation is equal to its width: E=2B. The case of surface foundation, E/B=0, has been shown in Figure 1. The discretisations used for the cases E/B=2/3 and E/B=2 are shown in Figure 3.

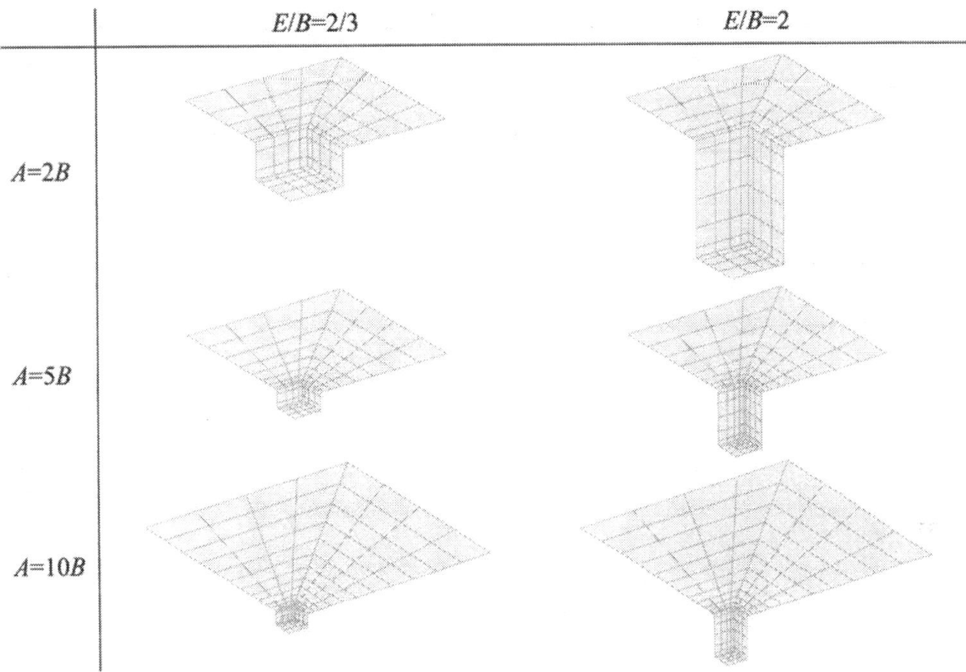

Figure 3. Discretisations used for embedded foundations; left E/B=2/3, right E/B=2. The three cases refer to increasing extensions of discretised free surface: A=2B, A=5B, A=10B.

The number of elements considered for each discretisation is given in Table 2. The results, again specified as direct terms of the dynamic stiffness matrix, are shown in Figure 4.

Table 2. Number of elements considered for three different extensions of the free surface.

Free Surface Estension A	Number of elements				
	Foundation-Soil Interface	Foundation-Soil Lateral Interface			Free Surface
		E/B=0	E/B=2/3	E/B=2	
2B	16	-	24	48	40
5B	16	-	24	48	64
10B	16	-	24	48	96

It is evident from Figure 4 how the real and imaginary parts of the coefficients considered obtained by different discretisations tend to separate more and more as the frequency and the foundation depth both increase.

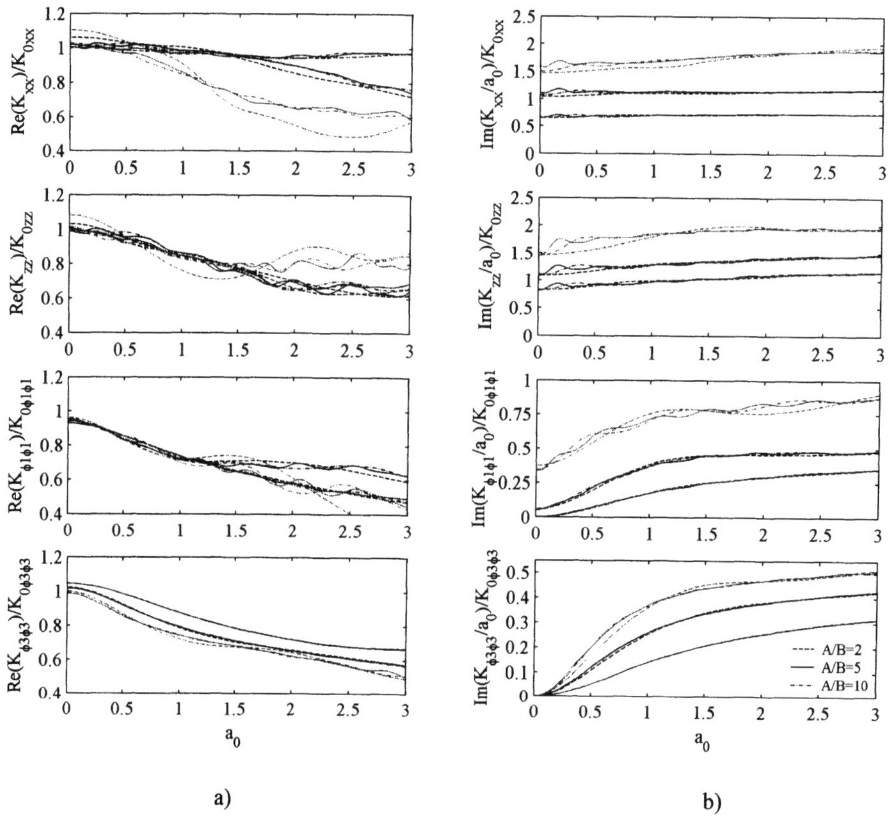

Figure 4. Dynamic stiffness coefficients for surface and embedded foundations; a) real part, b) imaginary part. —: surface foundation; —: E/B=2/3; —: E/B=2.

It is therefore evident that both the frequency and the foundation depth play a dominant role in the definition of the most appropriate discretisation for the problem in hand. In order to show in more detail the divergence of results as a function of the discretisation, the results relative to the extreme case E/B=2 are shown in Figure 5.

From the discussion above it follows that a particular care is needed when modelling extended plane foundations whether surface or embedded ones, depending on the problem in hand and of the result being sought. Particular attention must be devoted to the definition of the frequency range to be investigated, to the maximum length of the elements considered and to the extension of the free surface to be discretised. Insufficient consideration of such aspects may result in poor, if not completely unreliable, analyses.

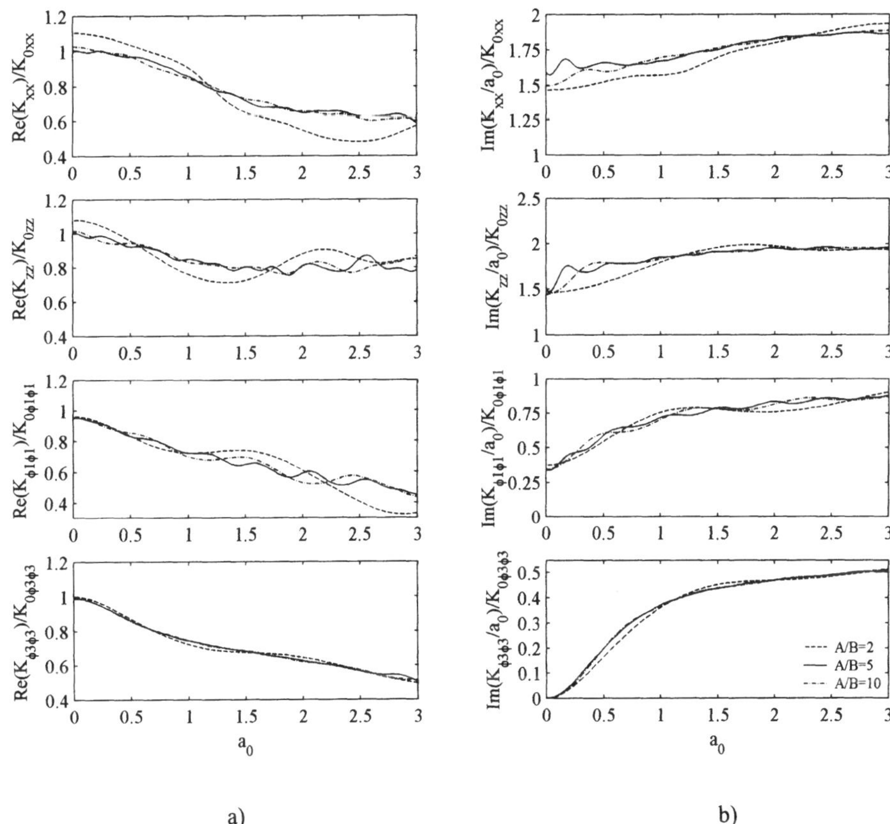

Figure 5. Dynamic stiffness coefficients for an embedded foundation with E/B=2; a) real part, b) imaginary part.

5.2 Deep Foundations.

The use of the term *deep foundations* in the following relates essentially to foundations on piles. In particular the pile-soil system will be considered and in some cases a system of three aligned piles in a testing configuration will also be considered. When the single pile-soil system is being studied it is necessary to describe with sufficient accuracy the behaviour of both pile and surrounding soil. In such a case problems of discretisation of the pile surface and of the free surface of the halfspace clearly arise. Because in many cases the foundation soil is layered rather than homogeneous, the various layer interfaces need also to be discretised. As previously done with plane extended foundations, the frequency range to be investigated needs to be defined, and within such a range the actual pile behaviour must be correctly described.

Pile Discretisation. In order to define minimum requirements for the pile discretisation, reference here is made to a fixed base pile of length L and radius R. It is assumed that the first 5 vibration modes of the system must be described with sufficient accuracy, separately for axial and bending behaviour. The limit frequencies to be considered in the two cases are given below:

$$\omega_{nf} = k_n^2 \sqrt{\frac{EI}{mL^4}} \, ; \qquad \omega_{ne} = (2n-1)\frac{\pi}{2}\sqrt{\frac{EA}{mL^2}} \, .$$

Given the longitudinal and transversal wave velocities, on the basis of the maximum element length, being a function of the wavelength, the minimum number of elements to be considered in the radial (n_R), circumferential (n_ϑ) and longitudinal (n_z) directions is established in the two cases. Such minimum values are shown in Table 3.

Table 3. Minimum numbers of elements to be used in the discretisation of the pile.

Flexural vibrations	Extensional vibrations
$n_R = k_n^2 \frac{R^2}{\pi L^2}\sqrt{\frac{1+\nu}{2}}$	$n_R = \frac{R}{L}\frac{2n-1}{2}\sqrt{\frac{(1+\nu)(1-2\nu)}{(1-\nu)}}$
$n_\vartheta = k_n^2 \frac{R^2}{2L^2}\sqrt{\frac{1+\nu}{2}}$	$n_\vartheta = (2n-1)\frac{\pi}{4}\frac{R}{L}\sqrt{\frac{(1+\nu)(1-2\nu)}{(1-\nu)}}$
$n_z = k_n^2 \frac{R}{\pi L}\sqrt{\frac{1+\nu}{2}}$	$n_z = \frac{2n-1}{2}\sqrt{\frac{(1+\nu)(1-2\nu)}{(1-\nu)}}$

It may be noticed from Table 3 that such minimum values depend essentially on two parameters, one geometrical, the ratio *L/R*, and the other mechanical associated to Poisson's ratio. The geometrical ratio depends on the design and technological characteristics of the pile. For current technologies it may be expected that such a ratio should not exceed the value 100, which corresponds to a length of 50 *m* for a diameter of 1 *m*. As it is well known, Poisson's ratio may vary in the range $0 \le \nu \le 1/2$. The following tables were compiled for the following values of Poisson's ratio: ν=0, ν=1/4, ν=1/2. Examination of the results collected in Tables 4, 5 and 6 shows that for extensional vibration the value ν=0 is dominant, while for flexural vibration it is the value ν=1/2. Therefore the limit values arising from those choices will be adopted. In this way the minimum number of elements will depend only on the geometrical ratio.

Table 4. Limit values (minima) for the number of elements required in pile discretisation for ν=0.

L/R	Extensional			Flexural		
	n_ϑ	n_z	n_R	n_ϑ	n_z	n_R
4	1.767	4.500	1.125	4.416	11.246	2.812
6	1.178	4.500	0.750	1.963	7.497	1.250
8	0.884	4.500	0.563	1.104	5.623	0.703
10	0.707	4.500	0.450	0.707	4.498	0.450
12	0.589	4.500	0.375	0.491	3.749	0.312
14	0.505	4.500	0.321	0.361	3.213	0.230
16	0.442	4.500	0.281	0.276	2.812	0.176
18	0.393	4.500	0.250	0.218	2.499	0.139
20	0.353	4.500	0.225	0.177	2.249	0.112
22	0.321	4.500	0.205	0.146	2.045	0.093
24	0.295	4.500	0.188	0.123	1.874	0.078
26	0.272	4.500	0.173	0.105	1.730	0.067
28	0.252	4.500	0.161	0.090	1.607	0.057
30	0.236	4.500	0.150	0.079	1.499	0.050
35	0.202	4.500	0.129	0.058	1.285	0.037
40	0.177	4.500	0.113	0.044	1.125	0.028
45	0.157	4.500	0.100	0.035	1.000	0.022
50	0.141	4.500	0.090	0.028	0.900	0.018
55	0.129	4.500	0.082	0.023	0.818	0.015
60	0.118	4.500	0.075	0.020	0.750	0.012
65	0.109	4.500	0.069	0.017	0.692	0.011
70	0.101	4.500	0.064	0.014	0.643	0.009
75	0.094	4.500	0.060	0.013	0.600	0.008
80	0.088	4.500	0.056	0.011	0.562	0.007
85	0.083	4.500	0.053	0.010	0.529	0.006

Table 5. Limit values (minima) for the number of elements required in pile discretisation for ν=1/4.

L/R	Extensional			Flexural		
	n_ϑ	n_z	n_R	n_ϑ	n_z	n_R
4	1.613	4.108	1.027	4.938	12.573	3.143
6	1.075	4.108	0.685	2.194	8.382	1.397
8	0.807	4.108	0.513	1.234	6.287	0.786
10	0.645	4.108	0.411	0.790	5.029	0.503
12	0.538	4.108	0.342	0.549	4.191	0.349
14	0.461	4.108	0.293	0.403	3.592	0.257
16	0.403	4.108	0.257	0.309	3.143	0.196
18	0.358	4.108	0.228	0.244	2.794	0.155
20	0.323	4.108	0.205	0.198	2.515	0.126
22	0.293	4.108	0.187	0.163	2.286	0.104
24	0.269	4.108	0.171	0.137	2.096	0.087
26	0.248	4.108	0.158	0.117	1.934	0.074
28	0.230	4.108	0.147	0.101	1.796	0.064
30	0.215	4.108	0.137	0.088	1.676	0.056
35	0.184	4.108	0.117	0.064	1.437	0.041
40	0.161	4.108	0.103	0.049	1.257	0.031
45	0.143	4.108	0.091	0.039	1.118	0.025
50	0.129	4.108	0.082	0.032	1.006	0.020
55	0.117	4.108	0.075	0.026	0.914	0.017
60	0.108	4.108	0.068	0.022	0.838	0.014
65	0.099	4.108	0.063	0.019	0.774	0.012
70	0.092	4.108	0.059	0.016	0.718	0.010
75	0.086	4.108	0.055	0.014	0.671	0.009
80	0.081	4.108	0.051	0.012	0.629	0.008
85	0.076	4.108	0.048	0.011	0.592	0.007

Table 6. Limit values (minima) for the number of elements required in pile discretisation for ν=1/2.

L/R	Extensional			Flexural		
	n_ϑ	n_z	n_R	n_ϑ	n_z	n_R
4	0.000	0.000	0.000	5.409	13.774	3.443
6	0.000	0.000	0.000	2.404	9.182	1.530
8	0.000	0.000	0.000	1.352	6.887	0.861
10	0.000	0.000	0.000	0.865	5.509	0.551
12	0.000	0.000	0.000	0.601	4.591	0.383
14	0.000	0.000	0.000	0.442	3.935	0.281
16	0.000	0.000	0.000	0.338	3.443	0.215
18	0.000	0.000	0.000	0.267	3.061	0.170
20	0.000	0.000	0.000	0.216	2.755	0.138
22	0.000	0.000	0.000	0.179	2.504	0.114
24	0.000	0.000	0.000	0.150	2.296	0.096
26	0.000	0.000	0.000	0.128	2.119	0.082
28	0.000	0.000	0.000	0.110	1.968	0.070
30	0.000	0.000	0.000	0.096	1.836	0.061
35	0.000	0.000	0.000	0.071	1.574	0.045
40	0.000	0.000	0.000	0.054	1.377	0.034
45	0.000	0.000	0.000	0.043	1.224	0.027
50	0.000	0.000	0.000	0.035	1.102	0.022
55	0.000	0.000	0.000	0.029	1.002	0.018
60	0.000	0.000	0.000	0.024	0.918	0.015
65	0.000	0.000	0.000	0.020	0.848	0.013
70	0.000	0.000	0.000	0.018	0.787	0.011
75	0.000	0.000	0.000	0.015	0.735	0.010
80	0.000	0.000	0.000	0.014	0.689	0.009
85	0.000	0.000	0.000	0.012	0.648	0.008

The minimum number of elements to be considered for each direction and evaluated according to previous limitations is shown in Table 7 as a function of the geometrical ratio.

Table 7. Limit values (minima) for the number of elements required in pile discretisation.

L/R	Extensional			Flexural		
	n_ϑ	n_z	n_R	n_ϑ	n_z	n_R
4	1.767	4.500	1.125	5.409	13.774	3.443
6	1.178	4.500	0.750	2.404	9.182	1.530
8	0.884	4.500	0.563	1.352	6.887	0.861
10	0.707	4.500	0.450	0.865	5.509	0.551
12	0.589	4.500	0.375	0.601	4.591	0.383
14	0.505	4.500	0.321	0.442	3.935	0.281
16	0.442	4.500	0.281	0.338	3.443	0.215
18	0.393	4.500	0.250	0.267	3.061	0.170
20	0.353	4.500	0.225	0.216	2.755	0.138
22	0.321	4.500	0.205	0.179	2.504	0.114
24	0.295	4.500	0.188	0.150	2.296	0.096
26	0.272	4.500	0.173	0.128	2.119	0.082
28	0.252	4.500	0.161	0.110	1.968	0.070
30	0.236	4.500	0.150	0.096	1.836	0.061
35	0.202	4.500	0.129	0.071	1.574	0.045
40	0.177	4.500	0.113	0.054	1.377	0.034
45	0.157	4.500	0.100	0.043	1.224	0.027
50	0.141	4.500	0.090	0.035	1.102	0.022
55	0.129	4.500	0.082	0.029	1.002	0.018
60	0.118	4.500	0.075	0.024	0.918	0.015
65	0.109	4.500	0.069	0.020	0.848	0.013
70	0.101	4.500	0.064	0.018	0.787	0.011
75	0.094	4.500	0.060	0.015	0.735	0.010
80	0.088	4.500	0.056	0.014	0.689	0.009
85	0.083	4.500	0.053	0.012	0.648	0.008

Two problems arise at this point. The first one comes from the need to make a whole number the minimum value reported in Table 7 as a function of the geometrical ratio L/R. The second one arises from the need not to have excessively elongated elements. These, in fact, generally lead to numerical instability problems in the integration over the same elements. In order to have good results, the ratio between maximum and minimum element size should be within the range $1 \le d_{max}/d_{min} \le 2$. In those cases in which, on a purely theoretical basis, it would be sufficient to use only one element the choice of using at least two elements has been made. The results shown in Table 8 have been derived by taking into account the above considerations.

Table 8. Minimum number of elements as a function of the geometrical ratio *L*/*R* for *n*=5.

L/*R*	Extensional			Flexural		
	n_ϑ	n_z	n_R	n_ϑ	n_z	n_R
4	3	4	2	5	6	3
6	3	6	2	3	6	2
8	3	8	2	3	8	2
10	3	10	2	3	10	2
12	3	11	2	3	11	2
14	3	13	2	3	13	2
16	3	15	2	3	15	2
18	3	17	2	3	17	2
20	3	19	2	3	19	2
22	3	21	2	3	21	2
24	3	23	2	3	23	2
26	3	25	2	3	25	2
28	3	27	2	3	27	2
30	3	29	2	3	29	2
35	3	33	2	3	33	2
40	3	38	2	3	38	2
45	3	43	2	3	43	2
50	3	48	2	3	48	2
55	3	53	2	3	53	2
60	3	57	2	3	57	2
65	3	62	2	3	62	2
70	3	67	2	3	67	2
75	3	72	2	3	72	2
80	3	76	2	3	76	2
85	3	81	2	3	81	2

It may be interesting to notice how the results shown in Table 8 are strictly related to the maximum extensional and flexural vibration frequencies considered in the analyses. In fact the results of Table 8 refer to case *n*=5. If it were appropriate or necessary to take *n*=10 the results of Table 9 would be obtained.

Table 9. Minimum number of elements as a function of the geometrical ratio L/R for n=10.

L/R	Extensional			Flexural		
	n_ϑ	n_z	n_R	n_ϑ	n_z	n_R
4	4	5	2	24	31	15
6	3	6	2	11	21	7
8	3	8	2	6	15	4
10	3	10	2	4	13	2
12	3	11	2	3	11	2
14	3	13	2	3	13	2
16	3	15	2	3	15	2
18	3	17	2	3	17	2
20	3	19	2	3	19	2
22	3	21	2	3	21	2
24	3	23	2	3	23	2
26	3	25	2	3	25	2
28	3	27	2	3	27	2
30	3	29	2	3	29	2
35	3	33	2	3	33	2
40	3	38	2	3	38	2
45	3	43	2	3	43	2
50	3	48	2	3	48	2
55	3	53	2	3	53	2
60	3	57	2	3	57	2
65	3	62	2	3	62	2
70	3	67	2	3	67	2
75	3	72	2	3	72	2
80	3	76	2	3	76	2
85	3	81	2	3	81	2

Numerical Applications. Some numerical applications have been carried out to illustrate the theoretical results previously presented. Figures 6 and 7 show the frequency response functions of the displacement at the top of the cantilever for a motion imparted to the base obtained by considering n=5 and $d_{\max}/d_{\min}=2$. The real and imaginary parts and the amplitude of the above functions are shown. It may be seen that corresponding to the natural frequencies, real and imaginary parts and amplitude of the functions become singular.

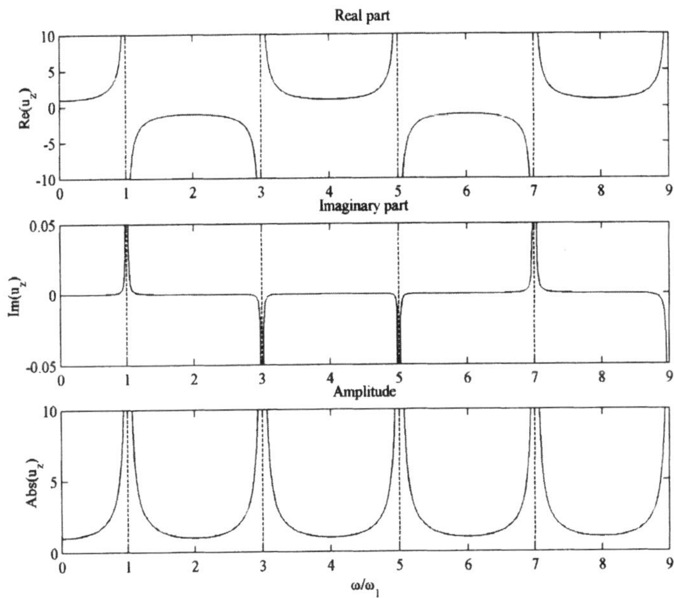

Figure 6. Real part, imaginary part and amplitude of the displacement at the tip of a cantilever beam subject to an axial harmonic motion imparted at the base, as a function of frequency; L/R=80. Broken straight lines represent the frequencies of vibration of the cantilever beam considered as a one-dimensional system.

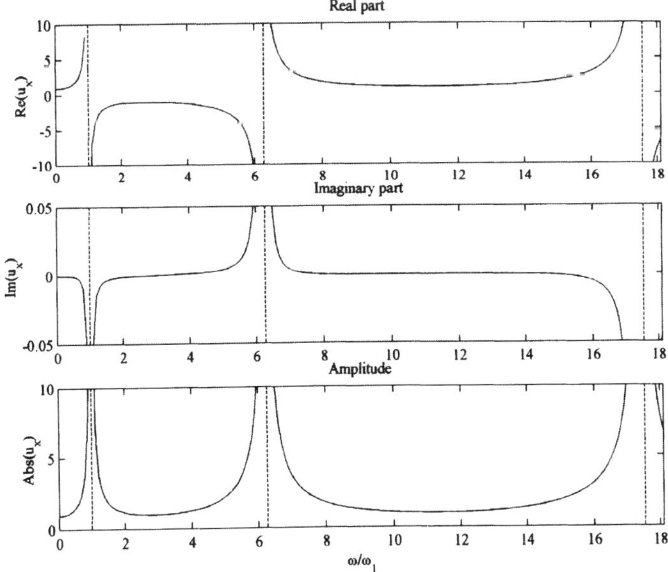

Figure 7. Real part, imaginary part and amplitude of the displacement at the tip of a cantilever beam subject to a transverse harmonic motion imparted at the base, as a function of frequency; L/R=80. Broken straight lines refer to the frequencies of vibration of the cantilever considered as a Bernoulli beam.

In order to show the importance of the size or element elongation ratio d_{max}/d_{min}, the previous numerical applications have been repeated using values of such a ratio equal to 1, 1.5, 2.5, 3 and 4 (Figure 8). It may be seen as for ratios up to 2.5 the results are in good agreement amongst themselves and with those corresponding to $d_{max}/d_{min} = 2$. For values of the elongation ratio equal to 3 and 4 there exist frequency ranges where the results are significantly different from those obtained for elongation ratios ≤ 2 pointing to a substantial numerical instability. For elongation ratios $d_{max}/d_{min} > 3$ the results show a clear-cut overestimate of the first few frequencies of vibration as one could have expected anyway.

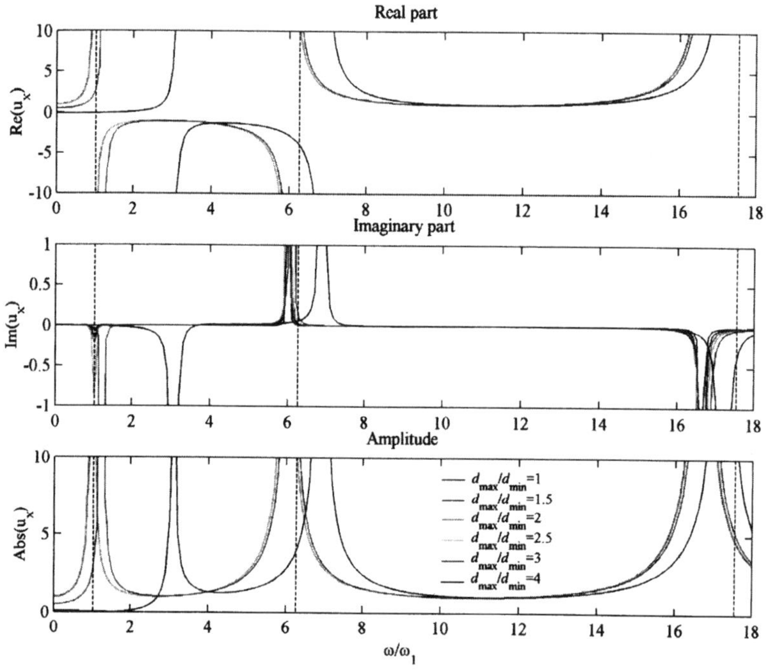

Figure 8. Real part, imaginary part and amplitude of the displacement at the tip of a cantilever beam subject to a transverse harmonic motion imparted at the base, as a function of frequency. *L*/*R*=40. Different discretisations corresponding to d_{max}/d_{min} = 1, 1.5, 2, 2.5, 3, 4.

Another important aspect to be considered in the discretisation of a pile is the number of elements to be used over ¼ of the circumference. Some numerical investigations have been performed by assuming an elongation ratio $d_{max}/d_{min} = 2$ and varying the number of elements n_ϑ from 1 to 4. The results are shown in Figure 9. It is immediately seen that results are practically invariant for $n_\vartheta \geq 2$. It follows that $n_\vartheta = 2$ represents the optimal number of elements from a point of view of computational efficiency without affecting the accuracy of the results.

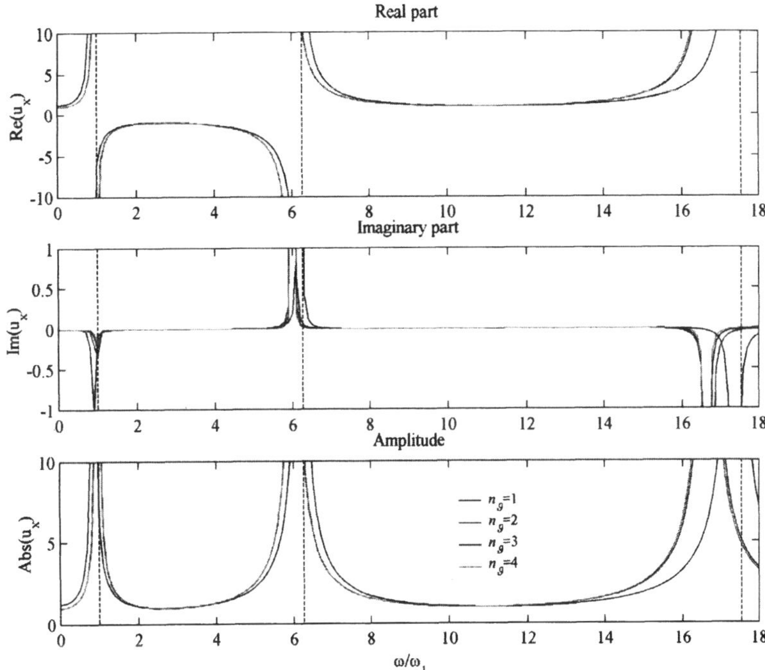

Figure 9. Real part, imaginary part and amplitude of the displacement at the tip of a cantilever beam subject to a transverse harmonic motion at the base, as a function of frequency; L/R=40. Different discretisations corresponding to $n_\vartheta = 1, 2, 3, 4$.

Discretisation of the free surface. As has been already observed for surface and embedded plane foundations the extension of the free surface that must be discretised is highly dependant on the depth of the foundation plane. By extrapolation it may be thought that in the case of a pile such an extension must depend in a significant way on the length of the pile itself. From a purely intuitive viewpoint it does not appear reasonable to consider free surface extensions of radius R_S smaller that the pile length (L). For this reason the minimum extension considered has been taken with radius equal to the pile length, $R_S/L = 1$. Extensions characterised by the ratios R_S/L equal to 2 and 3 have also been considered. In spite of the fact that the extension of the discretised free surface appears to be a significant parameter, it may also be that the density of the discretisation considered is of significance. To test this, two types of investigation have been carried out. In one the discretisation emerging from the pile surface has been maintained also on the free surface keeping unaltered the limits on the maximum element length and on the elongation ratio. In the other one, the extension of the discretised free surface has been kept as in the previous case, but the density of the discretisation has been increased by reducing the element size and increasing their number. The discretisations for the first case are shown in Figure 10 while those for the second case are shown in Figure 11.

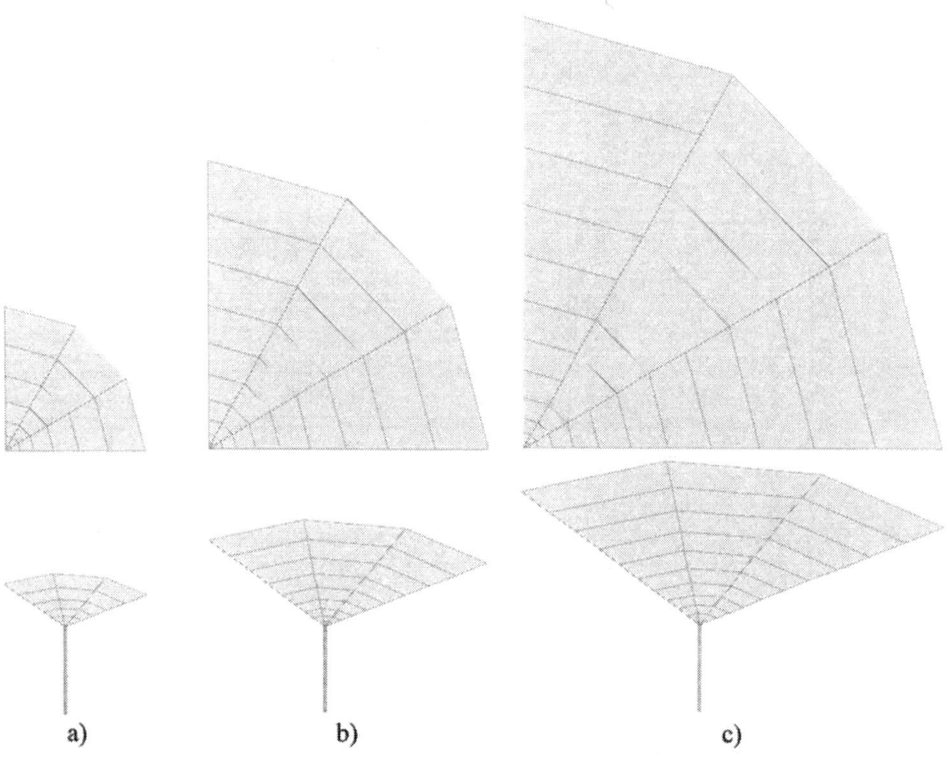

Figure 10. Three different discretisations of the free surface. a) $R_S = L$, b) $R_S = 2L$, c) $R_S = 3L$.

The number of elements considered for each discretisation is reported in Table 10.

Table 10. Number of elements on the free surface for different discretisations.

R_S/L	Discretisations of Figure 10	Discretisations of Figure 11
1	21	84
2	30	120
3	36	192

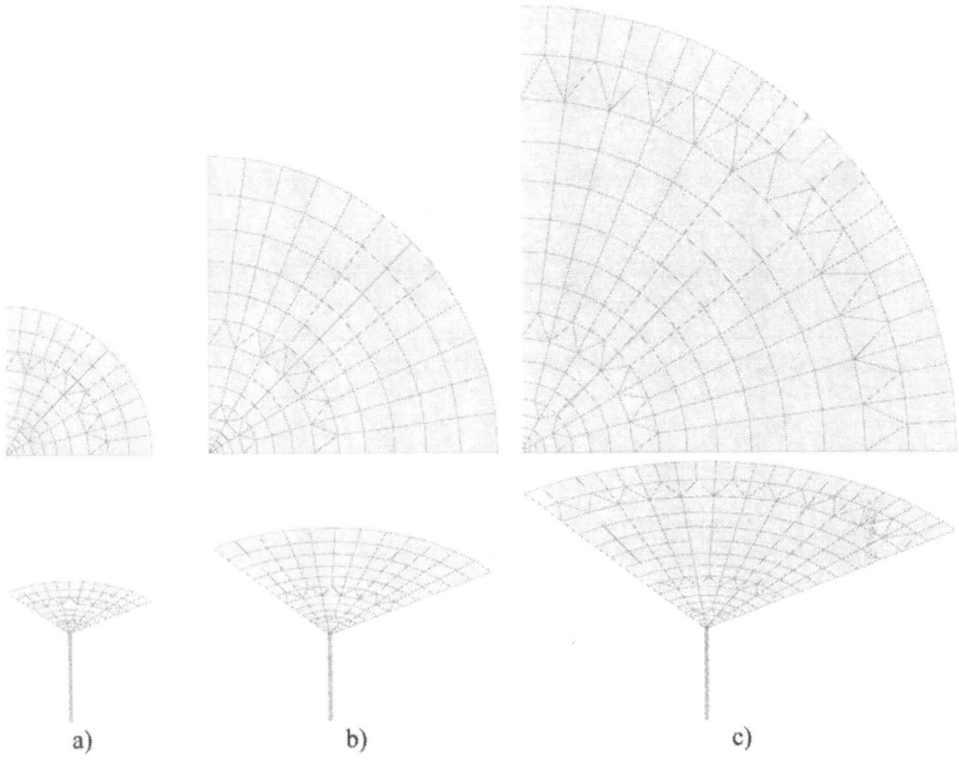

Figure 11. Three different discretisations of the free surface. a) $R_S = L$, b) $R_S = 2L$, c) $R_S = 3L$.

Comparison between the three discretisations has been carried out for the direct terms of the dynamic stiffness matrix (diagonal terms) referred to the pile head. The results for the first investigation (first case), normalised with respect to the corresponding static stiffness, are shown in Figure 12. With reference to the real part of such coefficients (stiffness) it may be seen that the results do not differ significantly, especially at low frequencies. At high frequencies the results show a tendency to diverge, implying that for higher frequencies it may be appropriate to increase the extension of the discretised free surface. Similar comments apply for the imaginary part (radiation damping). In the latter case the behaviour at low frequencies should be examined further as the damping coefficients should vanish as the frequency approaches zero. In Figures 13, 14 and 15 the previous results are compared for the same free surface extension, R_S/L, with those for denser discretisations. Overall the results seem to imply that no significant advantage may be gathered from a denser discretisation nor from a larger extension of the discretised free surface.

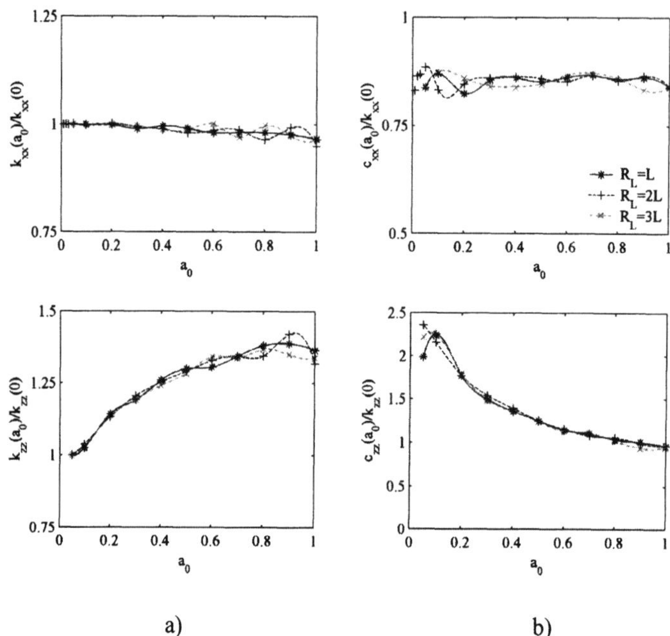

Figure 12. Direct dynamic stiffness coefficients for a pile: *L*/*R*=40, $E_s/E_p = 10^{-2}$, $\rho_s/\rho_p = 0.7$. a) real part, b) imaginary part.

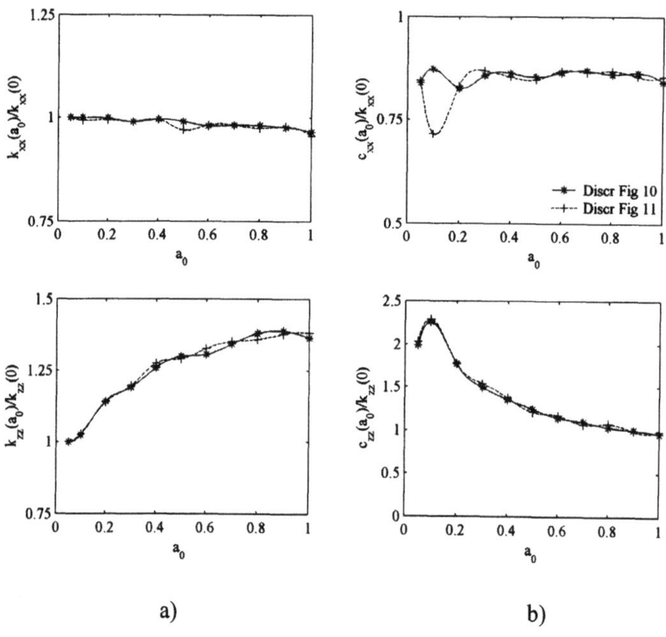

Figure 13. Direct dynamic stiffness coefficients for a pile: *L*/*R*=40, $E_s/E_p = 10^{-2}$, $\rho_s/\rho_p = 0.7$, $R_S = L$. a) real part, b) imaginary part.

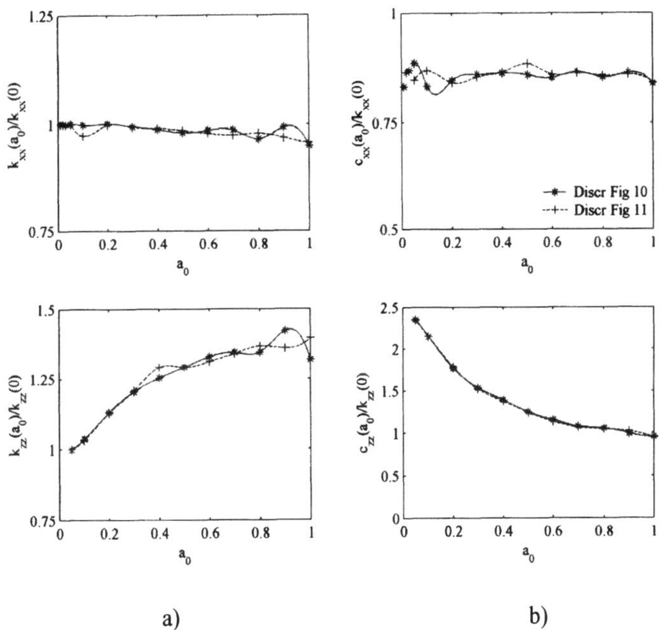

Figure 14. Direct dynamic stiffness coefficients for a pile: *L*/*R*=40, $E_s/E_p = 10^{-2}$, $\rho_s/\rho_p = 0.7$, $R_S = 2L$. a) real part, b) imaginary part.

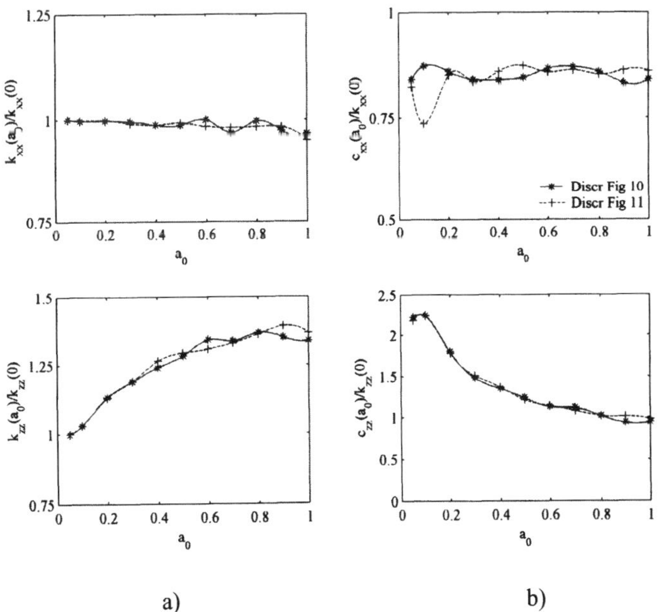

Figure 15. Direct dynamic stiffness coefficients for a pile: *L*/*R*=40, $E_s/E_p = 10^{-2}$, $\rho_s/\rho_p = 0.7$, $R_S = 3L$. a) real part, b) imaginary part.

Pile-soil-pile interaction. The dynamic interaction among piles has considerable importance when, as often occurs, the dynamic behaviour of foundations on pile groups must be studied. In the absence of interaction the results for a single pile could be realistically used for the group. In the presence of interaction the system must be studied as a whole. In the present study the case of three aligned piles will be considered in order to obtain information on the behaviour of a similar system undergoing experimental static and dynamic testing. The latter is described in a companion paper appearing in the present volume. The cases studied and results are graphically shown in Figure 16.

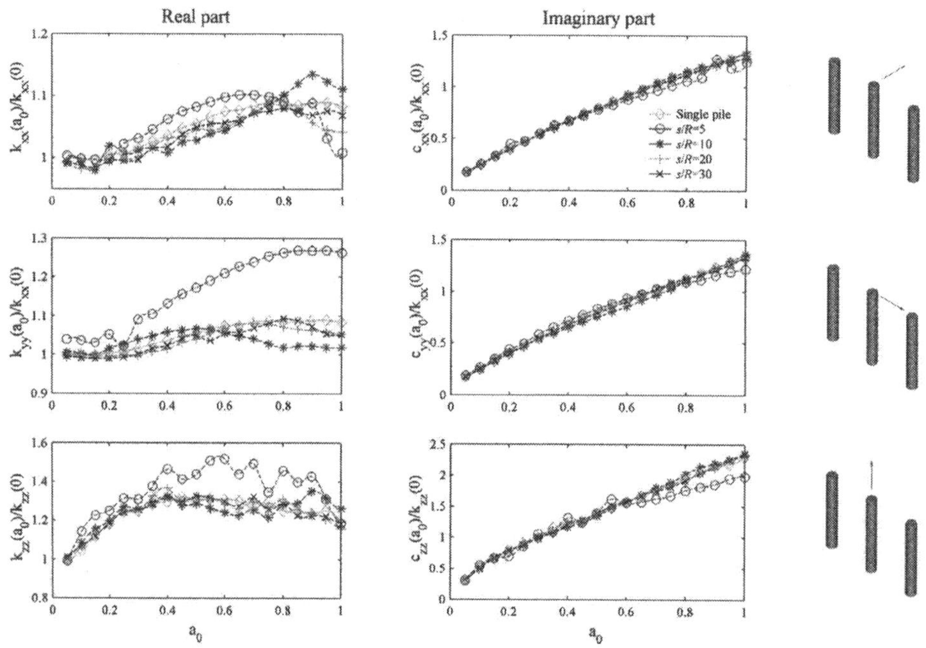

Figure 16. Dynamic stiffness coefficients for a 1×3 pile group: *L*/*R*=40, $E_s/E_p = 5 \cdot 10^{-4}$, $\rho_s/\rho_p = 0.8$, $R_S/R = 80$.

The loading conditions considered are shown in Figure 16 already referred to. It should be noted that the loaded pile is always the central one while the presence of the lateral piles determines the interaction. The response of the single pile is shown for comparison together with those for several separation distances. The cases considered are respectively: *s*/*R*=5, *s*/*R*=10, *s*/*R*=20, *s*/*R*=30 and the single pile (*s*/*R*=∞). It may be seen immediately that, independent of the loading condition, the interaction is more significant for the real part (stiffness) than for the imaginary part (radiation damping) of the impedance function. Moreover, in the case of transversal loading (horizontal) the interaction is more marked for the loading acting parallel to the pile alignment than for the loading acting perpendicularly. Moreover it is clear that the interaction effect becomes the more significant for increasing frequencies.

Nevertheless it is particularly relevant for the aim of the present study to see that, within the investigated frequency range, results do not differ significantly from those of the single pile for values of the separation parameter s/R greater than 10. In the case of the previously mentioned testing programme the minimum s/R ratio is equal to 15. Therefore, also in view of the uncertainties in the mechanical parameters, it appears reasonable in the present case to limit the study to the single pile. The results obtained should provide useful information for the planning of the testing programme. The complete modelling should subsequently provide a basis for the dynamical identification of the mechanical parameters of the soil-pile system.

Dynamical testing on three aligned piles. Within the present research project a device for static and dynamical testing on three aligned piles has been designed and a prototype has been constructed (Figure 17). The device is arranged in such a way that each pile, while providing contrast for the other two, is subject at the same time to a testing procedure. The dynamical tests are conducted by imposing two horizontal forces to the end piles, a horizontal force and a couple to the central pile. According to the original design two equal and opposite horizontal forces were to be applied to the end piles and a pure couple to the central one. Unavoidable friction forces have resulted in unequal forces being applied to the end piles and in a small horizontal force being applied to the central pile in parallel to the couple. Forces are however accurately measured by the testing apparatus at every stage of the loading process. The forces previously mentioned are initially applied in a pseudo-static fashion which allows for the conduct of static tests, including creep tests.

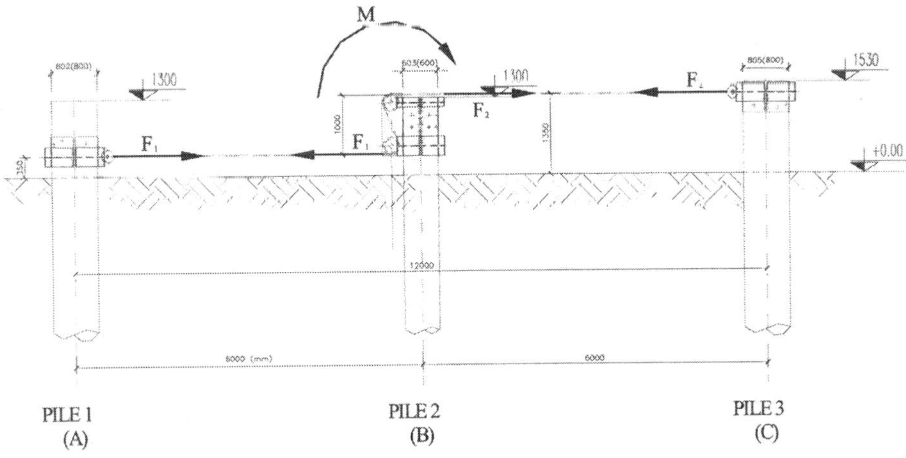

Figure 17. Schematic description of the testing apparatus on three aligned piles.

The dynamical test is initiated by the sudden release of the applied forces. For the preparation of the testing programme and the predisposition of the data acquisition apparatus considerable importance has acquired the prediction of the expected displacements at the various stages of the loading programme. On the basis of the analysis presented in the previous paragraph the study has been conducted with reference to a single pile. The description of the pile surface and of the free surface has been carried out according to the criteria established in paragraph 5.2. The mechanical characteristics of the soils were derived on the basis of the wave propagation velocities (longitudinal and transversal) acquired by means of down-hole and cross-hole seismic tests and of additional information provided by laboratory tests. Uncertainties about the wave velocity profiles have resulted in a poor characterisation of the soil properties. Two limiting profiles have been established for the initial elastic shear modulus of the soil as shown in Figure 18. Those correspond to values of Poisson's ratio respectively equal to 0.48 and 0.49.

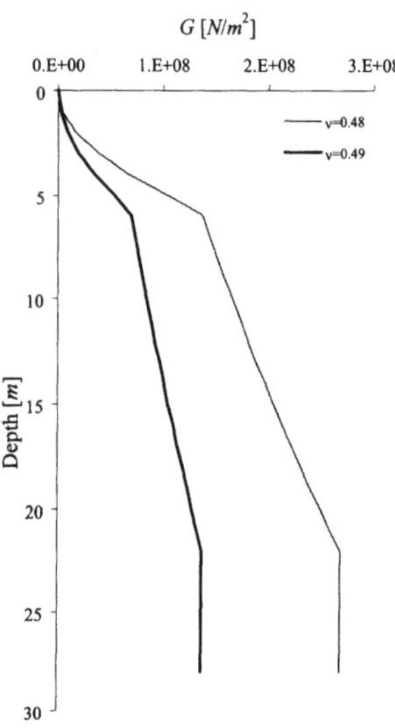

Figure 18. Extreme profiles for the initial elastic shear modulus for the soil.

It is obvious that to each profile will correspond different values of the static and/or dynamic stiffness coefficients and consequently different predictions of the displacements expected at the pile head.

Table 11. Attributed values of the shear modulus G for each soil layer.

Depth [m]		G [MN/m^2]	
from	to	ν=0.49	ν=0.48
0	6	2.83	11.25
6	22	86	167
22	∞	137	269

The shear modulus variability with depth produces considerable modelling problems. In the present study a layered soil model with the G values specified in Figure 19 and in Table 11 has been adopted. The Poisson's ratio values have been specified correspondingly.

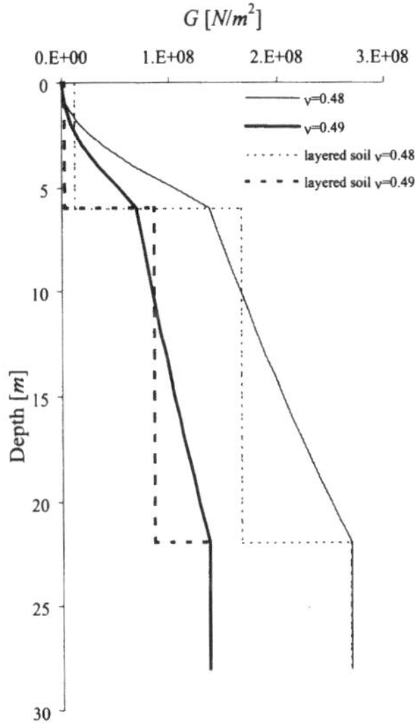

Figure 19. Values of the shear modulus G attributed to each soil layer.

As an equivalence criterion the equality of the displacement at the top of the layer considered has been enforced for the actual distribution of G and for the equivalent case of constant G. For this a rigid model on distributed springs has been used with spring stiffness proportional to G. Two limit models of layered soil are obtained in this way which hopefully may provide upper and lower bounds to the expected displacement at the pile head. It is important to realise that in the case of a layered soil model, besides the discretisation of the free surface, the soil layer interfaces need also to be discretised. In this way equilibrium and compatibility conditions can be enforced at layer interfaces. In the present study the same discretisation has been used for the free surface and for the soil interfaces.

Results. The solution is sought by using the following approach. A unit horizontal displacement is enforced at the nodal points at the pile head while keeping the vertical component equal to zero. The numerical model leads to the system of equations (6) which, once solved, provides the nodal parameters needed to evaluate the stress distribution on the pile head. The stress and moment resultants are then evaluated by statics. The stress resultant represents the direct horizontal translation stiffness while the moment resultant represents the cross translation-rotation stiffness or cross term. In a similar way, by enforcing a unit rotation at the pile head, the direct rotational stiffness and the cross rotation-translation stiffness are evaluated. The latter, by the reciprocity theorem of linear elasticity, must equal the cross translation-rotation stiffness previously evaluated. The results obtained for the two soil models considered are collected in

Tables 13 and 14 having assumed the geometrical and mechanical characteristics of Table 12 for the piles.

Table 12. Geometrical and mechanical characteristics of piles.

Piles	G $[MN/m^2]$	ν	ρ $[kg/m^3]$	L $[m]$	R $[m]$
End piles (A, C)	11875	0.25	2500	25	0.4
Central pile (B)	11875	0.25	2500	25	0.3

Specifically the results of Table 13 refer to the two end piles of diameter 80 *cm* while the results of Table 14 refer to the central pile of diameter 60 *cm*.

Table 13. Stiffness coefficients for the piles of length L=25 *m* and diameter 2R=0.8 *m*.

Soil			k_{hh} $[MN/m]$	$k_{\phi h}$ $[MNm/m]$	$k_{h\phi}$ $[MN]$	$k_{\phi\phi}$ $[MNm]$
G $[MN/m^2]$	ν	ρ $[kg/m^3]$				
2.83 86 137	0.49	1900 2000 2100	48.15	90.25	107.2	414.6
11.25 167 269	0.48	1900 2000 2100	125.9	152.9	179	522.8

Table 14. Stiffness coefficients for the pile of length L=25 *m* and diameter 2R=0.6 *m*.

Soil			k_{hh} $[MN/m]$	$k_{\phi h}$ $[MNm/m]$	$k_{h\phi}$ $[MN]$	$k_{\phi\phi}$ $[MNm]$
G $[MN/m^2]$	ν	ρ $[kg/m^3]$				
2.83 86 137	0.49	1900 2000 2100	31.18	41.65	51.16	157.9
11.25 167 269	0.48	1900 2000 2100	89.39	79.52	94.88	212.7

Once the stiffness matrix relative to the pile head is available for the pile-soil system, an estimate of the expected displacements during the static tests may be obtained. Displacements and rotations at the pile head for different amplitudes of the applied load are collected in Table 15 according to the sign conventions shown in Figure 17. From a graphical viewpoint the results are shown in Figure 20 for the end piles, subjected to horizontal forces, and in Figure 21 for the central pile under the action of a couple.

Table 15. Displacements and rotations on pile heads.

ν	End piles (A, C)						Central pile (B)					
	Displacement $[mm]$			Rotation $[rad \times 10^{-3}]$			Displacement $[mm]$			Rotation $[rad \times 10^{-3}]$		
	50 *kN*	100 *kN*	150 *kN*	50 *kN*	100 *kN*	150 *kN*	50 *kN*	100 *kN*	150 *kN*	50 *kN*	100 *kN*	150 *kN*
0.49	1.977	3.953	5.930	-0.430	-0.861	-1.291	-0.464	-0.928	-1.392	0.283	0.565	0.848
0.48	0.667	1.334	2.001	-0.195	-0.390	-0.585	-0.209	-0.419	-0.628	0.197	0.395	0.592

The performance of the experimental tests should permit the assessment of the reliability of the hypotheses at the basis of the present modelling. Hopefully the results should at least fall within the fuzzy domains of Figures 20 and 21.

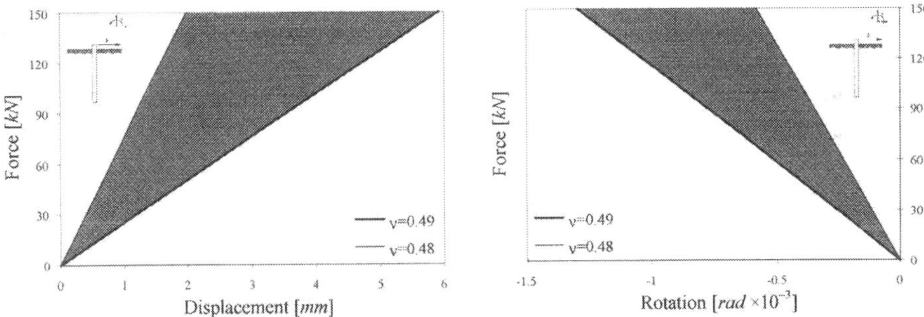

Figure 20. Force-Displacement and Force-Rotation diagrams for the end pile A (Opposite displacements and rotations for pile C at the opposite end).

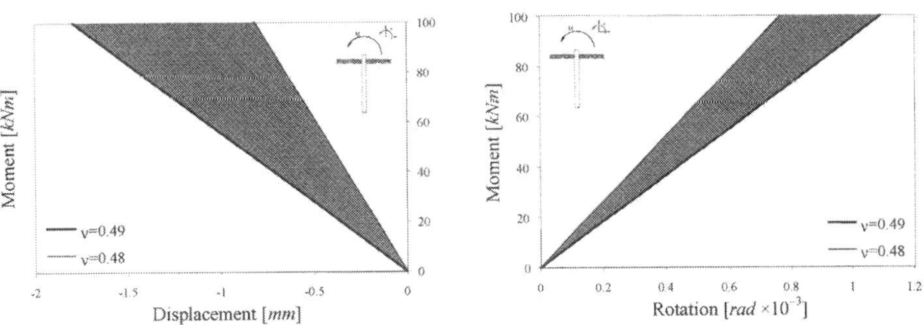

Figure 21. Moment-Displacement and Moment-Rotation diagrams for the central pile (B).

6 Conclusions

In this study some discretisation problems for the pile-soil system have been considered and partially solved in both the static and the dynamic field. More specifically, important problems concerning the accuracy of the model and of the results have been addressed. The problem of the definition of the number of elements necessary for an adequate modelling of the pile surface has been dealt with in detail and the parameters which determine it have been singled out. The problem of the element elongation ratio, which is dominant when modelling oblong bodies, has been treated in depth and a solution has been proposed. Finally particular attention has been devoted to the modelling of the free surface and of the layer interfaces.

Numerical applications have been directed to the specific situation of an experimental test field, in a way as to directly compare the results of the numerical modelling with those of the experimental tests. This has highlighted important situations that occur frequently in practice and which should be accounted for if numerical analyses are to be of practical value. It is well known from geotechnical science that the mechanical characteristics of clay, as well as those of other geotechnical materials, are highly dependant on the deformation level. This has not been taken into account for in the present work and the results of the experimental tests should be interpreted in the light of this behaviour. Moreover clay deposits occur in the form of heterogeneous formations, sometimes layered, but more often with mechanical characteristics gradually varying with depth, especially near the surface. The considerable variation of the mechanical properties with depth poses serious limitations to the half-space model for the soil.

This very important aspect has been dealt with in an approximate manner within the present research. Hopefully fundamental solutions for an heterogeneous half-space, at least in the sense specified above, will become available in the near future. In that direction efforts are being carried out in the literature and some results are already appearing (Manolis, G. D, 1997; Manolis, G. D. Shaw, R. P. 2000; Manolis, G. D. Rangelov, T. V. Shaw, R. P. 2002).

In the present study all the preparatory work for the conduction of dynamical analyses has been performed. Even in this case the treatment is limited to linear behaviour. The experimental tests will show the limitations of linear models and hopefully provide hints on the necessary corrections required to obtain through such analyses satisfactory estimates of the real behaviour.

References.

Aliabadi, M. H. BEM for crack dynamics. In *Hall, W. S., Oliveto, G., Editors* (2003). *Boundary Element Methods for Soil-Structure Interaction.*

Bonnet, M., Maier, G., Polizzotto, C., (1998). Symmetric Galerkin Boundary Element Method. *Applied Mechanics Reviews*, 51:669-704

Buda, G., Oliveto, G. (2001), Dynamic identification of a rectangular foundation raft on piles, *Minisymposium on Interaction Problems in Structural Mechanics*, AIMETA 2001.

Buda, G. Sciacca, P., Oliveto, G. (2003), A device for static and/or dynamic identification tests on foundation piles, *Parallel contribution to the present volume.*

Domínguez, J. (1993). Boundary Elements in Dynamics. *Computational Mechanics Publications*, Elsevier Applied Science, London New York.

Euromech Colloquium 414 *Boundary Element Methods for Soil-Structure Interaction*, University of Catania, Italy, 21-23 June 2000. Hall, W. S. (University of Teesside), Oliveto, G. (University of Catania), Chairmen.

Frangi, A., Maier, G., Novati, G., Springhetti, R., A general boundary element tool for the analysis of 3D linear elastic fracture mechanics. In *Hall, W. S., Oliveto, G., Editors* (2003). *Boundary Element Methods for Soil-Structure Interaction.*

Hall, W. S., Oliveto, G., Editors (2001), Boundary Element Methods in Soil-Structure Interaction and other applications. *Special Issue of the International Journal Meccanica*, Kluwer Academic.

Hall, W. S., Oliveto, G., Editors (2003), *Boundary Element Methods for Soil-Structure Interaction*, Kluwer Academic Publishers, Dordrecht, Hardbound, ISBN 1-4020-1300-0.

Interaction Problems in Structural Mechanics, Mini-symposium, AIMETA 2001, Oliveto, G., Hall, W. S., Chairmen.

Manolis, G. D., (1997). Green's Function for the Vector Wave Equation in a Mildly Heterogeneous Medium. *Solid Mechanics*, 87:163-170

Manolis, G. D., Shaw, R. P., (2000). Fundamental Solutions for variable density two-dimensional elastodynamic problems. *Engineering Analysis with Boundary Elements*, 24:739-750

Manolis, G. D., Rangelov, T. V., Shaw, R. P., (2002). Conformal mapping methods for variable parameter elastodynamics. *Wave Motion*, 36:185-202

Wrobel, L. C., Aliabadi, M. H. (2002). The Boundary Element Method. Wiley, Chichester, Vol.1: ISBN: 0-471-72039-9. Vol.2: ISBN: 0-470-84298-9

A device for static and/or dynamic identification tests on foundation piles

Giuseppe Oliveto, Giuseppe Buda and Pietro Sciacca

Department of Civil and Environmental Engineering, Structural division, University of Catania, Catania, Italy

Abstract. This work describes the design and construction process of an experimental device for static and/or dynamic identification tests on foundation piles. The aim is to provide information on the behaviour of soil-pile systems subject to seismic action. The primary objective is the validation of numerical models currently available for research purposes pointing out inevitable shortcomings and outlining aspects that should be investigated in detail in the near future. In this respect this paper is closely related to a companion one on numerical aspects published in this same volume. The device is specifically designed for the application of horizontal forces and couples to pile heads. Besides the evaluation of the limiting values of dynamic parameters, generally speaking stiffness and flexibility coefficients, static tests have the objective of identifying the viscoelastic behaviour of the system through creep tests. The experimental tests carried out by means of this prototype device will provide an opportunity to gather information not only on the global behaviour of the soil-pile system but also on the strain and stress state within the pile stem and indirectly within the soil.

1 Introduction

The experience gained by the research group in Catania on static and dynamic testing of soil-structure interacting systems, such as full scale buildings and foundation rafts on piles, has shown to the complexity of the problem and has suggested the design and execution of a simpler research programme addressing the simplest of the structural systems. Within the field of foundations on piles this may be identified as the single pile-soil system.

In such a case, even only for the static field, the behaviour under horizontal loads is still little known. For this reason our choice has been on the identification of the static and dynamic behaviour of the single pile-soil system under horizontal loads. Therefore it has appeared appropriate to design and construct an experimental device with the capability of applying horizontal forces and/or couples to pile heads.

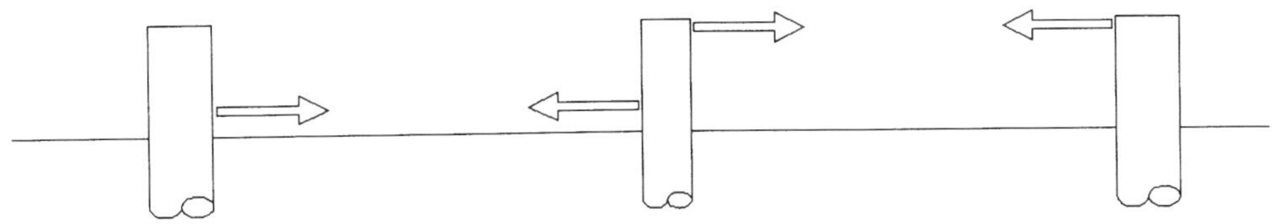

Figure 1. Schematic view of the test system.

For reasons of economy a device which could apply at the same time horizontal forces on some piles and couples on others has been conceived. In this way the single piles, besides constituting test specimens, provide contrast to the other piles. The basic scheme is shown in Figure 1.

Three aligned piles are set at such a distance that the pile-soil-pile interaction turns out to be negligible. The end piles are subject to horizontal forces applied at different heights; the central pile, through appropriate gearing, is subjected to a pure couple. It is evident that the testing system is a closed one and that each pile provides contrast to the other two. The above description illustrates the basic idea. In the following, the various elements that have had to be designed and constructed to realize the basic idea will be presented. At this point it should be emphasised that the above description explains readily only the static part of the experiments and perhaps the part concerning time delayed effects (viscous behaviour, creep tests). However nothing has been said about the dynamic nature of the test. This is activated at the end of each static test by the sudden release of the applied forces by the breaking of a properly designed weak link inserted within the testing devise. Because the test results are to be used to assess the reliability of numerical algorithms for the description of the static and/or dynamical behaviour of the soil-pile system it has seemed appropriate to include in the experimental programme a series of tests intended to find the characterisation of the pile and of the soil. Therefore, besides the description of the basic experimental device, in the following the tests and instrumentation used to characterise the pile and the soil will also be briefly illustrated.

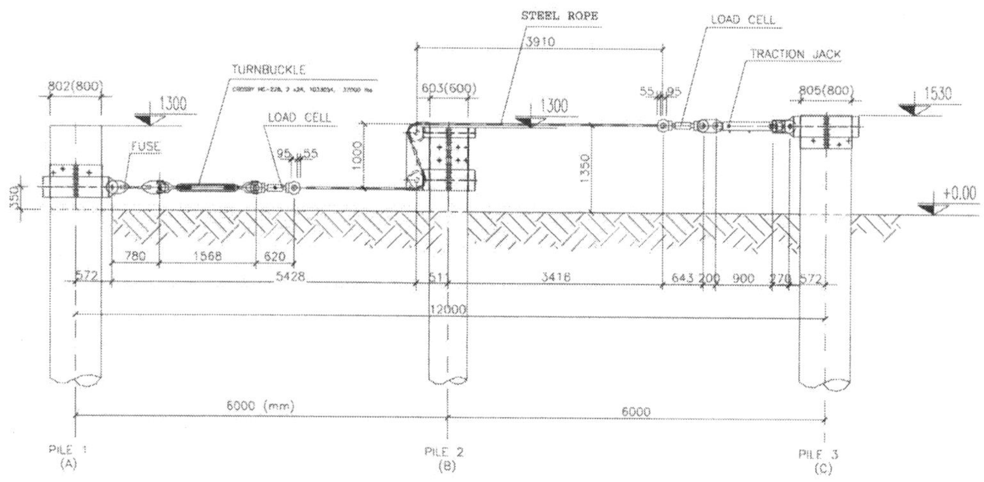

Figure 2. Components of the testing device.

2 Description of the test device

The elements which related to the constitution of the test device are shown schematically in Figure 2. They are listed below:

1. three collars to attach to the pile heads;
2. one steel rope for load transmission;
3. two pulleys to change load height and direction for couple application;
4. one traction jack for application of horizontal load;
5. two load cells to measure load in each branch of the load transmission rope;
6. One turnbuckle for fine tuning initial configuration (rope in horizontal position with no applied load);
7. Quick release device (fuse).

The device is useless in the absence of reaction and test piles. In the present case these have been built as test piles within the construction of the Teaching Activities Building of the Engineering Faculty at the University of Catania. The research group is grateful to the University of Catania for providing the test site, the piles and for allowing the testing activities. The description of the piles will be part of a following paragraph.

2.1 Pile head collars

The collars needed to engage pile heads have different characteristics for each pile. The differences are essentially related to the type of load to be applied and to the elements to be connected. With reference to Figure 2, the end piles are subject to horizontal forces applied at different heights. In principle there should be no need for any difference between the two collars; however they appear with the characteristics shown in Figure 3 and in Figure 4.

Figure 3. Collar on pile n°1.

Figure 4. Collar on pile n° 3.

The collar for pile n°1 in Figure 2 is shown in Figure 3, while the collar for pile n°3 is shown in Figure 4. It may be observed how the collars are essentially identical, although for loading requirements they are mounted in a symmetrical configuration. Owing to the fact that the force must be applied at a lower height, through collar n°1 it is mounted as in Figure 3, while an equal and opposite force must be applied at a greater height through collar n°3; therefore it is mounted as in Figure 4. Both are built as half-cylinders properly stiffened and provided with attachment devices. After positioning by means of a site fork-lift the two half-cylinders are tied together by means of high strength bolts.

As may be seen from Figure 4, the traction jack for the load generation is attached to the collar for pile n°3 . At the other end the quick release device (fuse) is attached on one side to the collar on pile n°1 and on the other to a CROSBY turnbuckle model HG-228, as may be seen in Figures 3, 9 and 10.

The collar for the central pile is shown in Figure 5. Special characteristics of this collar are the supports for the two pulleys used to change the steel rope alignment. This has resulted in a larger height of this collar with respect to the other two, especially in order to provide a suitable lever arm for the application of the couple. Additional dimensional differences are apparent because the central pile has a diameter of 600 mm while the end piles have a diameter of 800mm. The choice of dimensions arose from the need to test both types of piles used in the foundation of the TA building of the Faculty of Engineering at the University of Catania. As was the case for the other two collars the one for the central pile consists of two half-cylinders tied together by means of high strength bolts. The fastening system and the rope alignment system are both clearly visible in Figure 5.

Figure 5. Collar on pile n°2.

2.2 The steel rope

The steel rope is a fundamental component of the testing device because of its function of load transmission. The size and mechanical characteristics of the rope have been evaluated as a function of the maximum testing load. The evaluation of the latter forms the object of an appropriate paragraph. The characteristics of the steel rope, as provided by the manufacturer, are:

- nominal diameter 32 mm;
- length 7048 mm;
- operational load 172 kN;
- safety margin 6;
- tested in our laboratory up to T = 344 kN.

The rope was manufactured and assembled by CERTEX UK Ltd in the required design length with end fittings G416 CROSBY for the connection to the load cells for the load measurement.

Part of the rope is visible in Figure 7 when passing on top of the central pile, while one end with end fitting and eye connector to a load cell is visible in Figure 6.

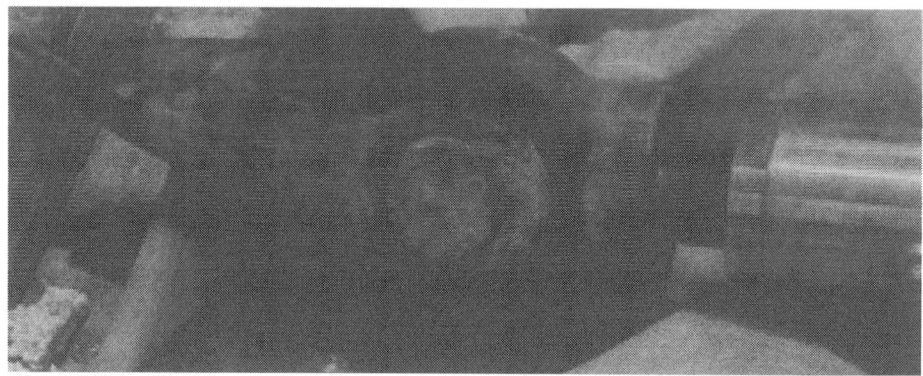

Figure 6. Rope end with fitting.

2.3 The pulleys

The function of the pulleys is to displace the steel rope path is such a way as to apply a couple to the central pile. Especially important for that purpose is the pulley rotation so that the rope may slide freely and the same traction be applied at both ends of the rope. The inevitable friction and mounting defects makes this aim difficult to achieve. A first prototype of the pulley system displayed an unacceptable load difference at the two ends of the rope. Successive mounting improvements have lead to a diminution in the traction unbalance but the improvement obtained has not been considered sufficient for the aims of the experiment. For this reason a new pair of pulleys has been constructed with the inclusion of suitable bearings to minimise friction and traction unbalance.

Figure 7. Change of the steel rope path via pulleys.

2.4 The traction jack

The traction jack type COT30N300X, built by EURO PRESS PACK s.r.l. (Genoa – Italy), is visible in Figure 4. This is a fundamental component of the test system; besides providing the force to be applied to the pile head, it must have a long enough stroke to accommodate the displacement at the pile heads. The chosen jack can apply a maximum load of 324 kN at a pressure 700 bar for a maximum stroke of 300 mm. These values cover very well the theoretical range of the test parameters which requires a maximum load of 150 kN for a maximum stroke not larger than 15 mm. On the theoretical prediction of the maximum displacement at pile heads and therefore on the required jack stroke, see the parallel article in this volume [Vinciprova, Maeso Fortuny, Aznárez and Oliveto].

2.5 The load cells

Two traction load cells for static and dynamic measurements have been built on the design specifications of the research group by NOVATECH Measurements Ltd (England). One of the two load cells is shown in Figure 8. The purpose of the cells is the correct evaluation of the forces applied to the pile heads both in static and dynamic conditions. In particular it may occur that is applied only a couple or a couple plus a horizontal force to the central pile. For that reason the two load cells have been placed at the two ends of the steel rope respectively in the section between piles 1 and 2 and in the section between piles 2 and 3. Particularly important, in view of the size of the cells and of the transmitted load, are the connectors which link the cells to the steel rope and other components of the test system: the turnbuckle in the vicinity of pile n°1 and the traction jack close to pile n°3. Such connections are clearly visible in Figure 8.

Figure 8. Load cell type F204 built on request by NOVATECH Ltd.

The characteristics of the load cells are as follows:

- load range 0 – 500 kN;
- non-linearity 0.15% maximum load;
- repeatability 0.02% maximum load.

The characteristics of the eye connectors M64x3 mm type BS4278 are:

- operational load 200 kN;
- safety margin 5.

2.6 The turnbuckle

In order to facilitate the assembling and mounting operations the length of the steel rope is slightly larger than the minimum required. Therefore the total length of the system chain exceeds by a small amount the distance between the heads of the end piles. The resulting sag is removed by means of the CROSBY turnbuckle type HG-228 supplied by CERTEX UK Ltd, Figure 9. As provided by the manufacturer, the characteristics of the turnbuckle are:

- weight 420 N;
- operational load 168 kN;
- safety margin 5;
- tested in our laboratory at T = 420 kN;
- length 1940/1340 mm.

The presence of such a component is fundamental in order to preserve the jack stroke for the loading operations. The nominal length which can be recovered by the turnbuckle is $\pm$ 300 mm. The type of turnbuckle was selected on the basis of its maximum regulation length and of the maximum transmissible load. In the assembly phase the turnbuckle is inserted in the configuration of maximum extension allowing for a maximum recoverable length of 600 mm.

Figure 9. CROSBY turnbuckle type HG-228 for sag removal.

2.7 The quick release device (fuse)

The quick release device show in Figure 10 is fundamental for the application of dynamic actions. In fact it allows the execution of free vibration tests under different amplitudes of the imposed displacement. Such a device is built starting from a high strength steel bar, therefore having a limited rupture strain, by reducing the cross section until the required rupture load is obtained. On the basis of the mechanical characteristics of the material and ad-hoc laboratory tests, the strength is regulated in such a way that rupture occurs at not more than 10% above the maximum load specified for the corresponding static test. Once the static test has been completed, the load is increased rapidly until the rupture of the device occurs with the sudden load release and subsequent free oscillation of the system. The fuse is clearly visible in the photograph on the right of Figure 10 while in the photograph on the left is visible in the assembled configuration.

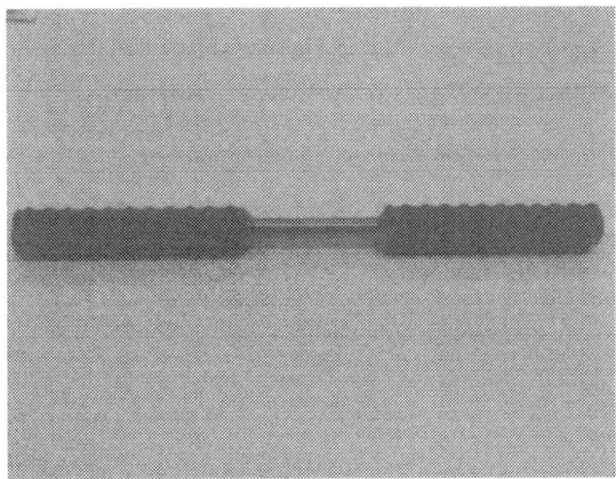

Figure 10. Quick release device designed and calibrated by laboratory staff. Right: calibrated fuse; left: fuse in assembled configuration.

Figure 11. Components of the test device between pile n° 1 and pile n° 2; starting from pile n° 1 are visible the quick release device, the CROSBY turnbuckle, the load cell and the steel rope passing over pile n°2. The support structure for the displacement transducers is also visible in the vicinity of pile n°1.

An overall view of the test device is visible in the photographs of Figures 11 and 12. In Figure 11 the components between pile n° 1 are visible, and pile n°2 to the left in the photograph,. In Figure 12, instead the elements between pile n°3 are visible, and pile n°2 to the right on the photograph,; in the background pile n°1 is also visible.

Figure 12. Components of the test device between pile n°2 and pile n°3. Starting from pile n°3, to the right in the photograph, the traction jack, the load cell and the steel rope passing on pile n°2 are visible. In the vicinity of pile n°3 the support structure for the displacement transducers is also visible. In the background pile n°1 is visible.

3 The test piles

As was mentioned earlier, the construction of the TA building of the Engineering Faculty at the University of Catania provided the opportunity for the execution of the present test series; in fact, the test piles and the test site were provided by the University of Catania within the previously mentioned contract. The test results will be used within the building certification procedure. The test piles were built in the immediate proximity of the building which is founded on piles. As it has been previously mentioned, the test piles were built along one alignment at a relative distance, centre to centre, of 6.00 m. The end piles have a diameter of 800 mm while the central pile has a diameter of 600 mm. In this way the two types of piles used in the construction of the building foundation will be both tested. The design length of the two piles is 25.00 m. The reinforcement bars and other design details are shown in the drawings of Figure 13 for the 800 mm pile and in the drawings of Figure 14 for the 600 mm pile. The piles are of the type bored and cast on site. The longitudinal reinforcement is uniform over all the length of the pile apart from a terminal section near the pile head, where the reinforcement area is doubled. As shown in Figure 13, the 800 mm piles are reinforced with 10 bars of nominal diameter 20 mm while the 600 mm piles are reinforced with 8 bars of nominal diameter 16 mm, see Figure 14. In both cases the number of bars doubles in the terminal section near the pile heads, as is clearly shown in the drawings of Figures 13 and 14. For both pile types the transversal reinforcement is helical with 30 cm pitch and 8 mm diameter.

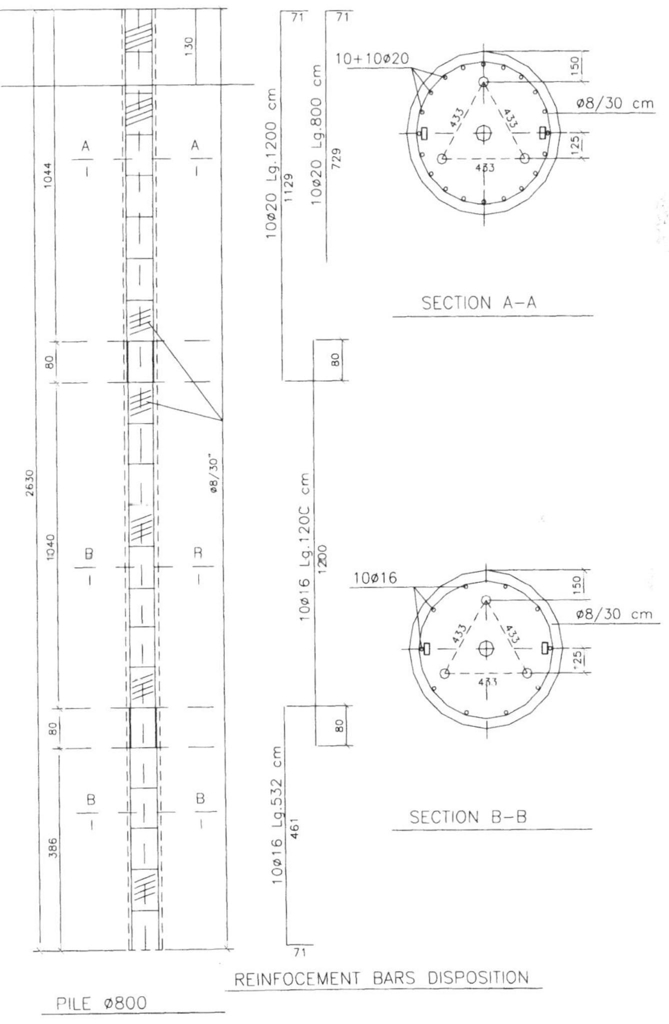

Figure 13. Reinforcement details for the piles of diameter 800 mm.

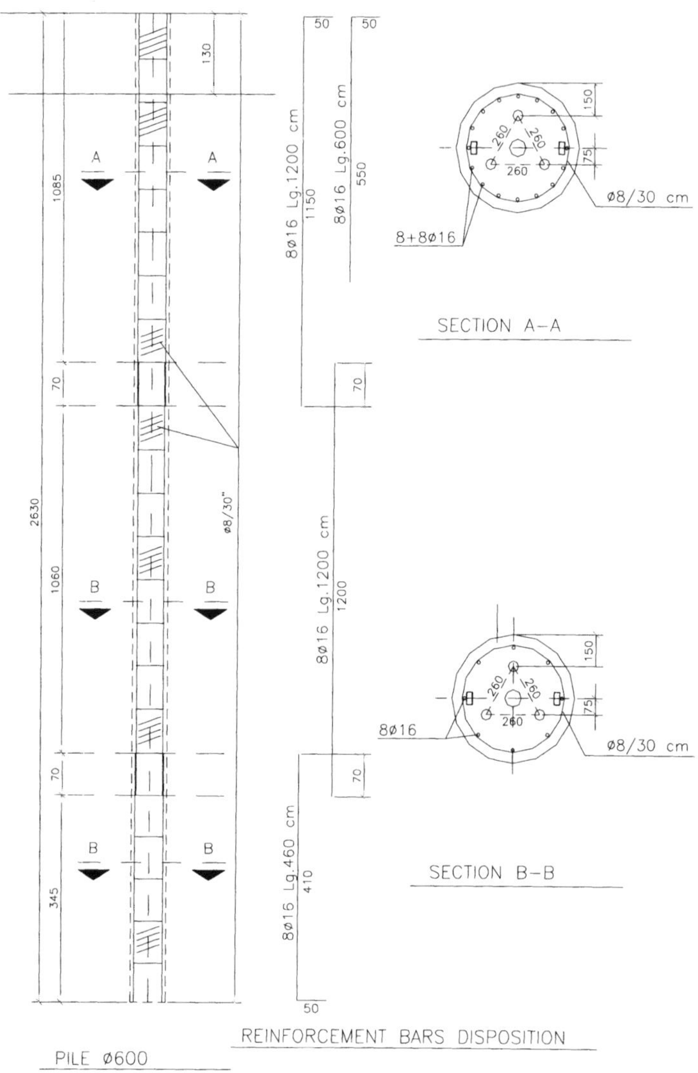

Figure 14. Reinforcement details for the piles of diameter 600 mm.

3.1 Materials testing

Within the certification procedure for the TA building of the Faculty of Engineering, concrete and reinforcement steel samples have been collected on site and tested according to present regulations. The concrete strength, evaluated on the basis of 20 samples, has exhibited a mean value of 29 N/mm^2 and a standard deviation 5 N/mm^2. The results of the tests conducted on the steel samples allow the classification of the material as Fe 44k according to the current Italian classification. In addition to the previously quoted laboratory tests, ultrasonic tests have been performed in order to verify casting quality and pile integrity. For this purpose each test pile has been fitted with 3 PVC pipes of diameter 50 mm and positioned as shown in the drawing of Figure 15. In the same figure there is a photograph where the longitudinal disposition of the three pipes is shown. The central pipe of diameter 80 mm should have allowed cross-hole seismic tests from pile to pile. Unfortunately, due to the contractor's negligence, these pipes have been found obstructed and not usable for the original purpose.

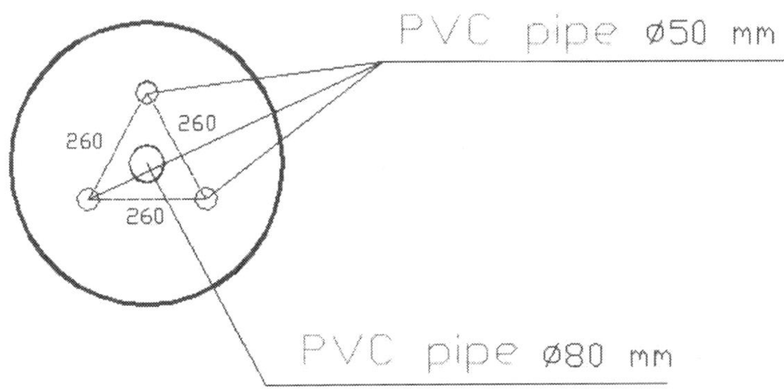

Figure 15. Configuration of PVC pipes for cross-hole ultrasonic tests.

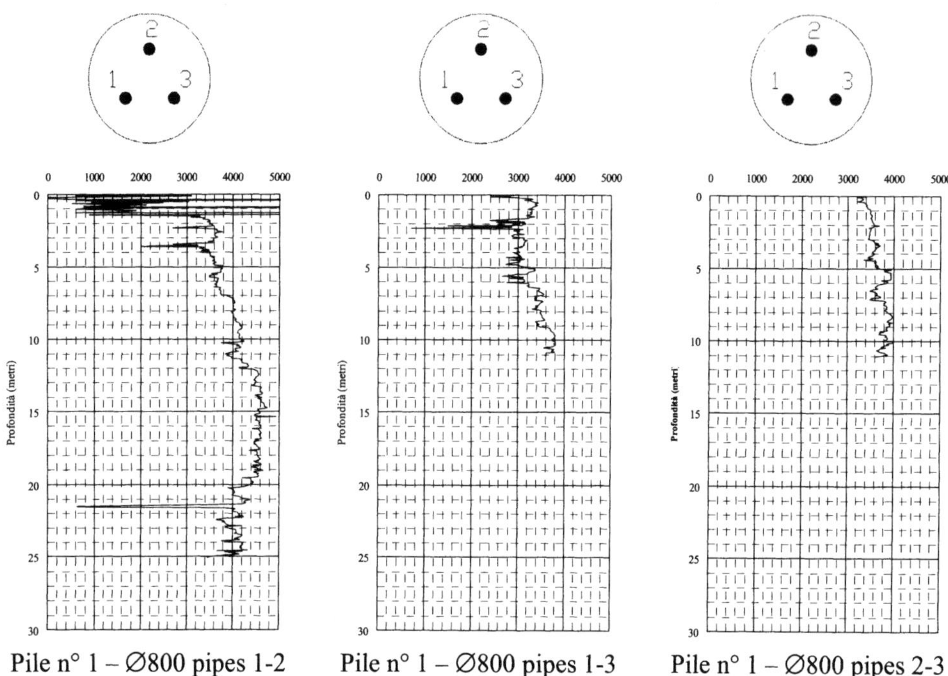

Pile n° 1 – Ø800 pipes 1-2 Pile n° 1 – Ø800 pipes 1-3 Pile n° 1 – Ø800 pipes 2-3

Figure 16. Ultrasonic tests on pile n° 1, diameter 800 mm - P wave velocity in m/s.

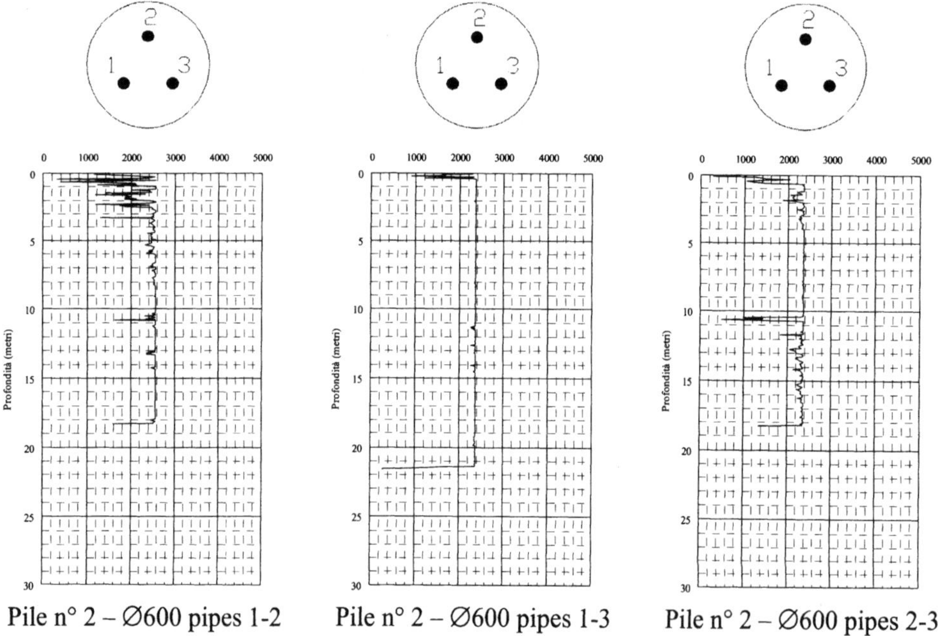

Pile n° 2 – Ø600 pipes 1-2 Pile n° 2 – Ø600 pipes 1-3 Pile n° 2 – Ø600 pipes 2-3

Figure 17. Ultrasonic tests on pile n° 2, diameter 600 mm - P wave velocity in m/s.

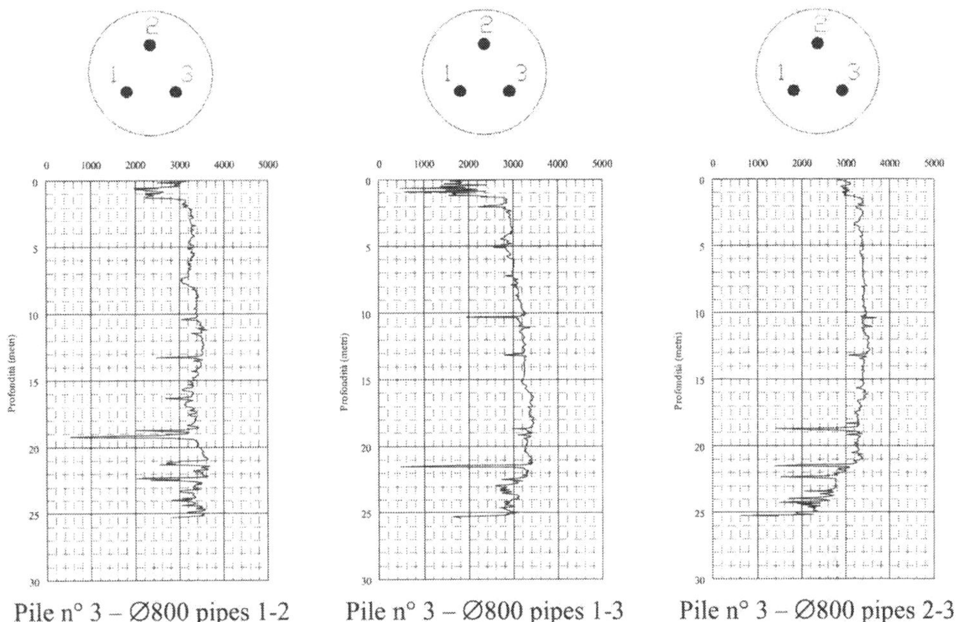

Figure 18. Ultrasonic tests on pile n° 3, diameter 800 mm - P wave velocity in m/s.

Pile casting occurred on 15 July 2002 while the ultrasonic integrity tests were performed on 9 September 2002. The results are shown in Figure 16 for pile n°1, in Figure 17 for pile n°2 and in Figure 18 for pile n°3. Again because of the contractor's negligence, pipe n°3 for the execution of the ultrasonic tests in pile n°1 showed an obstruction at the depth of 11.05 m. Nevertheless the ultrasonic tests have provided invaluable information on the casting quality and on the mechanical properties of concrete. The irregularities in the velocity profiles are more likely attributable to measurement defects than to casting vagaries. In fact, inevitable deformations of the PVC pipes, possibly responsible for the observed obstruction, may have determined significant variations in the distance between couples of pipes. This distance is obviously reflected in the determined wave velocity. At any rate a longitudinal wave velocity in the range 3000 m/s ÷4000 m/s may be considered acceptable for the considered piles.

With reference to pile n°2 an obstruction has been reported at the depth of 18.30m in pipe n°2, while an obstruction was found at the depth of m 21.80 in pipe n°3. For the length of pile that was possible to examine the wave velocity was found in the range 2400÷2600 m/s. These values are rather out of range, especially if compared with those for piles n°1 and n°3; therefore a repetition of the tests is advisable in order to confirm or disprove the presently available results.

With reference to pile n°3 it is worth noticing that it was possible to perform the tests for the complete length. In this case the wave velocity was in the range 3000÷3400 m/s. At any rate the ultrasonic tests will soon be repeated on all the test piles also in order to detect possible variations in the wave velocity eventually due to concrete maturing.

3.2 Curvature measurements

The test piles have been instrumented in such a way as to measure curvatures at discrete points along the pile length. Two channel shaped profiles have been placed in the pile as shown in the drawing of Figure 19. These profiles were fitted with spearheads to ensure better adherence to the concrete. The channel shape is useful on one hand to bind the strain gauges on the inner side of the web and on the other to channel the electric cables that carry the signals to the data acquisition system. Once all the strain gauges and corresponding electrical cables have been fixed, the channel is filled with polyurethane foam and then covered with a lid which is welded to the channel profile. In this way the open channel profile becomes a closed one which protects the strain gauges and electric cables during casting operations and later on. A schematic representation of the apparatus described is shown in Figure 20. In Figure 20, two photographs show the channel profiles, the spearheads, the welded lid and the polyurethane foam. From one of the two channels the electric cables and a connector are seen to emerge. In the photograph on the right of Figure 20 the bundles of cables, emerging from the two channel profiles, and connectors are visible.

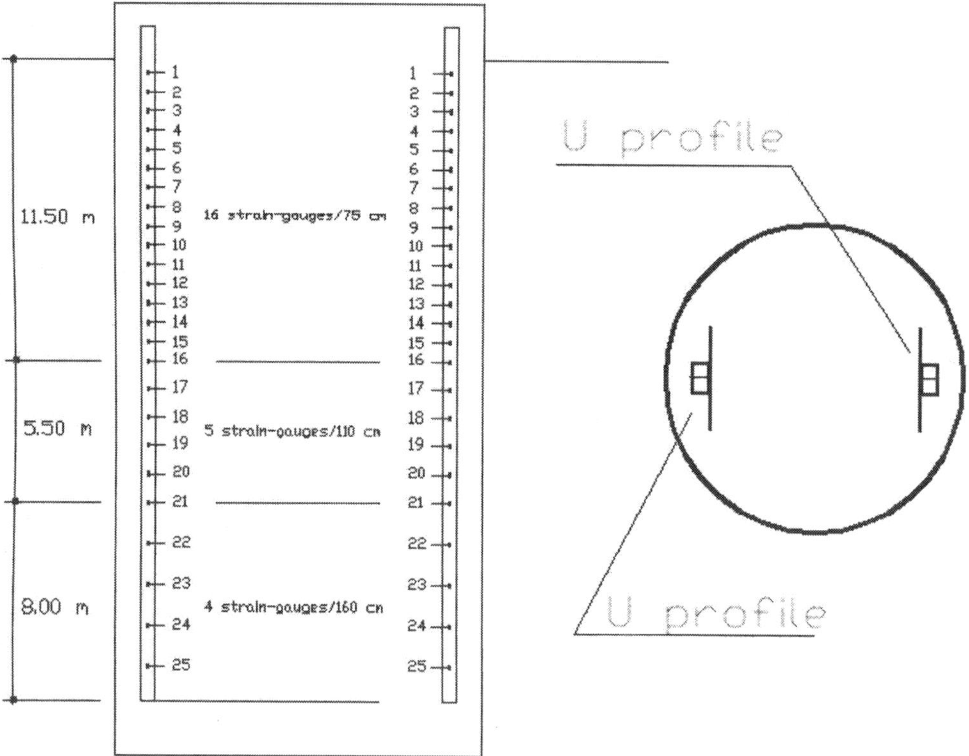

Figure 19. Strain-gauge positioning along the pile length for curvature readings.

Figure 20. Channel profiles for strain-gauges and electric cables.

4 Soil Characterisation

For the soil characterisation the preparatory geological and geotechnical studies for the design of the TA building of the Engineering Faculty at the University of Catania were made available. These studies contain 8 borehole log-plots and the results of some laboratory tests on soil samples drawn during the borehole operations.

For the reasons previously explained it was not possible to evaluate the mechanical characteristics of the soil through seismic down-hole and cross-hole tests conducted through the central holes in the piles. Thus three new boreholes were drilled as shown in Figure 21 and denoted respectively with BH n°1, BH n°2 and BH n°3. The boreholes of the diameter of 100 mm where perforated up to a depth of 28 m from the ground surface.

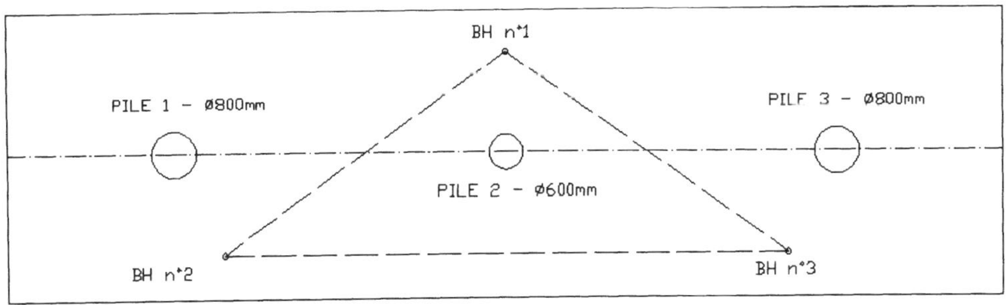

Figure 21. Relative position of piles and boreholes for geophysics tests.

The down-hole and cross-hole tests provided the wave velocity profiles shown in Figure 22. In Figure 22a the longitudinal wave velocity profiles obtained from two different down-hole and cross-hole tests are compared against each other. Down-hole tests usually show coherent values and so do cross-hole tests. Overall the results of cross-hole and down-hole tests are comparable only at depths larger than about 5 m. For lower depths the longitudinal wave velocities measured by the down-hole and the cross-hole tests differ dramatically. It is not clear which of the two methods provides more accurate results.

The results for the transverse wave velocities are shown in Figure 22b. Once again the velocity profiles of two different down-hole and cross-hole tests are reported for comparison against each other. The results are generally scattered and the coherence between the results of the down-hole and cross-hole tests is rather low. The tests were conducted by GEOCHECK – Catania. It is under consideration whether to require a repeat of the tests by the same operator or others. At the moment, in order to perform numerical analyses, it has been decided to make reference only to the results of the down-hole tests.

During the borehole operations samples of materials as shown in Figure 23 were collected.

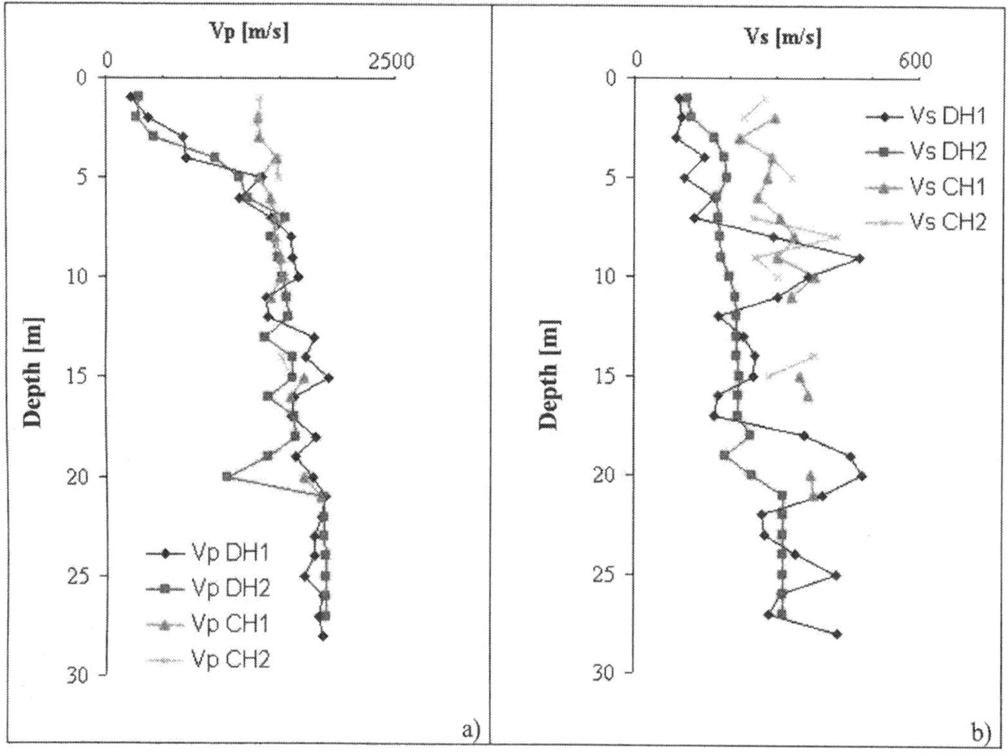

Figure 22. Body wave velocity profiles from down-hole (DH) and cross-hole (CH) tests.

Up to a depth of 7.50 m the presence of yellow clay has been detected; beyond this depth only blue clay has been found. Ultrasonic equipment has been used to measure the longitudinal wave velocity on some of the samples collected.

Figure 23. Clay samples from borehole operations.

The results obtained are collected in Table 1.

Table 1 . Measured parameters for soil samples.

Soil sample	Weight per unit mass [Nm^{-3}]	P- wave velocity [ms^{-1}]
Yellow clay – depth 6 m	19627.27	1454.00
Blue clay – depth 18 m	20717.15	1677.00

The above velocity values are consistent with those obtained at the same depths by down-hole seismic tests.

Table 2 . Parameters for the evaluation of the collapse load of the test system.

Cu : undrained cohesion	150 kNm^{-2}
Distance "a " (see Figure 21)	0.30 m
Plastic moment: 800 mm diameter - pile	823 kNm
Plastic moment: 600 mm diameter - pile	311 kNm

5 Evaluation of the maximum test load

During the preparation of the test programme intended to gather information on the behaviour of the pile-soil system and ultimately to the identification of the dynamic flexibility matrix, it has been seen to be of considerable importance that the piles should not be damaged during the tests as is usually the case under service load conditions and often even under ultimate load conditions. Therefore the collapse load of the test system has been evaluated and optimised in such a way that the most stressed end pile and the central pile collapse simultaneously, as shown in Figure 24. For that purpose Broms' theory [Viggiani, 1994] has been used together with the parameters collected in Table 2.

The solution is shown in the drawings of Figure 24, where the load H applied at the head of pile n°3 is shown as a function of the lever arm "b" of the couple applied to pile n°2. In Figure 25, one curve provides the collapse load for pile n°2 while the other provides the collapse load for pile n°3. As it may be seen from the figure, there exists only one lever arm for which the two piles collapse for the same value of the horizontal load H. The corresponding values of "H" and "b" provide the required solution. The values obtained by using the parameters of Table 2 are H=311.51 kN and b=1.00 m. These values have been used for the design of the collars.

To prevent pile damage a safety factor of 2 has been applied to this value to determine the maximum testing load which has been taken equal to 150 kN.

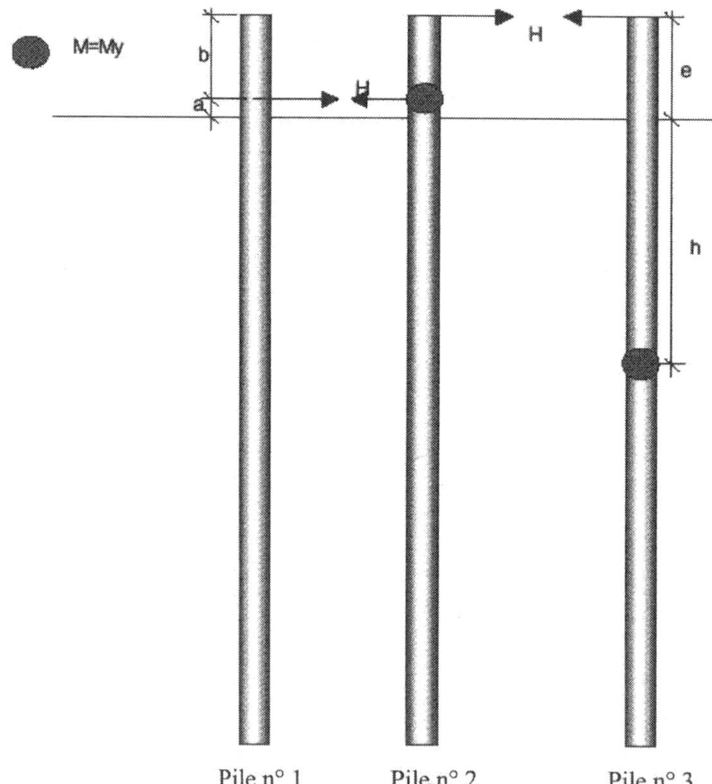

Figure 24. Collapse mechanism of the test system; The lever arm " b" is optimised so that piles n° 2 and n° 3 collapse simultaneously.

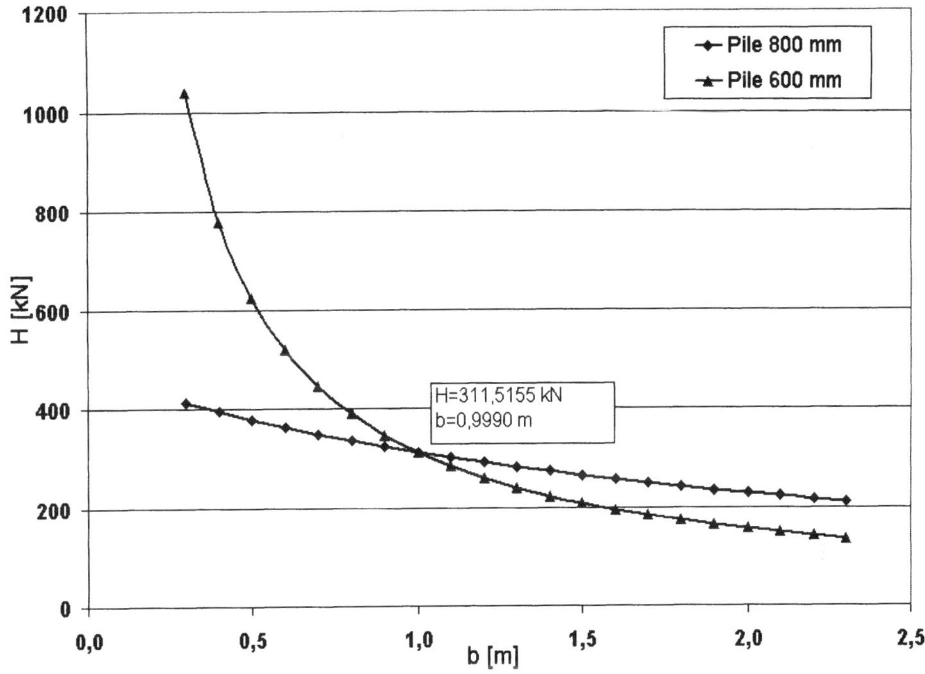

Figure 25. Evaluation of the collapse parameters of the system; Horizontal force "H" and couple "H b" .

6 Measurement and acquisition data systems

As it will be clear from the description of the loading programme, the experimental tests will have both a static and a dynamical character. The measurement and acquisition data systems will be different according to the loading phase considered, whether static or dynamic. In general the measurement devices for static parameters, such as displacements and rotations, will be removed at the beginning of any dynamic loading phase in order to avoid damaging them. In the following the measurement apparatus used for the measurement of static or dynamic parameters will be described separately.

6.1 Measurement and acquisition of static parameters

The physical parameters of static nature that needed to be measured and stored are essentially displacements and rotations at pile heads, and curvatures along the length of the pile stem. With reference to the latter, the distribution of strain-gauges along the length of the pile has already been illustrated when dealing with the test pile description, § 3.2. Here reference will be made to the instrumentation used for the measurement of displacements and rotations and the corresponding acquisition system.

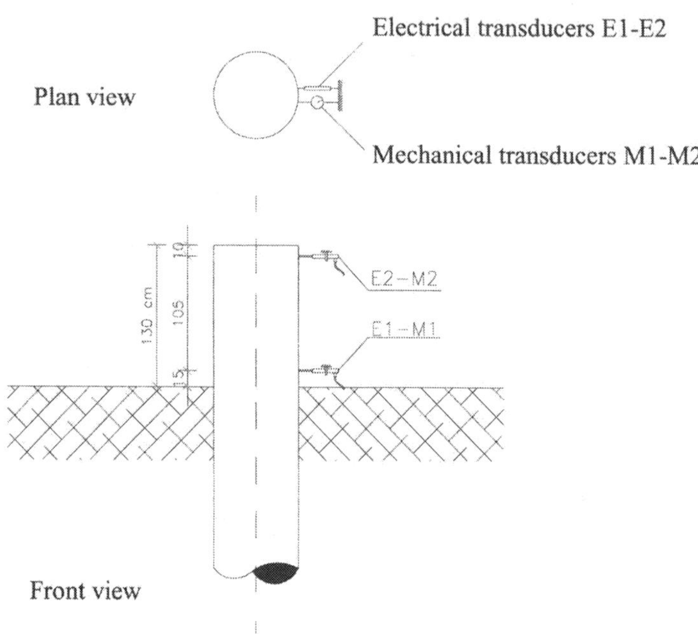

Figure 26. Instrument positioning for displacement and rotation measurements.

Experience gained within the testing laboratory for materials, models and structures at the University of Catania, points to the need for every measurement system to exhibit a certain degree of redundancy, because several instruments may malfunction when they are most needed during a test. In the present case two different measurement systems will be used. A system of electrical transducers with automatic acquisition and a system of mechanical transducers for direct reading and manual recording. For the case in hand the measuring instruments are positioned on pile heads as is shown in the schematic representations of Figure 26. The rotations are measured indirectly by the knowledge of the displacement differential and of the distance between measuring instruments.

Data relating to the mechanical displacement transducers is stored immediately after reading in pre-prepared data sheets. Data relative to the electrical transducers is acquired continuously by means of an HBM data acquisition system, type UPM 60.

The technical characteristics of the displacement transducers are shown in Table 3.

Table 3 . Technical characteristics of displacement transducers and strain-gauges.

Electrical transducers HBM WA - maximum range 20 – 50 mm	Resolution 10^{-3} mm
Mechanical transducers - maximum range 50 mm	Resolution 10^{-2} mm
Strain-gauge PFL10	Resolution 1μs

The strain-gauge signals are acquired by the same data acquisition system used for the electrical displacement transducers: HBM, type UPM 60. The technical characteristics of the strain-gauges are given in Table 3.

6.2 Measurement and acquisition of dynamical parameters

Two NOVATECH load cells, model F204, suitably modified for the present application, are used for the measurement of the traction in the two branches of the load transmission steel rope. Fourteen SETRA model 141B accelerometers are used for the measurement of the acceleration at the pile heads. Velocities and displacements are obtained by successive integrations. Rotations are obtained indirectly by the evaluation of differential displacements and the distance between measurement points. The positioning of the load cells for the measurement of the traction in the two branches of the steel rope has been dealt with in detail in § 2.5. The typical positioning of the accelerometers at pile heads is shown in the schematic representation of Figure 27. Accelerometers 1 and 2 are positioned to measure horizontal accelerations of the pile head and, after integration, velocities and displacements. The duplication obeys to the usual redundancy principle already mentioned when describing the static instrumentation. Accelerometers 3 and 4 are dedicated to the measurement of vertical accelerations and, indirectly, corresponding velocities and displacements.

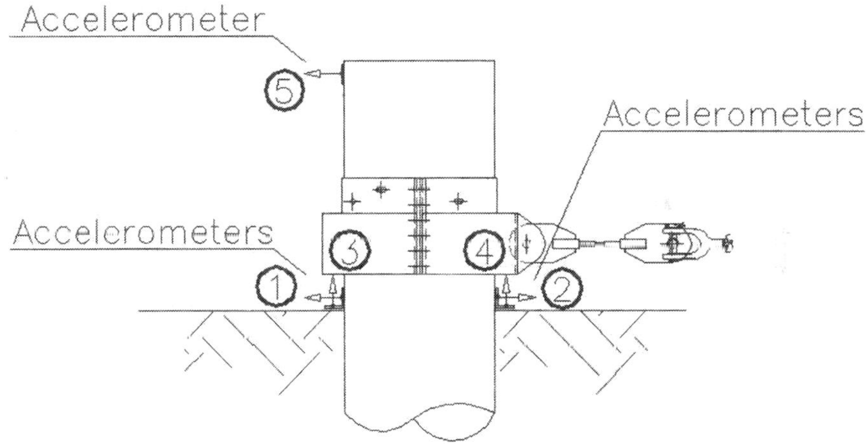

Figure 27. Positioning of accelerometers on pile heads.

Their main function is that of providing a differential measure of the rotation of the pile head. It is possible that the corresponding accelerations may turn out to be too low and there is, presently, no certainty that the results obtained will be sufficiently reliable. This explains the positioning of accelerometer 5, which together with accelerometers 1 and/or 2 should also provide an indirect measurement of the rotation of the pile head. This additional accelerometer positioning corresponds to the redundancy principle already quoted. A crosscheck of both horizontal displacements and rotations becomes in this way possible, provided that all accelerometers function properly and the acquired signals are sufficiently reliable. The depicted accelerometer disposition is the test design one. When the actual testing phase begins, the results of a few dynamical tests will be carefully analysed and any suitable correction to the present disposition will be adopted if required. The technical data for the load cells have already been given in § 2.5, those for the accelerometers are given in Table 4.

The signals produced by the accelerometers are acquired by an AT-MIO-16X NATIONAL INSTRUMENTS card mounted on personal computer.

Table 4 . Technical characteristics of SETRA 141B accelerometers.

Nominal sensitivity	0.5V/g
Utilisation range	±2 g
Resolution	10^{-4} g

7 Testing programme

The testing programme considers three maximum load levels respectively equal to 50 kN, 100 kN e 150 kN. For each load level a static test will be performed where the load (H force) will be increased in a pseudo-static fashion up to the maximum considered value. Once the maximum value has been reached, the load is maintained at that level for the next 24 hours in order to capture the time dependent characteristics of the soil-pile system (creep test). Once a static equilibrium configuration has been reached the load is suddenly increased to activate the quick release device (fuse). The subsequent phase of free vibration, together with the phase of rapid load growth, constitutes the dynamical aspect of the testing programme. Once the equilibrium configuration has been reached (rest condition of the system) the procedure is repeated until 8 test realisations are achieved. After the lower load level phase (maximum load of 50 kN) has been completed, the intermediate phase is executed (maximum load of 100kN) and finally the last phase with a maximum load of 150 kN ends the test programme.

8 Conclusions

An experimental device for static and/or dynamical tests on foundation piles has been described. The device allows for the application of horizontal forces to two test piles and of a couple to the third test pile. Besides static tests, impulsive-type dynamical tests may be performed with the acquisition of the dynamical response in the time domain. The tests performed for the mechanical characterisation of piles and soil have also been described. The instrumentation for the measurement of the response within the pile stem and at the pile heads has been discussed and the reasons for its positioning explained.

Overall the test programme should allow the gathering of information on the short and long term static behaviour of the pile-soil system and on the dynamical behaviour of the same system.
The experimental results will be useful in the validation and calibration of numerical algorithms for the analysis of the static and/or dynamical behaviour of the soil-pile system and of structures and buildings founded on piles (Finite Elements and/or Boundary Element based numerical codes).

The results may also be used in the formulation and/or validation of simplified models representing the pile-soil system.

References

Relazione Geologica e Geognostica (1998), Edificio per la Didattica della Facoltà di Ingneria nel N.C.U. di S. Sofia. Università degli Studi di Catania, Istituto di Geologia e Geofisica.

Viggiani C. (1994). *Fondazioni*. Ed. CUEN.

Vinciprova F., Maeso Fortuny O., Aznárez J. J. and Oliveto G., Interaction of BEM analysis and experimental testing on Pile-Soil Systems, in this volume.

Zeitfracht Medien GmbH
Ferdinand-Jühlke-Straße 7
99095 Erfurt, Deutschland
produktsicherheit@kolibri360.de